Bacteriophages for Plant Disease Control

Bacteriophages for Plant Disease Control

Edited by:

Dr Prashant Pant, Dr Pratibha Pant and
Dr Prerna Pandey

Delve PUBLISHING

www.delvepublishing.com

Bacteriophages for Plant Disease Control

Dr Prashant Pant, Dr Pratibha Pant and Dr Prerna Pandey

Delve Publishing
2010 Winston Park Drive,
2nd Floor
Oakville, ON L6H 5R7
Canada
www.delvepublishing.com
Tel: 001-289-291-7705
001-905-616-2116
Fax: 001-289-291-7601
Email: orders@arclereducation.com

ISBN: 978-1-77361-196-9 (Hardcover)

Library of Congress Cataloging-in-Publication Data

A CIP record for this book has been applied for and is available from Library and Archives Canada.

Delve Publishing publishes wide varicty of books and eBooks. For more information about Delve Publishing and its products, visit our website at www.dclvepublishing.com.

DECLARATION

ABOUT THE EDITORS

Dr. Prashant Pant

Dr. Prashant Pant, is an Assistant Professor of Botany in Dyal Singh College, University of Delhi, New Delhi, India. He has been teaching for more than 5 years now and his main areas have been Bioinformatics, Molecular Biology and Genomics at the college. His PhD was on understanding the Microbial Succession using a Metagenomic approach in a Barren mined-out site undergoing ecological restoration. For this, he deployed many molecular and computational Biology tool and techniques. He has several publications (articles and book chapters) in International Journals and International Publications. Currently, he is also Chief Editor of Bioinformatics Review and a co-founder/Executive Chief Editor of a Science Journal in Hindi Language "Swadeshi Vigyan" for the native language speaking communities. Both the Journals are a service to the scientific community as they are fully open access.

Dr. Pratibha Pant

Dr. Pratibha Pant, is an Assistant Professor of Botany in Daulat Ram College, University of Delhi, New Delhi, India and has a teaching experience of more than 6 years. She is a plant virologist and biotechnologist by training and has done her PhD in Characterization of Begomoviral genomes from Papaya from Department of Botany, University of Delhi. During her PhD, she utilized a combination of molecular biology and Computational Biology tools to analyze begomoviral genomes and to her credit, a couple of novel begomoviruses have been isolated, sequenced and reported in public databases.

Prerna Pandey

Dr. Prerna Pandey, a biotechnologist with several years of wet lab research experience, worked at International Center for Genetic Engineering and Biotechnology, New Delhi. Her field of research in PhD included isolation and molecular characterization of geminiviruses, genome sequencing, gene annotation, and gene silencing using the RNA interference technology. She has also worked at Transasia Biomedicals and Advance Enzyme Technologies as a scientist. Prerna has published papers in peer-reviewed journals, and has submitted a number of annotated Geminiviral genome sequences in the GenBank, including two novel ones. She has also completed her editing and proof-reading courses from Society for the promotion of Editors and Proof-readers, UK and now works as a freelance scientific editor and writer. When Prerna is not busy with her assignments, she loves to explore historical places.

TABLE OF CONTENTS

LIST OF CONTRIBUTORS

Dr Pratibha Pant

[1]Dr. Pratibha Pant, is an Assistant Professor of Botany in Daulat Ram College, University of Delhi, New Delhi, India and has a teaching experience of more than 6 years. She is a plant virologist and biotechnologist by training and has done her PhD in Characterization of Begomoviral genomes from Papaya from Department of Botany, University of Delhi. During her PhD, she utilized a combination of molecular biology and Computational Biology tools to analyze begomoviral genomes and to her credit, a couple of novel begomoviruses have been isolated, sequenced and reported in public databases

Dr Prashant Pant

[2]Dr. Prashant Pant, is an Assistant Professor of Botany in Dyal Singh College, University of Delhi, New Delhi, India. He has been teaching for more than 5 years now and his main areas have been Bioinformatics, Molecular Biology and Genomics at the college. His PhD was on understanding the Microbial Succession using a Metagenomic approach in a Barren mined-out site undergoing ecological restoration. For this, he deployed many molecular and computational Biology tool and techniques. He has several publications (articles and book chapters) in International Journals and International Publications. Currently, he is also Chief Editor of Bioinformatics Review and a co-founder/Executive Chief Editor of a Science Journal in Hindi Language "Swadeshi Vigyan" for the native language speaking communities. Both the Journals are a service to the scientific community as they are fully open access.

Dr Prerna Pandey

[3]Dr Prerna Pandey, a biotechnologist with several years of wet lab research experience, worked at International Center for Genetic Engineering and Biotechnology, New Delhi. Her field of research in PhD included isolation and molecular characterization of geminiviruses, genome sequencing, gene annotation, and gene silencing using the RNA interference technology. She has also worked at Transasia Biomedicals and Advance Enzyme Technologies as a scientist. Prerna has published papers in peer-reviewed journals, and has submitted a number of annotated Geminiviral genome sequences in the GenBank, including two novel ones. She has also completed her editing and proof-reading courses from Society for the promotion of Editors and Proof-readers, UK and now works as a freelance scientific editor and writer. When Prerna is not busy with her assignments, she loves to explore historical places.

Colin Buttimer
Department of Biological Sciences, Cork Institute of Technology, Cork, Ireland

Olivia McAuliffe
Moorepark Food Research Centre, Teagasc, Fermoy, Ireland

R. P. Ross
Alimentary Pharmabiotic Centre, University College, Cork, Ireland

Colin Hill
Alimentary Pharmabiotic Centre, University College, Cork, Ireland

Jim O'Mahony
Department of Biological Sciences, Cork Institute of Technology, Cork, Ireland

Aidan Coffey
Department of Biological Sciences, Cork Institute of Technology, Cork, Ireland

Alexandra Igorevna Klimenko
Institute of Cytology and Genetics SB RAS, Lavrentiev Avenue 10, Novosibirsk 630090, Russia
Novosibirsk State University, Pirogova st. 2, Novosibirsk 630090, Russia

Yury Georgievich Matushkin
Institute of Cytology and Genetics SB RAS, Lavrentiev Avenue 10, Novosibirsk 630090, Russia

Nikolay Alexandrovich Kolchanov
Institute of Cytology and Genetics SB RAS, Lavrentiev Avenue 10, Novosibirsk 630090, Russia
Novosibirsk State University, Pirogova st. 2, Novosibirsk 630090, Russia

Sergey Alexandrovich Lashin
Institute of Cytology and Genetics SB RAS, Lavrentiev Avenue 10, Novosibirsk 630090, Russia
Novosibirsk State University, Pirogova st. 2, Novosibirsk 630090, Russia

Andrew J. Curtright
Department of Microbiology, The Ohio State University, Mansfield, Ohio USA

Stephen T. Abedon
Department of Microbiology, The Ohio State University, Mansfield, Ohio USA

Rebekah A. Frampton
Department of Microbiology & Immunology, University of Otago, Dunedin 9054, New Zealand

Andrew R. Pitman
New Zealand Institute for Plant & Food Research, Christchurch 8140, New Zealand

Peter C. Fineran
Department of Microbiology & Immunology, University of Otago, Dunedin 9054, New Zealand

Sofie Rombouts
Laboratory of Gene Technology, Department of Biosystems, KU Leuven, Leuven, Belgium
Unit Plant— Crop Protection, Institute for Agricultural and Fisheries Research, Merelbeke, Belgium

Anneleen Volckaert
Vegetable Research Centre (PCG), Kruishoutem, Belgium

Sofie Venneman
Research Station for Vegetable Production, Sint-Katelijne-Waver, Belgium

Bart Declercq
Inagro, Rumbeke, Belgium

Dieter Vandenheuvel
Laboratory of Applied Microbiology and Biotechnology, Department of Bioscience Engineering, University of Antwerp, Antwerpen, Belgium

Camille N. Allonsius
Laboratory of Applied Microbiology and Biotechnology, Department of Bioscience Engineering, University of Antwerp, Antwerpen, Belgium

Cinzia Van Malderghem
Unit Plant— Crop Protection, Institute for Agricultural and Fisheries Research, Merelbeke, Belgium

Ho B. Jang
Laboratory of Gene Technology, Department of Biosystems, KU Leuven, Leuven, Belgium

Yves Briers
Laboratory of Gene Technology, Department of Biosystems, KU Leuven, Leuven, Belgium
Laboratory of Applied Biotechnology, Department of Applied Biosciences, Ghent University, Ghent, Belgium

Jean P. Noben
School of Life Sciences, Biomedical Research Institute and Transnational University Limburg, Hasselt University, Diepenbeek, Belgium

Jochen Klumpp
Institute of Food, Nutrition and Health, ETH Zurich, Zurich, Switzerland

Johan Van Vaerenbergh
Unit Plant— Crop Protection, Institute for Agricultural and Fisheries Research, Merelbeke, Belgium

Martine Maes
Unit Plant— Crop Protection, Institute for Agricultural and Fisheries Research, Merelbeke, Belgium
Lab. of Microbiology, Department of Biochemistry and Microbiology, Ghent University, Gent, Belgium

Rob Lavigne
Laboratory of Gene Technology, Department of Biosystems, KU Leuven, Leuven, Belgium

Mayukh Das
Department of Plant Pathology and Microbiology, Texas A&M University, College Station, Texas, United States of America
Center for Phage Technology, Texas A&M University, College Station, Texas, United States of America

Tushar Suvra Bhowmick
Department of Plant Pathology and Microbiology, Texas A&M University, College Station, Texas, United States of America
Center for Phage Technology, Texas A&M University, College Station, Texas, United States of America

Stephen J. Ahern
Department of Plant Pathology and Microbiology, Texas A&M University, College Station, Texas, United States of America
Center for Phage Technology, Texas A&M University, College Station, Texas, United States of America

Ry Young
Center for Phage Technology, Texas A&M University, College Station, Texas, United States of America
Department of Biochemistry and Biophysics, Texas A&M University, College Station, Texas, United States of America

Carlos F. Gonzalez
Department of Plant Pathology and Microbiology, Texas A&M University, College Station, Texas, United States of America
Center for Phage Technology, Texas A&M University, College Station, Texas, United States of America

Robert Czajkowski
Laboratory of Plant Protection and Biotechnology, Department of Biotechnology, Intercollegiate Faculty of Biotechnology, University of Gdansk and Medical University of Gdansk, Gdansk, Poland

Zofia Ozymko
Laboratory of Plant Protection and Biotechnology, Department of Biotechnology, Intercollegiate Faculty of Biotechnology, University of Gdansk and Medical University of Gdansk, Gdansk, Poland

Victor de Jager
Netherlands Institute of Ecology (NIOO-KNAW), Wageningen, the Netherlands,

Joanna Siwinska
Laboratory of Plant Protection and Biotechnology, Department of Biotechnology, Intercollegiate Faculty of Biotechnology, University of Gdansk and Medical University of Gdansk, Gdansk, Poland

Anna Smolarska
Laboratory of Plant Protection and Biotechnology, Department of Biotechnology, Intercollegiate Faculty of Biotechnology, University of Gdansk and Medical University of Gdansk, Gdansk, Poland

Adam Ossowicki
Laboratory of Biological Plant Protection, Department of Biotechnology, Intercollegiate Faculty of Biotechnology, University of Gdansk and Medical University of Gdansk, Gdansk, Poland

Magdalena Narajczyk
Laboratory of Electron Microscopy, Faculty of Biology, University of Gdansk, Gdansk, Poland

Ewa Lojkowska
Laboratory of Plant Protection and Biotechnology, Department of Biotechnology, Intercollegiate Faculty of Biotechnology, University of Gdansk and Medical University of Gdansk, Gdansk, Poland

PREFACE

The growing need for increased food production globally has led the farmers and scientists equally to devise means to increase plant productivity through several practices, including disease management. A number of plant diseases caused by bacteria have huge economic impact on global crop productivity. Although chemical control practices may be effective, they pose a threat to the environment and to the food chain. With the discovery of antibiotics in the 1940s, the concept of using bacteriophages was abandoned. However, its use has re-emerged, and is a rapidly growing area for controlling bacterial plant pathogens. In this book, we present the historical and recent developments, concepts, factors, challenges and concerns of using bacteriophages for plant disease control. We hope that the book will be a valuable reference for students, researchers, teachers as well as agriculturists while choosing environmentally safe and sustainable plant pest management techniques. We also look forward to your suggestions for improving the book.

Editor

CHAPTER

1

ORGANIZATION AND DIVERSITY OF BACTERIOPHAGES

Dr Pratibha Pant[1], Dr Prashant Pant[2], Dr Prerna Pandey[3]

[1]Dr. Pratibha Pant, is an Assistant Professor of Botany in Daulat Ram College, University of Delhi, New Delhi, India and has a teaching experience of more than 6 years. She is a plant virologist and biotechnologist by training and has done her PhD in Characterization of Begomoviral genomes from Papaya from Department of Botany, University of Delhi. During her PhD, she utilized a combination of molecular biology and Computational Biology tools to analyze begomoviral genomes and to her credit, a couple of novel begomoviruses have been isolated, sequenced and reported in public databases

[2]Dr. Prashant Pant, is an Assistant Professor of Botany in Dyal Singh College, University of Delhi, New Delhi, India. He has been teaching for more than 5 years now and his main areas have been Bioinformatics, Molecular Biology and Genomics at the college. His PhD was on understanding the Microbial Succession using a Metagenomic approach in a Barren mined-out site undergoing ecological restoration. For this, he deployed many molecular and computational Biology tool and techniques. He has several publications (articles and book chapters) in

International Journals and International Publications. Currently, he is also Chief Editor of Bioinformatics Review and a co-founder/Executive Chief Editor of a Science Journal in Hindi Language "Swadeshi Vigyan" for the native language speaking communities. Both the Journals are a service to the scientific community as they are fully open access.

[3]Dr Prerna Pandey, a biotechnologist with several years of wet lab research experience, worked at International Center for Genetic Engineering and Biotechnology, New Delhi. Her field of research in PhD included isolation and molecular characterization of geminiviruses, genome sequencing, gene annotation, and gene silencing using the RNA interference technology. She has also worked at Transasia Biomedicals and Advance Enzyme Technologies as a scientist. Prerna has published papers in peer-reviewed journals, and has submitted a number of annotated Geminiviral genome sequences in the GenBank, including two novel ones. She has also completed her editing and proof-reading courses from Society for the promotion of Editors and Proof-readers, UK and now works as a freelance scientific editor and writer. When Prerna is not busy with her assignments, she loves to explore historical places.

INTRODUCTION

With the advent of high resolution microscopy especially Transmission Electron Microscopy in the mid 20th Century, the existence, nature, structure of Bacteriophages came to be known. These are essentially viruses, in fact the largest viral group, that attack bacteria and thus the name Bacteriophages (meaning Bacteria devouring). Being a virus, these are similar in structure, function, host dependence for multiplication, ecology and pathology with viruses that infects plants and animals. They are known to be carrying both DNA as well RNA as the genetic material (Clark and March, 2006). However, they have some basic differences that makes them distinct from general viruses. During infection, they leave the protein coat (the capsid) outside only to rebuild it later inside the host, they are known to initiate a process called Lysin. Lysin is a process of initiating the synthesis of other phages during insertion of genetic material (Mahy and van Regenmortel, 2010).

Beginning with establishing DNA and not Protein as a genetic material, (Hershey and Chase Experiment), phages have been crucial in establishing the central dogma of molecular biology. Since then, they have been implicated in Molecular Biology and Biotechnology as model systems for many specific and general purposes viz. phage display systems, vectors for cloning. gene and cDNA libraries, vehicles for vaccine delivery, phage typing etc (Clark and March, 2006). They have also been used as a model system to understand the mechanisms by which bacterial genes can be transferred from one bacterium to another (Transduction). Sequencing of bacterial genomes has revealed that phage genome elements are an important source of sequence diversity and can potentially influence the bacterial evolution and virulence. Research in molecular genetics Research has led to several new aspects and understanding in the field of Phage Research such as the recent discovery and exploitation of CRISPR–CAS9 phage resistance systems.

DISCOVERY AND DIVERSITY OF BACTERIOPHAGES (TYPES)

Much before Penicillin was discovered for killing bacterial cells, Bacteriophages were discovered in 1915 by an English researcher Frederick William Twort at the University of London. Félix d'Herelle, a French-Canadian microbiologist at the Institute Pasteur in Paris in 1917, independently confirmed Twort's discovery and coined the term *bacteriophage (from* "bacteria" and "phagein" (to eat or devour, in Greek). Félix d'Herelle, observed that cultures of the dysentery bacteria disappear with the addition of a bacteria-free filtrate obtained from sewage containing "an invisible, antagonistic microbe of the dysentery bacillus" (d'Herelle, 1917). In 1923, the Eliava Institute was opened in Tbilisi, Georgia, to study bacteriophages and to develop phage therapy. Since then many scientists have been involved in developing techniques to study phages and using them for various purposes. By 1950s, the experiments on bacteriophages infecting strains leading to the "Host induced variation in bacteriophage ability to progressively become more and more infective to certain strains of bacteria" had kickstarted the revolutionary discovery of Restriction-Modification System (RMS) which later founded the very base of rDNA technology. It was for the restriction of the Bacteriophage DNA in bacterial systems that the restriction enzymes were discovered and characterized. In 1969,

Max Delbrück, Alfred Hershey and Salvador Luria were awarded the prestigious Nobel Prize in Physiology and Medicine for their discoveries of the replication of viruses and their genetic structure. Cyanophages a distinct class of bacteriophages attacking autotrophic hosts of the group Cyanobacteria, the primary producer population in marine and freshwater ecosystems were discovered in 1963 by Safferman and Morris. Much of the initial molecular and genetics work on Bacteriophages was done by members of the "phage group," founded by Max Delbruck, Salvador Luria and Hershey in 1940 and in the extended group were future scientist to the likes of James Watson who not only continued working with these remarkable systems, they also contributed to the field of Molecular Biology.

IMPORTANCE OF BACTERIOPHAGES

Bacteriophages have been recognized as an innovative alternative to the conventional antibiotic drug based treatment and it is safe too (European Food Safety Authority, 2009; USDA, 2006) and innovative alternatives to the use of chemotherapies (phage-therapy) (Connerton et al., 2011; Kutter et al., 2010; Mai et al., 2010). Using Bacteriophages will also help in preventing building up of antibiotic resistance in pathogenic bacteria which is a major challenge e.g. in the case of *Mycobacterium tuberculosis*, *Enterococcus faecalis*, *Staphylococcus aureus*, *Acinetobacter baumannii*and *Pseudomonas aeruginosa,* and methicillin-resistant *S.aureus* (MRSA) which has been responsible for increased mortalities from bacterial infections (Hanlon, 2007). Thus, phages act and do serve as an efficacious biocontrol agents against pathogens, and can also be used to regulate the microbes of gut environment of several organisms viz. insects etc. to manipulate their population levels (Hagens and Loessner, 2007, 2010).

LIFE CYCLE OF BACTERIOPHAGES-A CONCEPT NOTE

A phage unlike Eukaryotic viruses exhibit unique life cycle. A phage may undergo depending on several factors, Lytic cycle or Lysogenic cycle and a lesser known strategy called Chronic cycle.

Temperate phages: the phages which undergo Lysogenic cycle, while,

Virulent phages: the phages undergoing only lytic phase.

A temperate phage is much less harmful to the host genome as it becomes associated with the host genome, causing very little reshuffling of the host and upon infection it encodes repressor proteins which prevent expression of other phage genes. On the other hand, lytic phage encode host-lethal proteins that disrupt host machinery of replication, transcription and translation leading to major alteration in the host enzyme systems, and subsequent degradation of the host genome. In the ***lytic phase*** phages infect, multiply using the host machinery and come out of host cell bursting it out (Abedon, 2008; Weinbauer and Rassoulzadegan, 2004; Wilhelm and Suttle, 1999). The Lytic phages can act as a potential biological control agents against many bacterial species, some of which are economically important pathogenic ones (Jones et al., 2008). In contrast, the lysogenic life cycle entails integration of the Phage genome into the host DNA to become prophage or existing as plasmids, replicating along with host genome replication only and staying there in that form for one to many generations until they switch to the lytic phase (Little, 2005). Less number of host cells is one of the triggers to switch a phage from being temperate to a virulent one. Also, under some harsh conditions the prophage leaves the bacterial cell by undergoing Lytic cycle (Inal, 2003). Ultraviolet radiation, Antibiotics, Reactive Oxygen Species and other different stress factors cause the departure of the phage genome initiated (Allen, et al., 2011; DeMarini and Lawrence, 1992; Little and Mount, 1982).

Lysogenic lifecycle is highly evolved, that requires the coevolution of the bacteriophage and its host, leading to significant changes in the properties of their host bacteria and conferring a few traits to the bacterial host such as resistance against host immunity, metabolic advantage, faster growth, antibiotic resistance. Temperate bacteriophages can also help to protect their host bacterial host from the infections by other bacteriophages. Thus, lysogeny appears to be a good survival strategy for both the bacteria and the prophage or the lysogen (Canchaya et al., 2003; Barondess and Beckwith, 1990; Edlin et al., 1975; Lin et al., 1977). Thus, these two types of survival strategies i.e. one of infecting, multiplying and killing the host out or the Lytic Cycle and the other one of maintaining itself along with the bacterial host genome under transcriptional repressors have been well studied and documented for the past several decades.

Apart from these two, another lesser known survival strategy is now described under the head ***Chronic phase***. In this case, there is continuous, non-lethal production and dislodging of virions by protrusion through the bacterial membrane (Duerkop et al., 2014)

EFFECTS OF VARIOUS CHEMICAL AND PHYSICAL AGENTS ON BACTERIOPHAGES

Bacteriophages differ in their response to various physical and chemical agents. The studies on the effects of these agents have increased our understanding of their behaviour and nature of interaction *in vivo* conditions as well as *in vitro*. Some of the physical agents are UV radiation, high frequency vibration, and temperature. All phages show negative affect on the genome constitution upon prolonged UV exposure as expected. It is mostly UV radiation of wavelength 265 nm which is mostly DNA damaging (Winkler et al., 1962). Similarly, studies on Temperature as a factor has been done to show that beyond a particular temperature (ranging from 50°C to above, the inactivation of phage proteins and denaturation of Phage DNA initiates. Bacteriophages have been also subjected to High frequency vibration exposure and the reluts indicate their high vulnerability to the same (Krueger et al., 1941). In the case pf chemical; agnets affecting the phage systems, among the most prominent ones are pH, Urea, Detergents, Ascorbic Acid, chelators, mustard gas and many others (Donald, 1956; Adams, 1959 and Ackermann and Dubow, 1987). Briefly, most phages are stable between 4-10, a range highly appreciable for a viral system. Although this is for a very short duration that the virus may be able to thrive at extremes of pH. For some bacteriophages, as the pH rises the capsid begins to break up releasing the inner content of the head and the tail separates from it (Poglazov and Tikhonenko, 1963). Most chemical agents viz. mustard gas, Nitrogen gas work by inducing lysis in lysogenic culture, an effect similar to that of UV radiation (Smith, 1953).

ECOLOGY OF BACTERIOPHAGES

Bacteriophages are highly specific to their host bacteria. Though disputed, it appears that Bacteriophages represent the most abundant 'life' forms on Earth (Suttle, 2005; Pedulla, et al., 2003; Wommack and Colwell, 2000), they are very common in all natural environments including soil, oceans and in deserts and thus serve as good indicators of bacterial contamination (Atterbury et al., 2005; Prestel et al., 2008; Prigent et al., 2005; Srinivasiah et al., 2008; Wommack and Colwell, 2000) and are associated with nearly all known bacterial taxa. As phages have an obligate requirement for a host, their abundance and distribution is likely to be based on that of their host organisms. Phages are found either directly associated with their bacterial hosts or in large numbers as free virions in the environment. The diversity of phage is directly proportional to the associated bacterial population as a host in any environment or habitat.

CLASSIFICATION OF BACTERIOPHAGES

Not only they are abundant (more than 5000 reported by Ackermann (2007), they are extremely diverse and so less studied in terms of the kinds and types of them we have in our environments. Given that they can inhabit practically all environments, only a small set of the total phage diversity has been characterized so far. One reason could be the difficulty in culturing microbes and the other could be because of extremely fast rate of recombination and thus with recombinants forming it is almost impossible to completely type the length and breadth of the diversity that we may have (Hendrix et al., 1999). Phages are known to infect the Eubacterial and Archaeal world alike. Like all other viruses, phages are currently classified by the International Committee on Taxonomy of Viruses (ICTV) according to their morphology and nucleic acid composition, single or double stranded DNA or RNA. The most updated classification for viruses was given by Ackermann (2006, 2007) based on Morphology, Auxillary structures (tail or Envelope), host preference, nature and kind of the genetic material. The vast majority of phages contain double strand DNA (dsDNA), while there are small phage groups with ssRNA, dsRNA, or ssDNA (ss stands for single strand). As the classification is based on general morphology, Transmission Electron Microscopy (TEM) though a very old technique still finds good place in characterization of phages till today (Ackermann, 2012). The role of

TEM is so pronounced that many workers have reported the possibility of defining some genome characteristics like single or double DNA chains or single RNA chains from the ultrastructure (Bradley, 1967). Negative staining has also played an important role in visualization and characterization of phages (and Viruses).

Morphologically, there are *filamentous types, icosahedral types, phages with and without tails, phages with a lipid moieties envelope.* There are six morphological types (A-F) of bacteriophages (Bradley, 1967). Bacteriophages A-E are distinguished according to head shape (icosahedral or elongated) and tail (presence or absence, contractile or non-contractile, short or long compared to head diameter, and though rarely tail fibers in some). Group F, comprises of filamentous phages.

Briefly, Group A phages are characterized by the following features: Hexagonal Head (may or may not elongated), Tail with a contractile sheath, tail may have appendages like fibers or terminal structures. Group B phages are the ones which have hexagonal head, flexible long tail with or without appendages with no contractile appendages. Phages under Group C type are similar to Group B but with a shorter tail (shorter than head diameter). The fourth group, Group D has knob like large capsomeres on the apex of the symmetrical hexagonal head and no tail while rest of the characters are similar to Group B. Group C is similar to group D with the only difference that large capsomeres are absent and the virion has a simple regular hexagonal head. Group F are flexible filamentous type with no additional structure present.

METHODS TO STUDY BACTERIOPHAGES

Since the middle of the 20th century, the studies on Bacteriophages have exponentially gone up. Many methods have been developed to isolate, characterize, and type the phage obtained from a variety of sources. The source of isolation, to begin with, could be soil, water or infected (suitable) host system (Clokie and Kropinski, 2009; Svircev et al., 2011; Adriaenssens et al., 2012; Balogh et al., 2008). These methods have been roughly classified into traditional (or conventional, culture based) and modern methods (molecular methods). For phage isolation, either suitable hosts can be isolated from environment or a model host system could be used. Although in some cases, phages may not infect the model host strain. With so many methods and their variations reported, five methods are most commonly referred to classify, quantify, and estimate the abundance of viruses in any given sample. These are Most-Probable Number assays (MPNs), Plaque Assays (PAs), Transmission Electron Microscopy (TEM), Epifluorescence Microscopy (EfM), and Flow Cytometry (FC) (Brussaard, 2009; Ortmann and Suttle, 2009; Ackermann, 2009).

Most Probable Number (MPN)

MPN assays are one of the oldest and most popular methods of quantifying the abundance in dilution series of cultivable fraction of microbes (Suttle, 1993).

Plaque Forming Assay (PAs)

But, as the host cells are not always cultivable, PAs are another method for uncultured bacterial samples to estimate the titres of viruses that cause the lysis of bacteria. The number of plaque-forming units in a given sample per unit volume can be estimated using this method. The dilution replicates with no growth, or growth followed by cell lysis are assumed to contain at least one infectious virus. The number of replicates at each dilution in which lysis occurred can be used to calculate the number of infective units in the original sample. PAs and MPNs are the only methods that can be used to directly determine the abundance of infectious viruses, and they can also be used to obtain and purify specific viral isolates.

Transmission Electron Microscopy (TEM)

TEM provides data on both the abundance and morphology of virus-like particles (Borsheim et al., 1993). For abundance-based analysis, EFM has replaced TEM (Hara et al., 1991; Hennes and Suttle, 1995). Flow Cytometry is another recent technique for estimating viral abundance (Marie et al., 1999, Brussaard, 2004). FC provides high-throughput method, quantification of viral subpopulations differing by fluorescent characteristics and the technique is very useful for analyzing large number of samples.

The modern or the molecular approach work is based on the genome analysis using a plethora of wet lab and dry lab tools and techniques. This is the computational biology era where diversity is studied at molecular level using marker systems. It is more efficient, complex and extremely useful for phages due to their nature of high recombinant formation. Unlike bacteria, there are no universal marker systems available as there are no specific genes or sequences which are conserved across all the groups (Paul et al., 2002). That said, regions which are conserved in some smaller taxonomic groups have been used as phylogenetic markers viz. structural proteins (Coat proteins), DNA polymerases and gene encoding Portal protein which is located at the top of the neck of the phage through which DNA passes en route down the tail sheath (Marston and Sallee, 2003; Sullivan et al., 2008; Wang and Chen, 2004; Zhong et al., 2002). Apart from this, general marker systems targeting different regions of the genome spanning almost the entire genome are also used e.g. RFLP, DGGE,

Another modern approach implicated in bacteriophage diversity has been metagenomics. A new branch of genomics dealing with DNA sample from a community comprising of many different individual or cell or components belonging to diverse taxonomic groups (Handelsman, 2004). It can be used to identify phages or phage genes of environmental significance, such as those that are highly abundant or specific to particular niches (Angly et al., 2006; Reyes et al., 2010; Willner et al., 2009). With the increasing ease and low cost in DNA sequencing technology and with next generation sequencing technology (high throughput) this approach has been made possible (Kristensen et al., 2010; Willner et al., 2009; Rodriguez-Brito et al., 2010). The approach is useful as it circumvents the issue of culturing host, finding the suitable strain etc., as this method requires community sample such as Soil DNA or from water sample, infected tissue etc. (Paul et al., 2002). Thus, we can have information on phages that are not amenable to propagation, or of those whose host is unknown or its media is not standardized. We know that, more than 95% of bacteria are still uncultured and so phages which infect them cannot currently be studied (Leroy et al.,2008; Rappe and Giovannoni, 2003). As of date, several viral metagenomic projects have been successfully completed and the data has been made available in public domain from marine coastal waters (Breitbart et al., 2002), marine sediments (Breitbart et al., 2004), human feces (Breitbart et al., 2003) and equine feces (Cann et al., 2005).

Metagenomics can also potentially provide abundance information based on the amount of coverage of particular phages or phage genes present in sample sets. Clearly there may be amplification, or sequencing bias but over, or under representation of particular genes can yield useful information about phage biology. The major limitation one comes across is that because viral genome diversity is so high, a large proportion of predicted genes are 'unknown' or 'hypothetical', and therefore currently much of the information collected by this approach is not immediately useful. This situation will improve as additional genomes from isolated bacteriophages are sequenced and annotated, and as bioinformatic tools based on structural protein homologies are developed to assist amino acid or nucleotide sequence comparisons. Finally, it often is difficult to test hypotheses made from sequence data. Although genes of interest from metagenomes can be cloned, expressed and biochemically characterized, their relevance to specific phages can only be established if cultured phages with those genes are in existence. Another disadvantage is that the output is pure data, and no biological materials for further experimentation are recovered (Edwards and Rohwer, 2005).

A third approach is related with field of bioinformatics and computational biology wherein deeply embedded phage sequences are mined out using computational power from the bacterial whole genome sequence data which is available in public databases. More than 1500 completely sequenced whole genome data are available (Leplae et al., 2010; Casjens, 2003; Canchaya et al., 2004).

GENOME CHARACTERISTICS OF BACTERIOPHAGES

One of the earliest genome to be sequenced and published was a phage genome, that of Φx174 infecting Escherichia coli (Sanger et al., 1977). Studies of bacteriophages became an essential part of biology because of their association tightly linked to microbes. Genome analyses of bacteriophage provide the opportunity to identify basic principles of genome organization, co-evolution, as well as to model and modify their genome. With more and more phage genome sequencing projects (such as that of coliphages, mycobacteriophages, and dairy phages) and data availability, it is becoming increasingly clear that they are not only useful to understand genomics, adaptation, interaction rather they are also potential tools for various molecular biology tools and techniques as well as helping in tackling some of the biggest challenges concerning microbial multi-resistance problems (Hendrix et al., 1999; Orlova, 2009). One of the interesting conclusion from the initial phylogenetic studies done till date is that the relationship among phage orders being more related to the choice of host species and the relationships among the host bacteria rather than the possibilities of gene transfer within phages in a given space and time setup (Chibani-Chennoufi et al., 2004). Comparative Genomic analyses of phages has revealed enormous diversity with similarities mostly restricted to cases with a common history of hosts, or envirornment (Hatfull, 2008; Pope et al., 2011; Kwan et al., 2005, 2006).

REFERENCES

1. Abedon S. T. *Phages, Ecology, Evolution*. In: Abedon ST, editor. Bacteriophage Ecology: Population Growth, Evolution, and Impact of Bacterial Viruses. Cambridge University Press; 2008. pp. 1–28.
2. Ackermann H. *Basic Phage Electron Microscopy*. In: Clokie MRJ, Kropinski A, eds. Bacteriophages Methods and Protocols Volume 1: Isolation, Characterisation and Interactions. New York, NY: Springer 2009; 113-126.
3. Ackermann H. W. (2012). Bacteriophages. Part A, Section 1, Chapter 1 Bacteriophage Electron Microscopy. *Adv. Virus Res*. 84: 1-16.
4. Ackermann H. W., Dubow M. (1987). *Viruses of prokaryotes*. CRC Press, Boca Raton FL.
5. Ackermann H-W. (2007) 5500 Phages examined in the electron microscope. *Arch Virol*. 152: 227–243.
6. Adams M. H. (1959). *Bacteriophage*. Interscience publishers, New York 1987
7. Adriaenssens E. M., Van Vaerenbergh J., Vandenheuvel D. (2012).

T4-related bacteriophage Limestone isolates for the control of soft rot on potato caused by "Dickeya solani". *PLoS ONE.* 7(3):e33227

8. Allen H.K., Looft T., Bayles D.O., Humphrey S., Levine U. Y., Alt D. (2011). Antibiotics in feed induce prophages in swine fecal microbiomes. *MBio*. 2(6):e00260–e00211.
9. Angly F. E., Felts B., Breitbart M., Salamon P., Edwards R. A., Carlson C. (2006). The marine viromes of four oceanic regions. *PLoS Biol.* 4, 2121–2131.
10. Atterbury R.J., Dillon E., Swift C., Connerton P. L., Frost J. A., Dodd C. E. R., Rees C. E. D.,
11. Connerton I.F. (2005). Correlation of Campylobacter Bacteriophage with Reduced Presence of Hosts in Broiler Chicken Ceca. *Appl Environ Microbiol.* 71(8): 4885-7.
12. Balogh B., Canteros B. I., Stall R. E., Jones J. B. (2008). Control of citrus canker and citrus bacterial spot with bacteriophages. *Plant Dis*. 92(7): 1048–1052.
13. Barondess J. J., Beckwith J. (1990). A bacterial virulence determinant encoded by lysogenic coliphage lambda. *Nature.* 346:871-874.
14. Borsheim K. Y., Bratbak G., Heldal M. (1990). Enumeration and biomass estimation of planktonic bacteria and viruses by transmission electron microscopy. *Appl Environ Microbiol.* 56: 352–356.
15. Bradley D. (1967). Ultrastructure of Bacteriophages and Bacteriocins. *Bacteriological Reviews.* 31(4): 230-314.
16. Breitbart M., Salamon P., Andresen B., Mahaffy J. M., Segall A. M. (2002). Genomic analysis of uncultured marine viral communities. *Proc Natl Acad Sci U S A*. 99: 14250–14255.
17. Breitbart M., Felts B., Kelley S., Mahaffy J., Nulton J. (2004). Diversity and population structure of a nearshore marine sediment viral community. *Proc R Soc Lond Ser B Biol Sci*. 271: 565–574.M.
18. Breitbart M., Felts B., Kelley S., Mahaffy J. M., Nulton J., Salamon P., Rohwer F. (2004). Diversity and population structure of a near-shore marine- sediment viral community. *Proc Biol Sci*. 271:565-574.
19. Breitbart M., Hewson I., Felts B., Mahaffy J. M., Nulton J. (2003).

Metagenomic analyses of an uncultured viral community from human feces. *J Bacteriol.* 85: 6220–6223.

20. Breitbart M., Salamon P., Andresen B., Mahaffy J.M., Segall A.M., Mead D., et al. (2002). Genomic analysis of uncultured marine viral communities. *Proc Natl Acad Sci USA*. 99:14250-14255.
21. Brussaard C. P. D. (2004). Optimization of procedures for counting viruses by flow cytometry. *Appl Environ Microbiol.* 70: 1506–1513.
22. Brussaard C. *Ennumeration of bacteriophages using flow cytometry.* In: Clokie MRJ, Kropinski A, eds. Bacteriophages Methods and Protocols Volume 1: Isoation, Charaterisation and Interactions. New York, NY: Springer 2009; 97-112.
23. Canchaya C., Fournous G., Brussow H. (2004). The impact of prophages on bacterial chromosomes. *Mol Microbiol.* 53:9–18.
24. Canchaya C., Proux C., Fournous G., Bruttin A., Brussow H. (2003). Prophage genomics. Microbiol Mol Biol Rev. 67:238-276.
25. Cann A. J., Fandrich S. E., Heaphy S. (2005). Analysis of the virus population present in equine faeces indicates the presence of hundreds of uncharacterized virus genomes. *Virus Genes.* 30:151–156.
26. Casjens S. (2003). Prophages and bacterial genomics: what have we learned so far? *Mol Microbiol.* 49:277–300.
27. Chibani-Chennoufi S., Bruttin A., Dillmann M-L., Brüssow H. (2004). Phage-Host Interaction: an Ecological Perspective. *J Bacteriol.* 186(12):3677-3686.
28. Clark J. R., March J. B. (2006). Bacteriophages and biotechnology: vaccines, gene therapy and antibacterials. *Trends Biotechnol.* 24(5):212-218.
29. Clokie M. R. J., Kropinski A. M. Eds., Bacteriophages Methods and Protocols, Volume 1: Isolation, Characterization, and Interactions, Humana Press, New York, NY, USA, 2009.
30. Connerton P. L., Timms A. R., Connerton I. F. (2011). Campylobacter Bacteriophages and Bacteriophage Therapy. *J Appl Microbiol.* 111:255-265.
31. Culley A. I., Lang A. S., Suttle C. A. (2006). Metagenomic analysis of coastal RNA virus communities. *Science.* 312: 1795–1798.

32. DeMarini D. M., Lawrence B. K. (1992). Prophage induction by DNA topoisomerase II poisons and reactive-oxygen species: role of DNA breaks. *Mutat Res*. 267(1):1–17.

33. Donald H. (1956). The Effects of Various Physical and Chemical Agents on a Staphylococcus Bacteriophage. *Butler University Botanical Studies*. 13, Article 8.

34. Duerkop B. A., Palmer K. L., Horsburgh M. J. Enterococcal Bacteriophages and Genome Defense. 2014 Feb 11. In: Gilmore MS, Clewell DB, Ike Y, et al., editors. Enterococci: From Commensals to Leading Causes of Drug Resistant Infection [Internet]. Boston: Massachusetts Eye and Ear Infirmary; 2014.

35. Edlin G., Lin L., Kudrna R. (1975). Lambda lysogens of *E. coli*reproduce more rapidly than non-lysogens. *Nature*. 255:735-737.

36. Edwards R.A., Rohwer F. (2005). Viral metagenomics. *Nat Rev Microbiol*. 3:504–510.

37. EFSA (European Food Safety Authority). (2009). Scientific Opinion of the Panel on Biological Hazards on a Request from European Commission on the Use and Mode of Action of Bacteriophages in Food Production. *The EFSA Journal* 1076: 1-26.

38. Filee J., Tetart F., Suttle C. A, Krisch H. M. (2005). Marine T4-type bacteriophages, a ubiquitous component of the dark matter of the biosphere. *Proc Natl Acad Sci USA*. 102:12471-12476.

39. Hagens S., Loessner M. J. (2007). Application of Bacteriophages for Detection and Control of Foodborne Pathogens. *Appl. Microbiol. Biotec*. 76 (3): 513-9.

40. Hagens S., Loessner M. J. (2010). Bacteriophage for Biocontrol of Foodborne Pathogens: Calculations and Considerations. *Curr. Pharmac. Biotec*. 11: 58-68.

41. Handelsman J. (2004). Metagenomics: application of genomics to uncultured microorganisms. *Microbiol Mol Biol Rev*. 68: 669–685.

42. Hanlon G.W. (2007). Bacteriophages: an appraisal of their role in the treatment of bacterial infections.

43. *Int J Antimicrob Agents*. 30: 118–128.

44. Hara S., Terauchi K., Koike I. (1991). Abundance of viruses in

marine waters: assessment by epifluorescence and transmission electron microscopy. *Appl Environ Microbiol.* 57, 2731–2734.

45. Hatfull G.F. (2008). Bacteriophage genomics. *Curr Opin Microbiol.* 11:447–453.
46. Hendrix R. W., Smith M. C., Burns, R. N., Ford M. E., Hatfull G.F. (1999). Evolutionary relationships among diverse bacteriophages and prophages: all the world's a phage. *Proc Natl Acad Sci U S A.* 96(5):2192–2197.
47. Hendrix R.W., Smith M.C., Burns R.N., Ford M.E., Hatfull G.F. (1996). Evolutionary relationships among diverse bacteriophages and prophages: all the world's a phage. *Proc Natl Acad Sci U S A.* 96(5):2192–2197.
48. Hennes K. P., Suttle, C. A. (1995). Direct counts of viruses in natural waters and laboratory cultures by epifluorescence microscopy. *Limnol. Oceanogr.* 40: 1050–1055.
49. Inal J.M. (2003). Phage therapy: a reappraisal of bacteriophages as antibiotics. *Arch Immunol Ther Exp.* 51(4):237-244.
50. Jones J. B., Jackson L. E., Balogh B., Obradovic A., Iriarte F. B., Momol M. T. (2008). Bacteriophages for plant disease control. *Annu Rev Phytopathol.* 45:245–262.
51. Kristensen D. M., Mushegian A. R., Dolja V. V., Koonin E. V. (2010). New dimensions of the virus world discovered through metagenomics. *Trends Microbiol.* 18:11–19.
52. Krueger A. P., Brown B. B., Scribner E. J. (1941). The effect of sonic vibrations on phage, precursor and the bacterial substrate. *J Gen Physio.* 24(6):691-698.
53. Kutter E., De Vos D., Gvasalia G., Alavidze Z., Gogokhia L., Kuhl S., Abedon, S. (2010). Phage Therapy in Clinical Practice: Treatment of Human Infections. *Curr. Pharmac. Biotec.* 11(1): 69-86.
54. Kwan T., Liu J., Dubow M., Gros P., Pelletier J. (2005). The complete genomes and proteomes of 27 Staphylococcus aureus bacteriophages. *Proc Natl Acad Sci U S A.* 102:5174–5179.
55. Kwan T., Liu J., Dubow M., Gros P., Pelletier J. (2006). Comparative genomic analysis of 18 Pseudomonas aeruginosa bacteriophages. *J Bacteriol.* 188:1184–1187.

56. Labrie S. J., Samson J. E., Moineau S. (2010). Bacteriophage resistance mechanisms. *Nat Rev Microbiol.* 8:317-27

57. Lang A. S., Beatty J. T. (2007). Importance of widespread gene transfer agent genes in alpha-proteobacteria. *Trends Microbiol.* 15:54–62.

58. Lavigne R., Darius P., Summer E. J, Seto D., Mahadevan P. (2009) Classification of *Myoviridae* bacteriophages using protein sequence similarity. *BMC Microbiol.* 9: 224.

59. Leplae R., Lima-Mendez G., Toussaint A. (2010). ACLAME: a Classification of Mobile genetic Elements, update 2010. *Nucleic Acids Res.* 38: D57–D61.

60. Leroy M., Prigent M., Dutertre M., Confalonieri F., Dubow M. (2008). Bacteriophage morphotype and genome diversity in Seine River sediment. *Freshw Biol.* 53:1176-85

61. Lin L., Bitner R., Edlin G. (1977). Increased reproductive fitness of*Escherichia coli* lambda lysogens. *J Virol.* 21:554-559.

62. Little J. W. *Loysogeny, Prophage Induction, and Lysogenic Conversion.* In: Waldor MK, Friedman DI, Adhya S, editors. Phages Their Role in Bacterial Pathogenesis and Biotechnology. Washington DC: ASM Press; 2005. pp. 37–54

63. Little J. W., Mount D. W. (1982). The SOS regulatory system of Escherichia coli. *Cell.* 29(1):11–22.

64. Mahy B. W. J., van Regenmortel M. H. V. (2010). Desk Encyclopedia of Plant and Fungal Virology. Academic Press, Oxford, 613.

65. Mai V., Ukhanova M., Visone L., Abuladze T., Sulakvelidze A. (2010). Bacteriophage Administration Reduces the Concentration of *Listeria monocytogenes* in the Gastrointestinal Tract and Its translocation to Spleen and Liver in Experimentally Infected Mice. *Int J Microbiol.* 1-6.

66. Mandilara G. D., Smeti E. M., Mavridou A. T., Lambiri M. P., Vatopoulos A. C., Rigas F. P. (2006). Correlation between Bacterial Indicators and Bacteriophages in Sewage and Sludge. *FEMS Microbiol. Lett.* 263: 119-126.

67. Marie D., Brussaard C. P. D., Thyrhaug R., Bratbak G., Vaulot D. (1999). Enumeration of marine viruses in culture and natural

samples by flow cytometry. *Appl Environ Microbiol.* 65: 45–52.

68. Marston M. F., Sallee, J. L. (2003). Genetic diversity and temporal variation in the cyanophage community infecting marine Synechococcus species in Rhode Island's coastal waters. *Appl Environ Microbiol.* 69: 4639-4647.

69. Millard A.D. *Isolation of Cyanophages from Aquatic Environments.* In: Clokie MRJ, Kropinski A, eds. Bacteriophages Methods and Protocols Volume 1: Isolation, Characterisation and Interactions. New York, NY: Springer 2009; 33-42.

70. Nakayama K., Takashima K., Ishihara H., Shinomiya T., Kageyama M., Kanaya S., et al. (2000). The R-type pyocin of Pseudomonas aeruginosa is related to P2 phage, and the F-type is related to lambda phage. *Mol Microbiol.* 38:213–231.

71. Novick R.P., Christie G.E., Penades J.R. (2010). The phage-related chromosomal islands of Gram-positive bacteria. *Nat Rev Microbiol.* 8:541–551.

72. Orlova E. V. (2009). How viruses infect bacteria? *EMBO J.* 28(7): 797–798.

73. Ortmann A. C., Suttle C. A. *Determination of Virus Abundance by Epifluorescence Microscopy.* In: Clokie MRJ, Kropinski A, eds. Bacteriophages Methods and Protocols Volume 1: Isolation,

74. Paul J. H., Sullivan M. B., Segall A. M., Rohwer F. (2002). Marine phage genomics. *Comp Biochem Physiol B Biochem Mol Biol.*133: 463-476.

75. Paul J.H., Sullivan M.B., Segall A.M., Rohwer F. (2002). Marine phage genomics. *Comp Biochem Physiol B Biochem Mol Biol.* 133: 463-476.

76. Pedulla M. L., Ford M. E., Houtz J. M., Karthikeyan T., Wadsworth C., Lewis J.A. (2003). Origins of highly mosaic mycobacteriophage genomes. *Cell.* 113(2):171–182.

77. Poglazov B. F., Tikhonenko A.S. (1963) Study of T2 phage contractile protein properties. *Biokhimiya* 28: 888-894.

78. Pope W. H., Jacobs-Sera D., Russell, D. A., Peebles C. L., Al-Atrache Z., Alcoser T. A., Alexander L. M., Alfano M.., Alford S. T., Amy N. E. (2011). Expanding the Diversity of Mycobacteriophages: Insights

into Genome Architecture and Evolution. *PLoS ONE*. 6:e16329

79. Prestel E., Salamitou S., DuBow M. S. (2008). An examination of the bacteriophages and bacteria of the Namib desert. *J Microbiol.* 46(4):364–372.
80. Prigent M., Leroy M., Confalonieri F., Dutertre M., DuBow M.S. (2005). A diversity of bacteriophage forms and genomes can be isolated from the surface sands of the Sahara Desert. *Extremophiles.* 9(4):289–296.
81. Rappe M. S., Giovannoni S. J. (2003). The uncultured microbial majority. *Annu Rev Microbiol.* 57:369- 394.
82. Reyes A., Haynes M., Hanson N., Angly F. E., Heath A. C., Rohwer F., Gordon J. I. (2010). Viruses in the faecal microbiota of monozygotic twins and their mothers. *Nature.* 466(7304):334-338
83. Rodriguez-Brito B., Li L., Wegley L., Furlan M., Angly F., Breitbart M., Buchanan J., Desnues C., Dinsdale E., Edwards R. (2010). Viral and microbial community dynamics in four aquatic environments. *Isme J.* 4:739–751.
84. Rooks D. J., Yan Y. X., McDonald J. E., Woodward M. J., McCarthy A. J., Allison H.E. (2010). Development and validation of a qPCR-based method for quantifying Shiga toxin-encoding and other lambdoid bacteriophages. *Environ Microbiol.*12:1194- 1204.
85. Sails A. D., Wareing D. R., Bolton F. J., Fox A. J., Curry, A. (1998). Characterisation of 16 *Campulobacter jejuni* and *Campylobacter coli* Typing Bacteriophages. *J Med Microbiol.* 47 (2): 123-8.
86. Sandaa R. A., Clokie M., Mann N. H. (2008). Photosynthetic genes in viral populations with a large genomic size range from Norwegian coastal waters. *FEMS Microbiol Ecol.* 63:2-11.
87. Sanger F., Air G. M., Barrell B. G., Brown N. L., Coulson A. R., Fiddes J. C., et al. (1997). Nucleotide sequence of bacteriophage φX174 DNA. *Nature.* 265: 687 -695.
88. Smith H. W. (1953). The Effect of Physical and Chemical Changes on the Liberation of Phage Particles by Lysogenic Strains of Salmonella. *J Gen Microbiol.* 8: 116-134.
89. Srinivasiah S., Bhavsar J., Thapar K., Liles M., Schoenfeld T., Wommack K. E. (2008). Phages across the biosphere: contrasts of viruses in soil and aquatic environments. *Res Microbiol.* 159(5):349–

357.

90. Sturino J. M., Klaenhammer T. R. (2006). Engineered bacteriophage-defence systems in bioprocessing. *Nat Rev Microbiol.* 4:395-404.
91. Sullivan M. B., Coleman M. L., Quinlivan V., Rosenkrantz J. E., DeFrancesco A. S., Tan G. (2008). Portal protein diversity and phage ecology. *Environ Microbiol.* 10:2810-2823.
92. Sutter M., Boehringer D., Gutmann S., Gunther S., Prangishvili D., Loessner M. J., et al. (2008). Structural basis of enzyme encapsulation into a bacterial nanocompartment. *Nat Struct Mol Biol.* 15:939–947.
93. Suttle C. A. in *Handbook of Methods in Aquatic Microbial Ecology* (eds Kemp, P. F., Sherr, B. F., Sherr, E. B. & Cole, J. J.) 121–134 (Lewis, Boca Raton, 1993).
94. Suttle C. A. (2005). Viruses in the sea. *Nature.* 437: 356–361.
95. Svircev A. M., Lehman S. M., Sholberg P., Roach D., Castle A. J. *Phage biopesticides and soil bacteria: multilayered and complex interactions* in Biocommunication in Soil Microorganisms, G. Witzany, Ed., pp. 215–235, Springer, Berlin, Germany, 2011.
96. Tomaru Y., Nagasaki K. (2007). Flow cytometric detection and enumeration of DNA and RNA viruses infecting marine eukaryotic microalgae. *J Oceanogr.* 63: 215–221.
97. USDA (United State Department of Agriculture). 2006. "GRAS Notice 000218, GRAS Notification of LISTEXTM P100 Bacteriophage." Accessed July 10, 2013.
98. Vale P. F., Little T. J. (2010). CRISPR-mediated phage resistance and the ghost of coevolution past. *Proc Biol Sci.* 277:2097-2103.
99. Wang K., Chen F. (2004). Genetic diversity and population dynamics of cyanophage communities in the Chesapeake Bay. *Aquat Microb Ecol.* 34:105-116.
100. Weinbauer M. G., Rassoulzadegan F. (2004). Are viruses driving microbial diversification and diversity? *Environ Microbiol.* 6:1–11.
101. Wen K., Ortmann A. C., Suttle C. A. (2004). Accurate estimation of viral abundance by epifluorescence microscopy. *Appl Environ Microbiol.* 70, 3862–3867.
102. Wilhelm S. W., Suttle C. A. (1999). Viruses and nutrient cycles in

the sea: Viruses play critical roles in the structure and function of aquatic food webs. *Bioscience*. 49:781–788.

103. Willner D., Furlan M., Haynes M., Schmieder R., Angly F. E., Silva J. (2009). Metagenomic analysis of respiratory tract DNA viral communities in cystic fibrosis and non-cystic fibrosis individuals. *PLoS One* 4:e7370

104. Willner D., Thurber R. V., Rohwer F. (2009). Metagenomic signatures of 86 microbial and viral metagenomes. *Environ Microbiol*. 11:1752–1766.

105. Wilson W. H., Carr N. G., Mann N. H. (1996). The effect of phosphate status on the kinetics of cyanophage infection in the oceanic cyanobacterium Synechococcus sp. WH7803. *J Phycol*. 32:506–516.

106. Winkler U., Johns H., Kellenberger E. (1962). Comparative study of some properties of bacteriophage T4D irradiated, with monochromatic ultraviolet light. *Viorolgy*. 18:343.

107. Wommack K. E., Colwell R. R. (2000). Virioplankton: viruses in aquatic ecosystems. *Microbiol Mol Biol Rev*. 64(1):69–114.

108. Zhong Y., Chen F., Wilhelm S. W., Poorvin L., Hodson R.E. (2002). Phylogenetic diversity of marine cyanophage isolates and natural virus communities as revealed by sequences of viral capsid assembly protein gene g20. *Appl Environ Microbiol*. 68:1576-84.

CHAPTER

2

BACTERIOPHAGES AND BACTERIAL PLANT DISEASES

Colin Buttimer[1], Olivia McAuliffe[2], R. P. Ross[3], Colin Hill[3], Jim O'Mahony[1] and Aidan Coffey[1]

[1]Department of Biological Sciences, Cork Institute of Technology, Cork, Ireland

[2]Moorepark Food Research Centre, Teagasc, Fermoy, Ireland

[3]Alimentary Pharmabiotic Centre, University College, Cork, Ireland

Keywords: bacteriophages, plant diseases, biocontrol, biopesticides, phytopathogens

Losses in crop yields due to disease need to be reduced in order to meet increasing global food demands associated with growth in the human population. There is a well-recognized need to develop new environmentally friendly control strategies to combat bacterial crop disease. Current control measures involving the use of traditional chemicals or antibiotics are losing their efficacy due to the natural development of bacterial resistance to these agents. In addition, there is an increasing awareness that their use is environmentally unfriendly. Bacteriophages, the viruses of bacteria, have received increased research interest in recent years as a realistic environmentally friendly means of controlling bacterial diseases. Their use presents a viable control measure for a number of destructive bacterial crop diseases, with some phage-based products already becoming available on the market. Phage biocontrol possesses advantages over chemical controls in that tailor-made phage cocktails can be adapted to target specific disease-causing bacteria. Unlike chemical control measures, phage mixtures can be easily adapted for bacterial resistance which may develop over time. In this review, we will examine the progress and challenges for phage-based disease biocontrol in food crops.

IMPORTANCE OF CROP DISEASES

The human population is expected to reach 9.6 billion by 2050 and this will result in increased demands for food. It has been estimated that the global food supply may need to grow by as much as 70% in order to meet these demands (UN, 2013). For such growth, it has been predicted that crop supply may have to increase as much as 80–110% (Ray et al., 2013). To achieve these yields, the impact of crop disease has to be reduced. It has been estimated that at least 10% of global food production is lost to plant diseases (Strange and Scott, 2005). The major pathogens of plants are parasitic plants, oomycetes, nematodes, viruses, fungi and bacteria. Among the latter, there are over 200 plant pathogenic bacterial species (Considine and Considine, 1995). Those considered to be the most important belonging to the genera of *Pseudomonas, Ralstonia, Agrobacterium, Xanthomonas, Erwinia, Xylella, Pectobacterium*, and *Dickeya*(Mansfield et al., 2012).

BACTERIOPHAGES, THEIR LIFE CYCLES AND THEIR MORPHOLOGY

Bacteriophages (phages) are the most abundant biological entity in the biosphere with an estimated number of 10^{31}, as total prokaryotic cell numbers are understood to be around 10^{30} in the biosphere and phage numbers are believed to be at least 10 times greater than this value (Whitman et al., 1998; Wommack and Colwell, 2000). Phages are specific viruses of bacteria that subvert the metabolism of their bacterial hosts in order to replicate. Of the phages that have been identified, the majority belong to the tailed phages; and these form the Taxonomic Order: *Caudovirales* (Ackermann, 2007). These phages possess icosahedral heads containing genomes comprised of double stranded DNA. The order *Caudovirales* is made up of three phage families; *Myoviridae* which have rigid contractile tails, *Podoviridae* with short, non-contractile tails and *Siphoviridae* with long flexible tails. Phages belonging to other families have highly variable morphologies with genomes of varying nucleic acid composition.

HISTORY OF BACTERIOPHAGES AND THEIR USE AS ANTIBACTERIAL AGENTS TOWARD PLANT DISEASES

The discovery of bacteriophages is credited to Frederick Twort (Twort, 1915) and Felix d'Herelle (d'Herelle, 1917). Similar findings of antibacterial agents that hinted on the existence of phage had been made prior to that of Twort and d'Herelle (Abedon et al., 2011). However, they were the first to suggest this phenomenon as being viral in origin. The potential of phages as antibacterial agents was quickly recognized, with d'Herelle in 1919 demonstrating the capability of his phage preparations to treat dysentery patients in the Hôpital des Enfants-Malades in Paris (Wilkinson, 2001). Following this work, many early studies and attempts were made to use phages to treat staphylococcal infections, cholera and bubonic plaque of humans (Sulakvelidze et al., 2001). This pre-antibiotic era approach became known as bacteriophage therapy. Studies were also initiated with the aim of using phages to control plant diseases. Mallmann and Hemstreet (1924) showed that the filtrate of decomposing cabbage could be used to inhibit the "cabbage-rot organism" *Xanthomonas campestris pv. campestris*. In 1925, Kotila and Coons demonstrated with bioassays that they could use phage to prevent soft rot by *Pectobacterium atrosepticum* and *Pectobacterium carotovorum* ssp *carotovorum* on slices of potato tuber and carrot, respectively (Coons and Kotila, 1925; Kotila and Coons, 1925). The first field trials were also done by Thomas (1935), who showed that he could reduce the incidence of Stewart's wilt disease by treating seeds with phage against the phytopathogen *Pantoea stewartii* from 18% (untreated) to 1.5% (treated). However, this type of research became neglected as understanding of the nature of phage was poor at the time, and data on their efficacy was limited (Okabe and Goto, 1963).

BACTERIOPHAGE TYPES USED FOR THERAPY/ BIOCONTROL

From a terminology perspective, the term bacteriophage therapy is usually reserved for human and animal applications. For plants the term bacteriophage biocontrol is more often used. In recent years, several studies have been published on phage biocontrol on a number of important bacterial plant pathogens, with many very promising results (see Table 1). The main deciding factor whether a phage is applicable for biocontrol (and also therapy in humans or animals) is whether a phage is exclusively lytic (virulent) or instead temperate in nature. Virulent phages are those which conduct infections that ultimately result in lysis of their host bacterium with the release of progeny phage particles. Temperate phages can follow the lytic route of infection but also follow the route of lysogeny, where the phage genome integrates into the bacterial chromosome or persists as a plasmid. In this form the phage is known as a prophage (Łobocka et al., 2004). With this strategy, the phage genome replicates as part of the bacterial genome of its host until a trigger switches it into the lytic cycle. These triggers can be chemical or physical (UV light or heat) in nature (Brunner and Pootjes, 1969; Müller et al., 2012). It is interesting to note that certain plant extracts can also trigger these events (Sato, 1983). Often, prophage DNA can increase the fitness of the bacterial host due to genes present on prophage genome. For example, in the case of plant pathogens, the *P. atrosepticum* prophages ECA41 and ECA29 both improve the motility of the bacterial host (Evans et al., 2010). Prophages may also harbor genes for toxins, e.g., shiga, cholera, and diphtheria toxins (Abedon and Lejeune, 2005). Another concern with these phages is the spread of virulence genes by transduction, where these phages can excise themselves from their host genomes incorporating host DNA into their own genomes facilitating horizontal transfer of genetic material among bacteria (Griffiths et al., 2000). Also, some lytic bacteriophages are capable of transduction, where they accidently pack bacterial DNA into their own capsid heads during the later stages of lytic cycle (Klumpp et al., 2008). There is also a third mechanism of phage-host interaction identified in filamentous phages (*Inovirus* family). Here, phages form a non-lethal chronic infection with continuous production of progeny phages. However, suitability of these phages for biocontrol is questionable as their infection can have varying effects on host virulence, as shown with phytopathogen *Ralstonia solanacearum* with its phage ϕRSS1 causing increased virulence (Yamada, 2013), although it has been shown possible to isolate virulent filamentous phage (Kuo et al., 1994). Another undesirable property in a phage intended for biocontrol is the ability to bring about superinfection exclusion to its host during infection. This prevents secondary infection of the host by another phage (Lu and Henning, 1994).

Table 1: Summary of bacteriophage biocontrol experiments which have been conducted since the year 2000 to the present.

Pathogen	Host	Disease	Information	Reference
Pectobacterium carotovorum ssp. *carotovorum*, *Pectobacterium wasabiae*, *Dickeya solani*	Potato	Soft rot	Bioassays with phage ΦPD10.3 and ΦPD23.1 could reduce severity of soft rot of tubers by 80% on potato slices and 95% with whole tubers from a mixed pathogen infection.	Czajkowski et al., 2015
Dickeya solani	Potato	Soft rot/Blackleg	Phage vB_DsoM_LIMEstone1 and vB_DsoM_LIMEstone2 reduced soft rot of inoculated tubers in bioassays and in field trials which produced a potato crop with higher yields	Adriaenssens et al., 2012
Dickeya solani	Potato	Soft rot	Bioassays with phage ΦD1, ΦD2, ΦD3, ΦD4, ΦD5, ΦD7, ΦD9, ΦD10, ΦD11 could reduce incidence of soft rot by up to 30–70% on co-inoculated potato slices with pathogen and phage.	Czajkowski et al., 2014
Streptomyces scabies	Potato	Common scab	Seed tubers treated with phage ΦAS1 resulted in producing tuber progeny with reduced levels of surface lesion of scab (1.2%) compared with tubers harvested from non -treated seed tubers (23%)	McKenna et al., 2001
Ralstonia solanacearum	Tomato	Bacterial wilt	Tomato plants treated with phage ΦRSL1 showed no symptoms of bacterial wilt during the experimental period; whereas all untreated plants showed wilting 18 days post infection.	Fujiwara et al., 2011
Ralstonia solanacearum	Tomato	Bacteria wilt	Simultaneous treatment of phage PE204 with *R. solanacearum* of the rhizosphere of tomato completely inhibited bacterial wilt. However, pre-treatment with phage before the inoculation of pathogen was not effective with control of bacterial wilt, whereas post treatment of PE204 delayed disease development.	Bae et al., 2012
Xanthomonas campestris pv. vesicatoria	Tomato	Bacterial spot	Greenhouse experiments with formulated phage cocktails could reduce disease severity with formulated phage cocktails providing better protection in comparison to unformulated. A similar effect was found in three consecutive field trials.	Balogh et al., 2003
Xanthomonas campestris pv vesicatoria	Tomato	Bacterial spot	In field experiments phage treatment was comparable to disease control with copper-mancozeb. Combination of phage and plant activator (ASM) resulted in enhanced control.	Obradovic et al., 2004
Xylella fastidiosa	Grapevines	Pierce's Disease	*X. fastidiosa* levels in grapevines were significantly reduced on pre and post inoculation of a four phage (*Sano, Salvo, Prado and Paz*) cocktail. Pierce disease symptoms could be stopped using phage treatment post infection as well as applying phage prophylactically to grapevines.	Das et al., 2015

Xanthomonas axonopodis pv. *allii*	Onion	*Xanthomonas* leaf blight of onion	Field trial showed that weekly and biweekly applications of phage could reduce disease severity, a result which was comparable to treatments of weekly applications of copper-mancozeb.	Lang et al., 2007
Pectobacterium carotovorum ssp *carotovorum*	Lettuce	Soft rot	Green house trials showed that phage PP1 could significantly reduce disease development on lettuce plants.	Lim et al., 2013
Streptomyces scabies	Radish	Common scab	Phages Stsc1 and Stsc3 could prevent disease development by treating radish seedlings. Non-treated radishes had 30% less weight than negative control, with phage treated radishes having masses similar to negative control.	Goyer, 2005
Xanthomonas axonopodis pv. *citri*	Grapefruit	Asiatic citrus canker	Five greenhouse experiments utilizing phage treatment could reduce disease severity by 59%. However, using a skim milk formulation of phage did not have increased disease control. Phage treatment was also capable of reducing disease occurrence in a citrus nursery. Control was less effective than copper-mancozeb. Combination did not give increased disease control.	Balogh et al., 2008
Xanthomonas axonopodis pv *citrumelo*	Orange	Citrus bacterial spot	Phage treatments reduced citrus spot occurrence by 35 and 48% in two trails in commercial citrus nursery. Control was equal or less effective than copper-mancozeb. Combination did not give increased disease control	Balogh et al., 2008
Pseudomonas syringae pv. *porri*	Leek	Bacterial blight	Specific bio-assays demonstrated the *in planta* efficacy of phages vB_PsyM_KIL1, vB_PsyM_KIL2, vB_PsyM_KIL3, and vB_PsyM_KIL3b. However, phage cocktail of six phages (vB_PsyM_KIL1, vB_PsyM_KIL2, vB_PsyM_KIL3, vB_PsyM_KIL4, and vB_PsyM_KIL5 and vB_PsyM_KIL3b), were tested with two parallel field trial experiments in three locations which showed variable results. In one trial, symptom development was attenuated.	Rombouts et al., 2016
Pseudomonas tolaasii	Mushrooms	Brown blotch Disease	Surface of mushrooms were inoculated with pathogen. The formation of blotches was completely blocked by co-incubation of phages with pathogen.	Kim et al., 2011
Erwinia amylovora	Pear, apple trees	Fire blight	Phages ΦEa1337-26 and ΦEa 2345 reduced infection of detached pear tree blossoms by 84 and 96%, respectively, with *Pantoea agglomerans* as a carrier. Also, infection of potted apple tree blossoms could be reduced by 54% with phage ΦEa1337-26 and *P. agglomerans*. Control was comparable to streptomycin.	Boulé et al., 2011

Ideally a phage for biocontrol applications should be exclusively lytic and possess a host range which allows productive infection on all strains of the pathogen genus/species being targeted. Also, current opinion is that phages should be able to lyse the host quickly while producing high numbers of progeny phage and diffuse easily though the environment to which they are being applied. However, there was a report of a phage (ϕRSL1) of the phytopathogen *R. solanacearum* which was described as not highly lytic but still exhibited great biocontrol effect. The current standing theory of this phage's disease prevention ability is that it is capable of co-existing without complete removal of its host from the soil surrounding crop roots, forming an equilibrium of infection that maintains the phage's population but yet suppresses bacteria pathogenicity (Fujiwara et al., 2011).

While a given phage's infection properties may appear to have great potential with *in vitro* studies, this does not necessarily translate into biocontrol potential in the field. Balogh (2006) showed in a study of three phages of *X. citri* pv *citri* exhibiting lytic activity in overlay plate assays that two of these phages were unable to lyse their host bacterium on grapefruit leafs, and indeed were later shown to be ineffective for the suppression of citrus canker in greenhouse trials. Attention should also be paid to the receptors that a given phage recognizes on a bacterial target. This can aid in the creation of phage mixtures with a reduced likelihood of host resistance (Frampton et al., 2014), and as such can lead to the development of phage combinations where individual members work in synergy to eliminate the target bacterium (Born et al., 2011).

ADVANTAGES OF PHAGE BIOCONTROL OVER OTHER STRATEGIES

Unlike chemical biocides, phages occur naturally in the environment and humans are thus exposed to them on a daily basis without any harm. After application, their numbers increase if their target bacterial host species are accessible to them. However, they tend to persist in high numbers in any environment only long as the host is present (Iriarte et al., 2012). Thus, phages are unlike copper-based pesticides which can potentially accumulate in the soil (Hirst et al., 1961; Pietrzak and McPhail, 2004). Phages generally have a narrow host range, typically being limited to stains within a particular species of bacteria. This can allow the creation of phage mixtures which can target bacterial species within a given genus of bacteria only. This could be a specific bacterial phytopathogen or it could be a particular bacterium in a microbial community whose suppression could help improve crop growth. Basit et al. (1992) for example, isolated phage which was unable to infect a desired strain of *Bradyrhizobium japonicum* which could aid soy bean crop growth due its nitrogen fixation properties, but could inhibit competing bacteria which did not possess this feature, thus allowing enhanced nitrogen fixation to occur.

Biofilm formation is an important factor in the virulence of phytopathogens such a *E. amylovora* (Koczan et al., 2011; Li and Wang, 2014). It is an attribute which has been shown to be involved in bacterial phytopathogen resistance to copper bactericides (Rodrigues et al., 2008). Phages have evolved to overcome this biofilm barrier through the use of depolymerase enzymes on their capsids but can also be released on host lysis, which allows them to degrade biofilm material, allowing the phage anti-receptor to gain access to the receptors on the surface of their host bacterium (Born et al., 2014). There is a growing demand by consumers for food produce that is free from chemicals biocides and preservatives. This has resulted in the restricted use of chemicals to produce "organic label" crops. The requirements of such food require the absence of chemical residues in crop production and processing (Lohr, 2001). Since phages are naturally occurring in the environment, they can be registered as biopesticides, making them suitable for more consumer-friendly organic farming (OmniLytics, 2006).

POTENTIAL ISSUES CONCERNING THE USE OF PHAGE IN BIOCONTROL

The main limitation for the application of phages in biocontrol in most settings is bacterial host-range. While this can be an advantage in certain circumstances, developing a phage-biocide that eliminates every member of a particular bacterial genus or species can be a challenge. Frequently the development of phage mixtures (cocktails) overcomes this disadvantage. Occasionally (but nevertheless, rarely) a phage is isolated which has an unexpectedly broad host-range. One example of this is a phage isolated from sewage and shown to target *Pectobacterium* and also enteric bacteria associated with humans (Pirhonen and Palva, 1988). Thus, careful attention should be given to ensure full understanding of likely host-range of a phage to avoid inefficacy or indeed to avoid the elimination of non-target potentially beneficial bacteria. In the latter context, instances of phage infecting beneficial bacteria resulting in reduced crop yield have been reported (Basit et al., 1992; Ahmad and Morgan, 1994).

It is believed that phages do not directly interact with plants. However, a number of phage-like genes have been identified in wheat, corn and *Arabidopsis* cress (Hedtke et al., 1997; Chang et al., 1999; Ikeda and Gray, 1999) which would suggest incorporation of phage DNA into the genomes of these crops and thus a possible a role in their evolution.

ADVANTAGES OF PHAGES IN THE CONTEXT OF HOST RESISTANCE

Like antibiotics and copper sprays, for which resistance has been reported, there is also the possibility of bacteria becoming resistant to phage infection following constant exposure. However, unlike chemicals, phages are biological entities which can evolve and overcome these biological alterations in their hosts. There has always been a constant race between phage and bacteria in nature. This is indicated by the fact that 10–20% of bacterial populations in certain habitats are lysed daily because of phage infection (Suttle, 1994). In the context of phage resistance, Qiao et al. (2010) found that *Pseudomonas syringae* phage phi2954 was dependent on a host protein glutaredoxin 3 for successful infection. Mutant host strains without this protein were shown to be resistant to the phage. Nevertheless, these authors showed it was possible to isolate mutants of the phage that had become independent of this host protein for infection and this observation has been developed and employed in certain phages aimed at biocontrol. Flaherty et al. (2001) also showed that phages could evolve to overcome phage resistance in target bacteria and these were referred to as H-mutants. This allowed the development of phages with broader host ranges.

In addition to simple mutation-based phage resistance, bacterial phytopathogens can also possess other more complex resistance mechanisms such as the altruistic abortive infection (Abi) systems which give a bacterial host population immunity against a phage by causing phage-infected cells to commit suicide in order to prevent phage reproduction (Parma et al., 1992). While a number of these systems have been identified in *Lactococcus* starter culture strains found in dairy fermentations (Coffey and Ross, 2002; Chopin et al., 2005), recently such a system was identified in the phytopathogen *P. atrosepticum* and was termed ToxIN. This was characterized as a plasmid encoded Type III protein-RNA toxin-antitoxin system. The toxic protein ToxN is bound to RNA antitoxin ToxI in its inactive form. However, when phage infection occured, ToxI RNA antitoxin became unbound from ToxN causing death of the bacterial host cell (Fineran et al., 2009). Indeed, Blower et al. (2012) also showed using phage phiTE, that the phage was capable of creating mutants that could overcome this system by producing a pseudo ToxI RNA antitoxin preventing ToxN toxic activity.

Another mode of phage resistance is CRISPR/Cas systems, which are used by bacteria as well as archaea to form an immunity to protect from infection by foreign DNA such as phage. These systems are comprised of clustered regularly interspaced short palindromic repeat (CRISPR) arrays and CRISPR associated (Cas) proteins. In a recent study of 1,724 bacterial and archaeal genomes it was found that these systems were present in 10% of studied genomes. Previous studies had estimated CRISPR/Cas prevalence values of 40 and 80% of studied bacteria and archaeal genomes, respectively (Burstein et al., 2016). These have been detected in phytopathogens such as *P. atrosepticum* (Przybilski et al., 2011), *E. amylovora* (Rezzonico et al., 2011), and *Xantomonas oryzae* (Semenova et al., 2009). CRISPR arrays are comprised of short stretches of DNA (termed spacers), which are transcribed into short RNAs which interact with Cas proteins to detect and cut foreign DNA that match the sequence of the spacer (protospacer). Spacer sequences are acquired during exposure to foreign DNA in phage or plasmids, and thus they provide a genetic immunity from invasion by foreign DNA due to previous encounters (Marraffini and Sontheimer, 2008). However, it is also possible for phage to evolve to overcome these systems. Indeed, Semenova et al. (2009) detected a spacer in *X. oryzae* which matched a protospacer of phage Xop411. However, the phage was still able to infect this bacterium, due a mutation having occured in the protospacer sequence.

Bacteria developing resistance against phage infection is not necessarily a negative development in the context of phage biocontrol. Phage-resistance mutations in bacteria frequently are accompanied by a fitness cost, one example being a reduction in virulence, resulting in reduced disease severity. This results from the fact that molecules involved in phage attachment are frequently also involved in the virulence process. Examples include lipopolysaccharide (LPS) (Evans et al., 2010a), extracellular polysaccharide (EPS) (Ayers et al., 1979), flagella (Evans et al., 2010b; Addy et al., 2012) and pili (Ahern et al., 2014). Thus, mutations leading to resistance frequently compromise virulence. There, are however, a few examples where these mutations in bacteria surface structures did not lead to reduced virulence as seen with LPS production mutants of *Pectobacterium* and *Dickeya* (Schoonejans et al., 1987; Pirhonen et al., 1988).

Bacteriophage and Chemicals

Phage have been shown to be stable in certain agrichemicals (Ravensdale et al., 2010). However, precautions need to be taken with some chemicals being combined with phage. Chemical biocides typically contain a range of phage inactivating substances such as surfactants and chelators (Yamamoto et al., 1968; Chattopadhyay et al., 2002). Also, copper-based bacteriocides have been shown to inactivate phage, but this inactivation can be avoided with the delayed application of phage (4–7 days) after initial application of copper-based bactericide (Iriarte et al., 2007).

COMPLEXITY OF PHAGE INTERACTION WITH SOIL

The rhizosphere is the area of soil which is in close proximity to the roots of a plant. There are several factors which can affect phage activity in this environment such as pH, moisture levels, presence of organic matter and soil type. A number of these factors either individually or in combination can cause phage inactivation. Different soil types affect the survival of phage. For example, clay loam soils appear better at maintaining phage at low soil moisture levels and high soil temperatures than that of sandy loam soils (Straub et al., 1992) As well, low soil pH can also negativity affect phage survivability (Sykes et al., 1981).

Levels of adsorption of phage are affected differently in differing soil types, with levels of hindrance varying from one phage type to another (Goyal and Gerba, 1979). Phage can become bound to soil components such as clays (kaolinite and montmorillonite) as these minerals possess positively and negatively charged surfaces to which phage can adsorb (Schiffenbauer and Stotzky, 1982). Such adsorption can be influenced by pH (Goyal and Gerba, 1979; Loveland et al., 1996) as well as the presence of organic materials (Zhuang and Jin, 2003). Under favorable conditions, phages have been identified that persist at relatively stable concentrations for several weeks in soil (Fujiwara et al., 2011).

PHAGE IN THE PHYLLOSPHERE

The phyllosphere is the portion of the plant which is above the ground and phages can readily be isolated from this location. How phages get there naturally has not been defined precisely, although it is possible that they originate in the soil from which the plant germinated - or alternatively get deposited by insect vectors. Indeed, phages for the phytopathogens *Pantoea stewartii* and *Erwinia herbicola* var. *herbicola*have been isolated from corn flea beetles (Woods et al., 1981). Another route is the translocation of phage from the roots to leaves of plants through the plant vascular system. And it has been shown that phages of *R. solanacearum, Xanthomonas perforans*, and *Xanthomonas euvesicatoria* can translocate though tomatoes plants, phage of *Xanthomanas oryzae* though the rice seedlings and phages of *E. amylovora*though apple seedlings and fire thorn (Rao and Srivastava, 1973; Iriarte et al., 2012; Kolozsváriné Nagy et al., 2015). However, this translocation may be influenced by the phage type, plant age, plant size, plant species, plant health and possibly soil type in which the plant is growing (Iriarte et al., 2012). It has also been reported that *E. amylovora* phages could pass from the leaves to the roots of apple seedlings (Kolozsváriné Nagy et al., 2015). The phyllosphere is nevertheless a harsh environment for phages to survive and it has been reported that their numbers can rapidly decline during daylight hours (Balogh et al., 2003; Iriarte et al., 2007). The destructive influence of UV light from the sun has been reported to be a limiting factor for the application of phages for successful biocontrol. The radiation causes the formation of lesions in DNA which can block DNA replication and transcription. In an vivo study with phage phiXV3-16, Iriarte et al. (2007) demonstrated a direct relationship between phage reduction on tomato leaves and increasing UVA+B dose. They also showed in an *in vitro* study that UV was capable of inactivating phage used against *Xanthomonas campestris pv. vesicatoria*, preventing it from exerting a biocontrol effect. Phage sensitivity against UV light has been shown to occur also with phage of *Dickya solani and E. amylovora* phages (Czajkowski et al., 2014; Born et al., 2015). However, there have been phages isolated against the phytopathogen *Pseudomonas syringae pv. actinidiae* which can tolerate extended UV-B doses (Yu et al., 2015). Other potential factors that could cause phage decline on the phyllosphere are desiccation, temperature, pH as well as certain chemicals produced by plants (Erskine, 1973; Delitheos et al., 1997; Iriarte et al., 2007).

PHAGE APPLICATION METHODS FOR OPTIMAL BIOCONTROL PERFORMANCE ON PLANTS

One of the limitations to effective phage biocontrol on crops is the possibility of poor persistence on the phyllosphere due to the factors discussed in the previous section. However, several methods have been found to reduce this problem. Survival of phage can be improved in the phyllosphere and rhizosphere if they are accompanied by a viable host. This can be an avirulent strain of the pathogen being targeted or indeed another species of bacteria which occurs naturally in that environment (Svircev et al., 2006; Bae et al., 2012; Iriarte et al., 2012). It has also been found that avoiding daylight during application can improve phage-based biocontrol. Indeed, it has been demonstrated that applying phage to tomato leaves in the evening resulted in longer phage persistence in the phyllosphere, giving phage more time to infect and kill their bacterial targets (Balogh et al., 2003; Iriarte et al., 2007).

Born et al. (2015) conducted studies with number of substances to investigate if they gave phage protection against UV and reported that natural extracts from carrot, red pepper and beetroot all gave protection as did casein, soy peptone and also purfied aromatic amino acids, astaxathin and Tween 80. None of these subtances had a compromising effect on phage infection and stability (Born et al., 2015). Thus, it appears that a wide range of substances could enhance phage preformance in the phyllosphere with the main requirement being that they need to absorb UV thus limiting phage exposure. Biodegradable polymers have also been shown to give these protective effects (Khalil et al., 2016). In addition, Balogh et al. (2003)also showed an enhanced phage activity by combining the following preparations with phage, namely (i) 0.5% pregelatinized corn (PCF) and 0.5% sucrose, (ii) 0.5% Casecrete NH400, 0.5% sucrose and 0.25% PCF and (iii) 0.75% skim milk and 0.5% sucrose. These tests were performed in greenhouse trials and in field trials on tomato plants with phages against *Xanthomonas campestris* pv. *vesicatoria*. All formulations were used under a variety of different conditions, but generally demonstrated enhanced disease protection.

Soil based phage delivery is another approach that has been looked at to improve phage presistance in the phylosphere. Iriarte et al showed that a proprietary mixture of phage (OmniLytics Inc.) active against *X. perforans* strain 97-2 could translocate to the upper leaves of a tomato plant from its roots. They demonstrated that these phages which were applied to soil at levels of 10^8PFU/mL could be detected at titres of 10^4 PFU/g in leaf tissue for 7 days, whereas with a direct foliar application of the same phage mix, phage were undetectable 1 to 2 days after application (Iriarte et al., 2012). This work would suggest that the phage control of foliar plant diseases could be controlled by applying the phages to surrounding soil of a plant rather than by foliar spraying.

COMBINATION OF PROTECTIVE METHODS APPEAR TO BE THE BEST DIRECTION FOR PHAGE BIOCONTROL

There is evidence to support that combining phage with several methods used to control crop disease results in better control. The bacterium *Pantoea agglomerans* has been used as a biocontrol agent to suppress growth of the agent of fire blight *E. amylovora* and is being sold under the band name Bloomtime® (Mikiciński et al., 2016). However, it has been reported that combining this bacterium with phage biocontrol can give enhanced protection that is comparable to that achieved with the antibiotic streptomycin (Svircev et al., 2006; Boulé et al., 2011). A similar observation of enhanced control was seen using a bacteriocin-producing strain of *R. solanacearum* with a phage to combat tobacco bacterial wilt (Tanaka et al., 1990). In another study, combining phage with Acibenzolar-S-methyl (ASM) was shown to have improved protection against bacterial spot of tomato in the field (Obradovic et al., 2004). However, combinations of phage with copper based pesticides do not appear to produce synergistic effects. Treatment with copper-mancozeb as seen with citrus canker and bacterial spot of citrus fruits did not produce synergy against *Xanthomanas axonopodis pv.citri* or *Xanthomanas axonopodis* pv. *citrumelo*, respectively (Balogh et al., 2008). As mentioned previously, this could be due to phage sensitivity to the components of these copper based sprays.

IMPROVED UNDERSTANDING OF BACTERIAL HOST DIVERSITY SHOULD AID PHAGE BIOCONTROL AND IMPROVE ITS SUCCESS IN THE FUTURE

Recent years have seen recognition of the increasing diversity and complexity of bacterial phytopathogens mainly due to advances in molecular techniques (16S rRNA sequencing). For example, *X. campestris* pv. *vesicatoria*, which was previously a single species has since been divided into four (Jones et al., 2004). Another example is of the soft rot *Erwinia* group, which has undergone a significant taxonomic reshuffle with creation of novel species and genera (Hauben et al., 1998; Gardan, 2003; Samson et al., 2005). These developments are very important, as while the afflictions caused by these bacteria may appear identical on their respective crop targets, the phage sensitivities of the pathogens are likely to differ significantly, but nevertheless are likely to have some correlation with their taxonomic groupings. For example, the soft rot *Erwinia* group, which affects potato crops, has more recently been reclassified into two new bacterial genera (*Pectobacterium* and *Dickeya*), and these are relatively distinct from the point of view of phage susceptibilities (Czajkowski, 2016).

PHYTOPATHOGENS TARGETED FOR PHAGE BIOCONTROL AND HOW THEY ARE CURRENTLY MANAGED

There are a number of important bacterial plant pathogens that have received attention for phage biocontrol in recent years (Table 1) as existing approaches are having limited efficacy or their use is restricted in certain regions of the world. The following section discusses selected crop pathogens where phage biocontrol has been evaluated and is showing promise.

Dickeya and Pectobacterium

Both *Dickeya* and *Pectobacterium* belong to the family of *Enterobacteriacea*, which collectively can be referred to as the Soft Rot *Enterobacteriacea* (SRE). Both genera characteristically produce several cell-wall-degrading enzymes that allow them to infiltrate and macerate the plant tissue on which they feed (Pérombelon, 2002). The plant host range of both bacterial genera is very broad: species belonging to *Dickeya* have been reported to infect 10 monocot and 11 dicot families, while those of *Pectobacterium* are reported to infect eleven monocot and sixteen dicot families (Ma et al., 2007).

P. carotovorum ssp *carotovorum* has a wide host range and global distribution, while *P. atrosepticum* is primarily found in temperate climates with a host range mainly limited to the potato (Pérombelon, 2002). *P. wasabie* and *P. carotovurum* ssp. *brasilensis* are also found to infect potato in several regions worldwide (Waleron et al., 2013; Lee et al., 2014). In Europe, *Dickeya dianthicola* is reported to be very important in potato disease, although more recently, a new *Dickeya* species called *D. solani* is being more frequently identified. Both also cause disease in other regions of the world (Toth et al., 2011). The economic impact of these potato infections can be severe. In the Netherlands, they cause annual losses in the seed potato sector of as much €30 million per year and in Israel, potato yield losses due to *Dickeya*have been as much as 20–25% (Prins and Breukers, 2008; Tsror (Lahkim) et al., 2008).

With regard to the potato, there are no effective bactericides to protect against SRE and the most effective approach has been through careful culturing practices, involving avoidance of contamination and the removal of diseased plants and/or diseased tissue. Certification systems are also employed. These involve the propagation of seed plants using healthy tissue culture plantlets followed by propagation in greenhouses, and then open field grow-out production. It is accompanied by careful monitoring and removal of diseased plants before release for general production. The generation number of these crops is also kept low to limit bacterial build up. However, the success of these certification schemes has been variable and heavily weather dependant (De Boer, 2004; Czajkowski et al., 2011).

Erwinia Amylovora

Erwinia amylovora, a member of the family of *Enterobacteriacea*, is the causative agent of fire blight which is a destructive disease that occurs to species of the plant family *Rosaceae.* The disease has been reported in 40 countries across North America, Europe, the Pacific Rim, and the

Middle East (Bonn and van der Zwet, 2000). It heavily affects apple and pear production in several regions, with costs estimated as much as $100 million per year in the USA due to production losses and control measures (Norelli et al., 2003). It is considered to be a quarantine concern in countries belonging to plant protection agencies of APPPC (Asia and Pacific Plant Protection Commission), COSAVE (Comite Regional de Sanidad Vegetal para el Cono Sur), EPPO (Europe and Mediterranean Plant Protection Organisation) and IAPC (Inter-African Phytosanitary Council) (CABI, 2016).

Pathogenesis typically involves the bacterium entering a susceptible plant host though the nectarthodes of its flowers, but it may also enter the plant though other openings such as wounds (Bubán and Orosz-Kovács, 2003). Once in the plant, it is capable of moving though the intracellular space of parenchyma, where at the latter stages it may reach the xylem vessels. Under favorable conditions, disease can present itself as wilting, necrosis of tissue and dieback of the plant (Vanneste and Eden-Green, 2000). The bacterium does not produce cell-wall-degrading enzymes but the exopolysaccharide amylovoran, biofim formation capacity, motility, a type III secretion system, and quorum sensing are all understood to be features in its virulence (Piqué et al., 2015).

Traditionally, control of fire blight relies on cultural practices involving the removal of diseased tissue as well as preventative sprays containing copper or antibiotics (Norelli et al., 2003). However, issues with these chemical controls is copper tolerance of the pathogen and also the long term of use antibiotics (such as streptomycin) as a control strategy may be limited in the future, with growing concern of antibiotic resistance and the resulting restricted used of antibiotics for agriculture in certain regions of the world such as EU countries (Ordax et al., 2006; Russo et al., 2008; Mayerhofer et al., 2009; de León Door et al., 2013). As mentioned, biological controls using antagonistic bacteria have shown a capacity for controlling the disease (Mikiciński et al., 2016)

Ralstonia Solanacearum

Ralstonia solanacearum is a Gram negative soil-borne bacterium. It is considered to be one of most destructive phytopathogens with a host range of up to 200 plant species from over 50 families (Denny, 2007). The bacterium is highly heterogeneous, historically being divided into five races (based on plant host range) and five biovars (based on carbon utilization) (Denny, 2007). It causes diseases of economically important crops, such as bacterial wilt of tobacco, banana and tomato as well as brown rot of the potato (Sanchez Perez et al., 2008). The bacterium has global distribution (Sanchez Perez et al., 2008), and with regard to tomato and potato production, has quarantine status in the EU (Anonymous, 2000). The species has considerable economic impact: for example, brown rot of the potato has been estimated to exceed more than €950 million in losses per year worldwide (Scherf et al., 2010). Infection begins by the bacterium entering the host plant though its roots where it will then colonize the xylem. Infection typically leads to the development of yellowing of the plant, stunted growth, wilting and death, although the bacterium is also capable of asymptomatic infections (Sanchez Perez et al., 2008). Typical methods of control include the use of cultural practices such as selection of planting time, crop rotation, using clean seedlings and the use of resistant cultivars (Mariano et al., 1998). However, the use of such cultivars has shown a negative correlation between resistance and yields (Yuliar et al., 2015). Also, resistance possessed by these cultivars tends to be strain specific (Wang et al., 2000).

Pseudomonas Syringe

The bacterial phytopathogen *P. syringea* belongs to the class of *Gammaproteobacteria* (Hirano and Upper, 2000). The species is currently subdivided into more than 50 pathovars, with different pathovars representing different strains with differing plant host ranges (Hirano and Upper, 1990; Parkinson et al., 2011). Stains of most pathovars typically exhibit narrow host ranges, with pathovar P. *syringea* pv. *syringea* being an exception, having been reported to infect more than 80 plant species (Hirano and Upper, 2000).

Pseudomonas syringea pv. *tomato* causes necrotic lesions surrounded by a yellow chlorotic halos on tomato, a disease known as bacterial speck (Cruz et al., 2010). The pathovar can also infect members of genera of *Arabidopsis* and *Brassica* in laboratory setting (Elizabeth and Bender, 2007). The disease reduces yields while also affecting fruit quality (Fatmi, 2003). Pathogenesis by the bacterium involves the invasion of plant tissue from natural openings, such as stomata, where a type III secretion system plays a major role in its virulence with the release of effectors to overcome the plant immune system (Xin and He, 2013). It is spread by contaminated tomato seeds but can also survive as an epiphyte for extended periods on tomato plant surfaces and is dispersed in windblown rain (Smitley and McCarter, 1982; McCarter, 1983; Preston, 2000). Control of the organism typically involves the use of uncontaminated seeds and the used of bactericides (copper and streptomycin) to limit its spread (Preston, 2000; Fatmi, 2003). However, copper tolerant strains of the bacterium have been reported (Alexander et al., 1999).

Xanthomonas Species

Xathomonas is a large genus, which belongs to the class of *Gammaproteobacteria*, containing at least 27 official species, many of which also possess several pathovars. Collectively, the genus host range is broad: infecting around 400 plant hosts, a number of which are important crops such rice, banana, tomato, and citrus fruits. Species and pathovars of this genus typically exhibit a high degree of host- as well as tissue-specificity, invading either the xylem or intercellular spaces of the mesophyll parenchyma tissue (Ryan et al., 2011).

Xathomonas campestris pv. *vesicatoria* is the causitive agent of bacterial spot disease of tomato and pepper, with the disease having been identified in many countries worldwide (Jones et al., 2005). This tomato disease can be very severe with yield losses of up to 50% reported for tomatoes grown both in greenhouses and fields in the USA and Caribbean (Camesano, 2015). Disease is caused by the bacterium entering the plant though stomata or wounds. The bacteria then colonize the intercellular space of the plant, inducing water-soaked lesions that later become necrotic, which can result in defoliation and severely spotted fruit (Thieme et al., 2005). Control of the disease has involved preventative cultural practices such as avoiding unnecessary crop damage and using uncontaminated seed, but also includes use of resistant cultivars as well as chemical controls with copper or streptomycin (Goode and Sasser, 1980). However, the use of resistant cultivars has not always been successful and there have been reports of bacteria developing resistance to the above two agents (Goode and Sasser, 1980; Ritchie and Dittapongpitch, 1991; McDonald and Linde, 2002).

Xylella Fastidiosa

Xylella fastidiosa belongs to the of class of *Gammaproteobacteria.* It is a xylem-limited phytopathogen that requires insect vectors (such as sharpshooters) for its distribution and infection of its host plants (Chatterjee et al., 2008). It causes disease on a number of crops such as the grape, citrus, almond, peach and coffee (Hopkins and Purcell, 2002). While it has primarily been contained in the Americas, it has been indentified in Europe in recent years causing disease on olive trees (Hopkins and Purcell, 2002; Loconsole et al., 2014). Disease caused by the bacterium is believed to be induced by the formation of biofilm aggregates in the vascular system, which restricts the movement of nutrients and water throughout the plant (Chatterjee et al., 2008). It causes Pierce disease of the grapevine, a highly destructive infection, which heavily affects grape production in the USA, and has been estimated to cost as much as $104.4 million annually to the state of California (Tumber et al., 2014). Existing control methods have been limited in their management of the disease and include removal of infected plants and control of the infected insect vector populations with neonicotinoid-based insecticides (Janse and Obradovic, 2010). However, the use of these insecticides has seen restrictions in recent years due to their possible effects on honey bee populations (Anonymous, 2013; Lu et al., 2014).

CRITICAL SUMMARY OF RECENT PHAGE BIOCONTROL STUDIES ON CROPS

There is growing evidence showing that phage have promising biocontrol applications for number of plant diseases in different crops. The following section describes recent studies that have been conducted since the year 2000 and the findings from these is summarized in Table 1.

The most common crops that appear to benefit from the application of phages for biocontrol in recent scientific literature are the potato and the tomato, as both have been the focus of numerous recent studies. The bacterial pathogens in the case of the potato are predominantly the SRE. As mentioned above, one of the most important SREs in Europe is *D. solani;* and the potential of phage to control this phytopathogen have been assessed indicating strong potential for disease control. For example, Adriaenssens et al. (2012)conducted a bioassay and a field trial using phage (LIMEstone1). The bioassay involved the incubation of seed tubers (cultivar Bintje), which had either been inoculated with the bacteria or co-inoculated with the bacteria and the phage (MOI of 100). They showed that tubers inoculated with the bacteria alone would experience to 40% maceration of tuber tissue, while those co-inoculated with the phage and bacteria exhibited no more than 10% maceration of tuber tissue. Similar results were observed with the seed tuber cultivar Kondor. The field trial using the same phage against the same pathogen also suggested it was capable of exerting this biocontrol effect *in-planta*, as phage treated infected seed potatoes resulted in higher crop yields than those without phage treatment. Similar findings were reported by Czajkowski et al. (2014) who also isolated phages specific for *D. solani*. These workers conducted bioassays with tuber slices incubated with the bacterial pathogen with or without phages (MOI of 0.01) and showed that the application of phages could prevent potato tuber tissue maceration by up to 70%. SREs other than *D. solani* were also studied for their susceptibility to phages by the same group. They found that the application of phages (MOI of 0.01) to control *P. carotovorum ssp carotovorum* and *P. wasabie*destruction could prevent damage of up to 80% on tuber slices and up to 95% on whole tubers against tissue maceration from a mixed bacterial infection (Czajkowski et al., 2015). Such data is highly encouraging as many SRE infections tend to result for a mixture of genera/species. Aside from potato, SRE infections have also been controlled by phage in lettuce, with high levels of disease prevention being reported (Lim et al., 2013). Aside from the SRE problem, potato infections from the Gram-positive bacterium *Streptomyces scabies* results in the formation of a corky lesion (known as common scab) on the tuber and indeed other root vegetables also, as well as causing the reduced growth of seedlings (Lerat et al., 2009). This pathogen has also been successfully treated in potato by phage biocontrol and thus has implications for other crops also as demonstrated by Goyer (2005). In conclusion, the above studies indicate strong potential for phage based control of these diseases.

Another crop which has been the focus of several studies in the context of phage therapy is the tomato, which is commonly infected by *R. solanacearum* (also causes brown rot in the potato) and *X. campestris*pathovars. Again, phage biocontrol approaches have been demonstrated to give a significant reduction in bacterial wilt (*Ralstonia*) and leaf spot caused by *Xanthomonas*. Indeed, the successful trials against *R. solanacearum* reported by Mansfield et al. (2012) are significant considering the wide host range of the bacterium. Similarly, in the case of *Xanthomonas*, the observed beneficial effect of the application of phages can also be extrapolated to other plants affected by pathogens belonging to the same genus. Indeed, studies on elimination of *Xanthomonas* using phage have been conducted with successful outcomes on both grapefruit and orange (Balogh et al., 2008) as well as onion (Lang et al., 2007). A variety of other crop infections have also been reduced in severity in other phage biocontrol studies. These include *Pseudomonas* infections of mushrooms (brown blotch) and leeks (bacterial blight) and infection of the grapevine by *Xylella* (Das et al., 2015).

COMMERCIALIZATION OF PHAGE FOR BIOCONTROL IN CROP DISEASE

In recent years, several phage biocontrol products have reached the market. A USA based company Omnilytics was the first company to receive registration (from the US Environmental protection agency) for their phage based biopesticide product Agriphage. The product is designed for the control of bacterial spot or speck of tomatoes and peppers (specific for *X. campestris* pv. *vesicatoria*, or *P. syringae* pv. *tomato*). This product has also received an OMRI listing making it suitable for use by commercial organic growers (OmniLytics, 2006). A Hungarian company Enviroinvest was the second company to receive registration for their biopesticide named Erwiphage for the control of fire blight of apple trees (specific for *Erwinia amylovora*) (Enviroinvest, n.d.). There is also a Scottish company, APS biocontrol, which has developed a bacteriophage-based wash solution (Biolyse) for potatoes tubers, which is to be used for prevention of soft rot disease (specific against soft rot *Enterobacteriacea*) during storage (APS Biocontrol Ltd, n.d.). Interestingly this product has been reported to be used by the Tesco supermarket chain (Branston, 2012).

However, in some regions of the world there are delays that have to be overcome with regard to legislation allowing phage biocontrol approaches for the of control of bacterial plant diseases. A problem with phage-mediated biocontrol is that phage mixtures/cocktails need to be updated constantly in order to lyse as many newly emerging strains of the target bacterium as possible. This approach is used by Omnilytics (OmniLytics, 2004). This allows a phage cocktail to be adapted to the relevant disease-causing bacterial strains in a given situation, also facilitating counteraction of any phage resistance development during the phage application. However, EU regulations (1107/2009 EC) require that any change to one of the components of a phage cocktail would require reregistering which requires time and expense, making the US approach currently unfeasible in the EU (Doffkay et al., 2015). Legislation governing phage biocontrol may need to become more malleable in the EU for the best application and performance of phage products as biopesticides.

OTHER PHAGE APPLICATIONS OF THE PAST AND POSSIBLE FUTURE WITH REGARDS TO PHYTOPATHOGENS

Phage typing schemes have been employed for several phytopathogens for epidemiology studies (Toth et al., 1999; Ahmad et al., 2014). These systems allow the identification of a particular strain of a species based on their susceptibility to series of phage. The downfall of this method, however, is that it depends on the isolation of pure cultures for identification as well as the maintenance of stocks of typing phage as well as host strains for which to propagate them. Nowadays, studies of phytopathogens has moved away from phage typing due to its tendency to generate false positives and false negatives results as well as its low resolution and the development of new and improved molecular techniques (Czajkowski, 2016)

Several phage-based detection systems have been developed for human and animal pathogens (van der Merwe et al., 2014). Recently however, work has been published on the promising application of these methods for the detection of plant pathogens. Such a detection system has been developed for *R. solanacearum*, which is based on detection of the bacterium by phage propagation followed with quantitative PCR (qPCR). Samples that contain the bacterium will cause added phage titres to increase, these titre increases can then be detected using qPCR. This method was found to be faster than conventional methods with greater sensitivity allowing detection of 10^2 CFU/g of soil, 10^3 CFU/ml from drainage water from potted plants and 10^2 CFU/g in 0.1 g of leaf tissue. The method also does not require the destruction of a plant for the detection of bacterium unlike those currently used to detect *R. solanacearum* (Kutin et al., 2009). It is possible to engineer phage of phytobacteria into reporter systems that can emit a detectable bioluminescent signal during infection. A "*luxAB*-tagged" reporter phage was developed for *Pseudomonas cannabina pv. alisalensis* (agent of bacterial blight of crucifers) which was shown capable of detecting the bacteria within minutes. This phage was also capable of emitting a detectable signal during infection of both cultures and diseased plant samples (Schofield et al., 2013). Both the mentioned systems have advantages over other molecular detection methods, in that phage propagation requires active metabolism, conveniently limiting it to viable bacterial cells.

CONCLUSION

Effective control of plant disease typically calls for a disease management strategy that involves several integrated approaches. Currently, the use of phage biocontrol is an emerging, but as yet uncommon practice. However, phages do possess several properties which can add to the arsenal of controls for crop diseases. They are natural, making them suitable for organic farming. They can be used to create phage cocktails with tailored host ranges. Also, phages naturally have the potential to evolve to adapt to overcome phage-resistance or overcome new strains of bacteria. They can be combined with other chemical or biocontrol agents. A possible limitation to their use is their sensitivity to UV light and to certain soil conditions. However, approaches have been found to overcome some of these limitations with the use of UV protectant formulas and timing of the application of phage to crops to avoid interaction with chemical pesticides and exposure to UV light. In addition to biocontrol applications, there is also good potential for phage-based diagnostics for plant pathogenic bacteria with a high sensitivity aimed specifically at viable bacteria.

Many pesticide companies are moving away from investment in chemical pesticides and increasingly directing their attention to biopesticides. The pesticide market is worth \$56 billion with the biopesticides forming only \$2–3 billion of this. However, growth of the biopesticide sector is expected to outpace chemical pesticides in the future (Marrone, 2014). This change is believed to be due to an increasing customer demand for chemical residue free foods and increasing legalization on the use of synthetic pesticides in certain regions in the word. In addition, many biopesticide products are potentially cheaper to develop and quicker to bring to the market (Marrone, 2014). With this economic environment, one can expect to see increased activity in the development of phage biocontrol as a viable approach for crop disease control in the future.

REFERENCES

1. Abedon, S. T., and Lejeune, J. T. (2005). Why bacteriophage encode exotoxins and other virulence factors. *Evol. Bioinform. Online* 1, 97–110.
2. Abedon, S. T., Thomas-Abedon, C., Thomas, A., and Mazure, H. (2011). Bacteriophage prehistory: is or is not Hankin, 1896, a phage reference? *Bacteriophage* 1, 174–178. doi: 10.4161/bact.1.3.16591
3. Ackermann, H.-W. (2007). 5500 Phages examined in the electron microscope. *Arch. Virol.* 152, 227–243. doi: 10.1007/s00705-006-0849-1
4. Addy, H. S., Askora, A., Kawasaki, T., Fujie, M., and Yamada, T. (2012). Loss of virulence of the phytopathogen *Ralstonia solanacearum*through infection by φRSM filamentous phages. *Phytopathology* 102, 469–477. doi: 10.1094/PHYTO-11-11-0319-R
5. Adriaenssens, E. M., Van Vaerenbergh, J., Vandenheuvel, D., Dunon, V., Ceyssens, P.-J., De Proft, M., et al. (2012). T4-related bacteriophage LIMEstone isolates for the control of soft rot on potato caused by '*Dickeya solani*'. *PLoS ONE* 7:e33227. doi: 10.1371/journal.pone.0033227
6. Ahern, S. J., Das, M., Bhowmick, T. S., Young, R., and Gonzalez, C. F. (2014). Characterization of novel virulent broad-host-range phages of *Xylella fastidiosa* and *Xanthomonas*. *J. Bacteriol.* 196, 459–471. doi: 10.1128/JB.01080-13
7. Ahmad, A. A., Ogawa, M., Kawasaki, T., Fujie, M., and Yamada, T. (2014). Characterization of bacteriophages Cp1 and Cp2, the strain-typing agents for *Xanthomonas axonopodis* pv. citri. *Appl. Environ.*

Microbiol. 80, 77–85. doi: 10.1128/AEM.02310-13

8. Ahmad, M. H., and Morgan, V. (1994). Characterization of a cowpea (*Vigna unguiculata*) rhizobiophage and its effect on cowpea nodulation and growth. *Biol. Fertil. Soils* 18, 297–301. doi: 10.1007/BF00570632
9. Alexander, S. A., Kim, S. H., and Waldenmaier, C. M. (1999). First report of copper-tolerant *Pseudomonas syringae* pv. *tomato* in Virginia. *Plant Dis.* 83, 964–964. doi: 10.1094/PDIS.1999.83.10.964C
10. Anonymous (2000). Council Directive 2000/29/EC of 8 May 2000 on protective measure against the introduction into the Community of organisms harmful to plants or plant products and against their spread within the commnity. *Off. J. L* 169, 1–12.
11. Anonymous (2013). Commission implementing regulation (EU) No. 485/2013 of 24 May 2013 amending Implementing Regulation (EU) No. 540/2011, as regards the conditions of approval of the active substances clothianidin, thiamethoxam and imidacloprid, and prohibiting the use and sale of seeds treated with plant protection products containing those active substances. *Off. J. Eur. Union L* 129, 12–26.
12. APS Biocontrol Ltd (n.d.). *APS Biocontrol [WWW Document].* Available at: http://apsbiocontrol.com/ (accessed March 24, 2016).
13. Ayers, A. R., Ayers, S. B., and Goodman, R. N. (1979). Extracellular polysaccharide of *Erwinia amylovora*: a correlation with virulence. *Appl. Environ. Microbiol.* 38, 659–666.
14. Bae, J. Y., Wu, J., Lee, H. J., Jo, E. J., Murugaiyan, S., Chung, E., et al. (2012). Biocontrol potential of a lytic bacteriophage PE204 against bacterial wilt of tomato. *J. Microbiol. Biotechnol.* 22, 1613–1620. doi: 10.4014/jmb.1208.08072
15. Balogh, B. (2006). *Characterization and Use of Bacteriophages Assciated with Citrus Bacterial Pathogens for Disease Control.* Ph.D. Dissertation, Unversity of Florida, Gainesville, FL.
16. Balogh, B., Canteros, B. I., Stall, R. E., and Jones, J. B. (2008). Control of citrus canker and citrus bacterial spot with bacteriophages. *Plant Dis.* 92, 1048–1052. doi: 10.1094/PDIS-92-7-1048
17. Balogh, B., Jones, J. B., Momol, M. T., Olson, S. M., Obradovic,

A., King, P., et al. (2003). Improved efficacy of newly formulated bacteriophages for management of bacterial spot on tomato. *Plant Dis.* 87, 949–954. doi: 10.1094/PDIS.2003.87.8.949

18. Basit, H. A., Angle, J. S., Salem, S., and Gewaily, E. M. (1992). Phage coating of soybean seed reduces nodulation by indigenous soil bradyrhizobia. *Can. J. Microbiol.* 38, 1264–1269. doi: 10.1139/m92-208
19. Blower, T. R., Evans, T. J., Przybilski, R., Fineran, P. C., and Salmond, G. P. C. (2012). Viral evasion of a bacterial suicide system by RNA-based molecular mimicry enables infectious altruism. *PLoS Genet.* 8:e1003023. doi: 10.1371/journal.pgen.1003023
20. Bonn, W. G., and van der Zwet, T. (2000). "Distribution and economic importance of fire blight," in *Fire Blight: The Disease and Its Causative Agent Erwinia amylovora*, ed. J. L. Vanneste (Wallingford: CAB International), 37.
21. Born, Y., Bosshard, L., Duffy, B., Loessner, M. J., and Fieseler, L. (2015). Protection of *Erwinia amylovora* bacteriophage Y2 from UV-induced damage by natural compounds. *Bacteriophage* 5:e1074330. doi: 10.1080/21597081.2015.1074330
22. Born, Y., Fieseler, L., Klumpp, J., Eugster, M. R., Zurfluh, K., Duffy, B., et al. (2014). The tail-associated depolymerase of *Erwinia amylovora* phage L1 mediates host cell adsorption and enzymatic capsule removal, which can enhance infection by other phage. *Environ. Microbiol* 16, 2168–2180. doi: 10.1111/1462-2920.12212
23. Born, Y., Fieseler, L., Marazzi, J., Lurz, R., Duffy, B., and Loessner, M. J. (2011). Novel virulent and broad-host-range *Erwinia amylovora* bacteriophages reveal a high degree of mosaicism and a relationship to Enterobacteriaceae phages. *Appl. Environ. Microbiol.*77, 5945–5954. doi: 10.1128/AEM.03022-10
24. Boulé, J., Sholberg, P. L., Lehman, S. M., O'gorman, D. T., and Svircev, A. M. (2011). Isolation and characterization of eight bacteriophages infecting *Erwinia amylovora* and their potential as biological control agents in British Columbia, Canada. *Can. J. Plant Pathol.* 33, 308–317. doi: 10.1080/07060661.2011.588250
25. Branston. (2012). *A Natural Solution to Takle Potential Soft Rot*

| *Branston Limited [WWW Document]*. Available at: http://www.branston.com/news/a-natural-solution-to-takle-potential-soft-rot/ (accessed March 24, 2016).

26. Brunner, M., and Pootjes, C. F. (1969). Bacteriophage release in a lysogenic strain of *Agrobacterium tumefaciens*. *J. Virol.* 3, 181–186.
27. Bubán, T., and Orosz-Kovács, Z. (2003). The nectary as the primary site of infection by *Erwinia amylovora* (Burr.) Winslow et al.: a mini review. *Plant Syst. Evol.* 238, 183–194. doi: 10.1007/s00606-002-0266-1
28. Burstein, D., Sun, L., Brown, C., Sharon, I., Anantharaman, K., Probst, A., et al. (2016). Major bacterial lineages are essentially devoid of CRISPR-Cas viral defense systems. *Nat. Commun.* 7:10613. doi: 10.1038/ncomms10613
29. CABI (2016). *Erwinia amylovora (Fireblight) [WWW Document]*. Available: http://www.cabi.org/isc/datasheet/21908
30. Camesano, T. A. (2015). *Nanotechnology to Aid Chemical and Biological Defense*. New York, NY: Springer.
31. Chang, C. C., Sheen, J., Bligny, M., Niwa, Y., Lerbs-Mache, S., and Stern, D. B. (1999). Functional analysis of two maize cDNAs encoding T7-like RNA polymerases. *Plant Cell* 11, 911–926. doi: 10.2307/3870824
32. Chatterjee, S., Almeida, R. P. P., and Lindow, S. (2008). Living in two worlds: the plant and insect lifestyles of *Xylella fastidiosa*. *Annu. Rev. Phytopathol.* 46, 243–271. doi: 10.1146/annurev.phyto.45.062806.094342
33. Chattopadhyay, D., Chattopadhyay, S., Lyon, W. G., and Wilson, J. T. (2002). Effect of surfactants on the survival and sorption of viruses. *Environ. Sci. Technol.* 36, 4017–4024. doi: 10.1021/es0114097
34. Chopin, M.-C., Chopin, A., and Bidnenko, E. (2005). Phage abortive infection in lactococci: variations on a theme. *Curr. Opin. Microbiol.*8, 473–479. doi: 10.1016/j.mib.2005.06.006
35. Coffey, A., and Ross, R. P. (2002). Bacteriophage-resistance systems in dairy starter strains: molecular analysis to application. *Antonie Van Leeuwenhoek* 82, 303–321. doi: 10.1023/A:1020639717181

36. Considine, D. M., and Considine, G. D. (1995). *Foods and Food Production Encyclopedia*. New York, NY: Springer.

37. Coons, G., and Kotila, J. (1925). The transmissible lytic principle (bacteriophage) in relation to plant pathogens. *Phytopathology* 15, 357–370.

38. Cruz, L., Cruz, J., Eloy, M., Oliveira, H., Vaz, H., and Tenreiro, R. (2010). First report of bacterial speck of tomato caused by *Pseudomonas* syringae pv. *tomato* race 1 in Portugal. *Plant Dis*. 94, 1504–1504. doi: 10.1094/PDIS-06-10-0415

39. Czajkowski, R. (2016). Bacteriophages of soft rot Enterobacteriaceae—a minireview. *FEMS Microbiol. Lett.* 363, fnv230. doi: 10.1093/femsle/fnv230

40. Czajkowski, R., Ozymko, Z., de Jager, V., Siwinska, J., Smolarska, A., Ossowicki, A., et al. (2015). Genomic, proteomic and morphological characterization of two novel broad host lytic bacteriophages ΦPD10.3 and ΦPD23.1 infecting pectinolytic *Pectobacterium* spp. and *Dickeya* spp. *PLoS ONE* 10:e0119812. doi: 10.1371/journal.pone.0119812

41. Czajkowski, R., Ozymko, Z., and Lojkowska, E. (2014). Isolation and characterization of novel soilborne lytic bacteriophages infecting *Dickeya* spp. biovar 3 "*D. solani*." *Plant Pathol.* 63, 758–772. doi: 10.1111/ppa.12157

42. Czajkowski, R., Pérombelon, M. C. M., van Veen, J. A., and van der Wolf, J. M. (2011). Control of blackleg and tuber soft rot of potato caused by *Pectobacterium* and *Dickeya* species: a review. *Plant Pathol.* 60, 999–1013. doi: 10.1111/j.1365-3059.2011.02470.x

43. Das, M., Bhowmick, T. S., Ahern, S. J., Young, R., and Gonzalez, C. F. (2015). Control of pierce's disease by phage. *PLoS ONE*10:e0128902. doi: 10.1371/journal.pone.0128902

44. De Boer, S. H. (2004). Blackleg of potato. *Plant Heal. Instuctor* doi: 10.1094/PHI-I-2004-0712-01

45. de León Door, A. P., Romo Chacón, A., and Acosta Muñiz, C. (2013). Detection of streptomycin resistance in *Erwinia amylovora* strains isolated from apple orchards in Chihuahua, Mexico. *Eur. J. Plant Pathol.* 137, 223–229. doi: 10.1007/s10658-013-0241-4

46. Delitheos, A., Tiligada, E., Yannitsaros, A., and Bazos, I.

(1997). Antiphage activity in extracts of plants growing in Greece. *Phytomedicine*4, 117–124. doi: 10.1016/S0944-7113(97)80055-4

47. Denny, T. (2007). “Plant pathogenic *Ralstonia* species,” in *Plant-Associated Bacteria*, ed. S. S. Gnanamanickam (Dordrecht: Springer), 573–644. doi: 10.1007/978-1-4020-4538-7_16
48. d’Herelle, F. (1917). Sur un microbe invisible antagoniste des *Bacillies dysentériques*. *C. R. Acad. Sci.* 165, 373–375.
49. Doffkay, Z., Dömötör, D., Kovács, T., and Rákhely, G. (2015). Bacteriophage therapy against plant, animal and human pathogens. *Acta Biol. Szeged.* 59, 291–302.
50. Elizabeth, S. V., and Bender, C. L. (2007). The phytotoxin coronatine from *Pseudomonas syringae* pv. *tomato* DC3000 functions as a virulence factor and influences defence pathways in edible brassicas. *Mol. Plant Pathol.* 8, 83–92. doi: 10.1111/j.1364-3703.2006.00372.x
51. Enviroinvest (n.d.). *Business Activity [WWW Document]*. Available at: http://biotechnologia.enviroinvest.hu/business-activity.html(accessed March 24, 2016).
52. Erskine, J. M. (1973). Characteristics of *Erwinia amylovora* bacteriophage and its possible role in the epidemiology of fire blight. *Can. J. Microbiol.* 19, 837–845. doi: 10.1139/m73-134
53. Evans, T. J., Coulthurst, S. J., Komitopoulou, E., and Salmond, G. P. C. (2010). Two mobile *Pectobacterium atrosepticum* prophages modulate virulence. *FEMS Microbiol. Lett.* 304, 195–202. doi: 10.1111/j.1574-6968.2010.01901.x
54. Evans, T. J., Ind, A., Komitopoulou, E., and Salmond, G. P. C. (2010a). Phage-selected lipopolysaccharide mutants of *Pectobacterium atrosepticum* exhibit different impacts on virulence. *J. Appl. Microbiol.* 109, 505–514. doi: 10.1111/j.1365-2672.2010.04669.x
55. Evans, T. J., Trauner, A., Komitopoulou, E., and Salmond, G. P. C. (2010b). Exploitation of a new flagellatropic phage of *Erwinia* for positive selection of bacterial mutants attenuated in plant virulence: towards phage therapy. *J. Appl. Microbiol.* 108, 676–685. doi: 10.1111/j.1365-2672.2009.04462.x

56. Fatmi, M. (2003). "Use of oxos, a complex of hydrogen peroxide, acetic acid and silver ion, to control bacterial speck of tomato (*Pseudomonas* syringac pv. *tomato) and angular leaf spot of melon (P. s.* pv. *lachrymans*)," in *Pseudomonas Syringae and Related Pathogens*, eds N. S. Iacobellis, A. Collmer, S. W. Hutcheson, J. W. Mansfield, C. E. Morris, J. Murillo, et al. (Dordrecht: Springer), 459–466. doi: 10.1007/978-94-017-0133-4_50

57. Fineran, P. C., Blower, T. R., Foulds, I. J., Humphreys, D. P., Lilley, K. S., and Salmond, G. P. C. (2009). The phage abortive infection system, ToxIN, functions as a protein-RNA toxin-antitoxin pair. *Proc. Natl. Acad. Sci. U.S.A.* 106, 894–899. doi: 10.1073/pnas.0808832106

58. Flaherty, J. E., Harbaugh, B. K., Jones, J. B., Somodi, G. C., and Jackson, L. E. (2001). H-mutant bacteriophages as a potential biocontrol of bacterial blight of *Geranium*. *Hortscience* 36, 98–100.

59. Frampton, R. A., Taylor, C., Holguín Moreno, A. V., Visnovsky, S. B., Petty, N. K., Pitman, A. R., et al. (2014). Identification of bacteriophages for biocontrol of the kiwifruit canker phytopathogen *Pseudomonas syringae* pv. *actinidiae*. *Appl. Environ. Microbiol.* 80, 2216–2228. doi: 10.1128/AEM.00062-14

60. Fujiwara, A., Fujisawa, M., Hamasaki, R., Kawasaki, T., Fujie, M., and Yamada, T. (2011). Biocontrol of *Ralstonia solanacearum* by treatment with lytic bacteriophages. *Appl. Environ. Microbiol.* 77, 4155–4162. doi: 10.1128/AEM.02847-10

61. Gardan, L. (2003). Elevation of three subspecies of *Pectobacterium carotovorum* to species level: *Pectobacterium atrosepticum* sp. nov., *Pectobacterium betavasculorum* sp. nov. and *Pectobacterium wasabiae* sp. nov. *Int. J. Syst. Evol. Microbiol.* 53, 381–391. doi: 10.1099/ijs.0.02423-0

62. Goode, M. J., and Sasser, M. (1980). Prevention-the key to controlling bacterial spot and bacterial speck of tomato. *Plant Dis.* 64, 831–834. doi: 10.1094/PD-64-831

63. Goyal, S. M., and Gerba, C. P. (1979). Comparative adsorption of human enteroviruses, simian rotavirus, and selected bacteriophages to soils. *Appl. Environ. Microbiol.* 38, 241–247.

64. Goyer, C. (2005). Isolation and characterization of phages Stsc1

and Stsc3 infecting *Streptomyces scabiei* and their potential as biocontrol agents. *Can. J. Plant Pathol.* 27, 210–216. doi: 10.1080/07060660509507218

65. Griffiths, A. J., Miller, J. H., Suzuki, D. T., Lewontin, R. C., and Gelbart, W. M. (2000). "Transduction," in *An Introduction to Genetic Analysis*, ed. W. H. Freeman, 7th Edn (New York, NY: W. H. Freeman).
66. Hauben, L., Moore, E. R., Vauterin, L., Steenackers, M., Mergaert, J., Verdonck, L., et al. (1998). Phylogenetic position of phytopathogens within the *Enterobacteriaceae*. *Syst. Appl. Microbiol.* 21, 384–397. doi: 10.1016/S0723-2020(98)80048-9
67. Hedtke, B., Börner, T., and Weihe, A. (1997). Mitochondrial and chloroplast phage-type RNA polymerases in *Arabidopsis*. *Science* 277, 809–811. doi: 10.1126/science.277.5327.809
68. Hirano, S. S., and Upper, C. D. (1990). Population biology and epidemiology of *Pseudomonas syringae*. *Annu. Rev. Phytopathol.* 28, 155–177. doi: 10.1146/annurev.py.28.090190.001103
69. Hirano, S. S., and Upper, C. D. (2000). Bacteria in the leaf ecosystem with emphasis on *Pseudomonas syringae*-a pathogen, ice nucleus, and epiphyte. *Microbiol. Mol. Biol. Rev.* 64, 624–653. doi: 10.1128/MMBR.64.3.624-653.2000
70. Hirst, J. M., Riche, H. H., and Bascomb, C. L. (1961). Copper accumulation in the soils of apple orchards near Wisbech. *Plant Pathol.* 10, 105–108. doi: 10.1111/j.1365-3059.1961.tb00127.x
71. Hopkins, D. L., and Purcell, A. H. (2002). *Xylella fastidiosa*: cause of pierce's disease of grapevine and other emergent diseases. *Plant Dis.* 86, 1056–1066. doi: 10.1094/PDIS.2002.86.10.1056
72. Ikeda, T. M., and Gray, M. W. (1999). Identification and characterization of T3/T7 bacteriophage-like RNA polymerase sequences in wheat. *Plant Mol. Biol.* 40, 567–578. doi: 10.1023/A:1006203928189
73. Iriarte, F. B., Balogh, B., Momol, M. T., Smith, L. M., Wilson, M., and Jones, J. B. (2007). Factors affecting survival of bacteriophage on tomato leaf surfaces. *Appl. Environ. Microbiol.* 73, 1704–1711. doi: 10.1128/AEM.02118-06

74. Iriarte, F. B., Obradović, A., Wernsing, M. H., Jackson, L. E., Balogh, B., Hong, J. A., et al. (2012). Soil-based systemic delivery and phyllosphere in vivo propagation of bacteriophages: two possible strategies for improving bacteriophage persistence for plant disease control. *Bacteriophage* 2, 215–224. doi: 10.4161/bact.23530

75. Janse, J. D., and Obradovic, A. (2010). *Xylella fastidiosa*: its biology, diagnosis, control and risks. *J. Plant Pathol.* 92, 1–35–S1.48. doi: 10.4454/JPP.V92I1SUP.2504

76. Jones, J. B., Lacy, G. H., Bouzar, H., Minsavage, G. V., Stall, R. E., and Schaad, N. W. (2005). Bacterial spot – worldwide distribution, importance and review. *Acta Hortic.* 695, 27–34. doi: 10.17660/ActaHortic.2005.695.1

77. Jones, J. B., Lacy, G. H., Bouzar, H., Stall, R. E., and Schaad, N. W. (2004). Reclassification of the xanthomonads associated with bacterial spot disease of tomato and pepper. *Syst. Appl. Microbiol.* 27, 755–762. doi: 10.1078/0723202042369884

78. Khalil, I., Irorere, V., Radecka, I., Burns, A., Kowalczuk, M., Mason, J., et al. (2016). Poly-γ-glutamic acid: biodegradable polymer for potential protection of beneficial viruses. *Materials* 9:28. doi: 10.3390/ma9010028

79. Kim, M. H., Park, S. W., and Kim, Y. K. (2011). Bacteriophages of *Pseudomonas tolaasii* for the biological control of brown blotch disease. *J. Appl. Biol. Chem.* 54, 99–104. doi: 10.3839/jksabc.2011.014

80. Klumpp, J., Dorscht, J., Lurz, R., Bielmann, R., Wieland, M., Zimmer, M., et al. (2008). The terminally redundant, nonpermuted genome of *Listeria* bacteriophage A511: a model for the SPO1-like myoviruses of gram-positive bacteria. *J. Bacteriol.* 190, 5753–5765. doi: 10.1128/JB.00461-08

81. Koczan, J. M., Lenneman, B. R., McGrath, M. J., and Sundin, G. W. (2011). Cell surface attachment structures contribute to biofilm formation and xylem colonization by *Erwinia amylovora*. *Appl. Environ. Microbiol.* 77, 7031–7039. doi: 10.1128/AEM.05138-11

82. Kolozsváriné Nagy, J., Schwarczinger, I., Künstler, A., Pogány, M., and Király, L. (2015). Penetration and translocation of *Erwinia amylovora*-specific bacteriophages in apple - a possibility of

enhanced control of fire blight. *Eur. J. Plant Pathol.* 142, 815–827. doi: 10.1007/s10658-015-0654-3

83. Kotila, J., and Coons, G. (1925). Investigations on the blackleg disease of potato. *Michigan Agric. Exp. Stn. Tech. Bull.* 67, 3–29.
84. Kuo, T. T., Chiang, C. C., Chen, S. Y., Lin, J. H., and Kuo, J. L. (1994). A long lytic cycle in filamentous phage Cf1tv infecting *Xanthomonas campestris* pv. *citri*. *Arch. Virol.* 135, 253–264. doi: 10.1007/BF01310012
85. Kutin, R. K., Alvarez, A., and Jenkins, D. M. (2009). Detection of *Ralstonia solanacearum* in natural substrates using phage amplification integrated with real-time PCR assay. *J. Microbiol. Methods* 76, 241–246. doi: 10.1016/j.mimet.2008.11.008
86. Lang, J. M., Gent, D. H., and Schwartz, H. F. (2007). Management of *Xanthomonas* leaf blight of onion with bacteriophages and a plant activator. *Plant Dis.* 91, 871–878. doi: 10.1094/PDIS-91-7-0871
87. Lee, D. H., Kim, J.-B., Lim, J.-A., Han, S.-W., and Heu, S. (2014). Genetic diversity of *Pectobacterium carotovorum* subsp. *brasiliensis*Isolated in Korea. *Plant Pathol. J.* 30, 117–124. doi: 10.5423/PPJ.OA.12.2013.0117
88. Lerat, S., Simao-Beaunoir, A.-M., and Beaulieu, C. (2009). Genetic and physiological determinants of *Streptomyces scabies*pathogenicity. *Mol. Plant Pathol.* 10, 579–585. doi: 10.1111/j.1364-3703.2009.00561.x
89. Li, J., and Wang, N. (2014). Foliar application of biofilm formation-inhibiting compounds enhances control of citrus canker caused by *Xanthomonas citri* subsp. *citri*. *Phytopathology* 104, 134–142. doi: 10.1094/PHYTO-04-13-0100-R
90. Lim, J.-A., Jee, S., Lee, D. H., Roh, E., Jung, K., Oh, C., et al. (2013). Biocontrol of *Pectobacterium carotovorum* subsp. *carotovorum*using bacteriophage PP1. *J. Microbiol. Biotechnol.* 23, 1147–1153. doi: 10.4014/jmb.1304.04001
91. Łobocka, M. B., Rose, D. J., Plunkett, G., Rusin, M., Samojedny, A., Lehnherr, H., et al. (2004). Genome of bacteriophage P1. *J. Bacteriol.*186, 7032–7068. doi: 10.1128/JB.186.21.7032-7068.2004
92. Loconsole, G., Potere, O., Boscia, D., Altamura, G., Djelouah, K., Elbeaino, T., et al. (2014). Detection of *Xylella fastidiosa* in olive

trees by molecular and serological methods. *J. Plant Pathol.* 96, 7–14. doi: 10.4454/JPP.V96I1.041

93. Lohr, L. (2001). "Factors affecting international demand and trade in organic Food products," in *Changing Structure of Global Food Consumption and Trade*, eds A. Regmi and A. Regmi (Collingdale, PA: Diane Publishing Co.), 67.
94. Loveland, J. P., Ryan, J. N., Amy, G., and Harvey, R. W. (1996). The reversibility of virus attachment to mineral surfaces. *Colloid Surf. A*107, 205–221. doi: 10.1016/0927-7757(95)03373-4
95. Lu, C., Warchol, K. M., and Callahan, R. A. (2014). Sub-lethal exposure to neonicotinoids impaired honey bees winterization before proceeding to colony collapse disorder. *Bull. Insectol.* 67, 125–130.
96. Lu, M. J., and Henning, U. (1994). Superinfection exclusion by T-even-type coliphages. *Trends Microbiol.* 2, 137–139. doi: 10.1016/0966-842X(94)90601-7
97. Ma, B., Hibbing, M. E., Kim, H. S., Reedy, R. M., Yedidia, I., Breuer, J., et al. (2007). Host range and molecular phylogenies of the soft rot enterobacterial genera *Pectobacterium* and *Dickeya*. *Phytopathology* 97, 1150–1163. doi: 10.1094/PHYTO-97-9-1150
98. Mallmann, W., and Hemstreet, C. (1924). Isolation of an inhibitory substance from plants. *Agric. Res.* 28, 599–602.
99. Mansfield, J., Genin, S., Magori, S., Citovsky, V., Sriariyanum, M., Ronald, P., et al. (2012). Top 10 plant pathogenic bacteria in molecular plant pathology. *Mol. Plant Pathol.* 13, 614–629. doi: 10.1111/J.1364-3703.2012.00804.X
100. Mariano, R. L. R., Silveira, N. S. S., and Michereff, S. J. (1998). "Bacterial wilt in Brazil: current status and control methods," in *Bacterial Wilt Disease*, eds P. Prior, C. Allen, and J. Elphinstone (Berlin: Springer), 386–393. doi: 10.1007/978-3-662-03592-4_59
101. Marraffini, L. A., and Sontheimer, E. J. (2008). CRISPR interference limits horizontal targeting DNA. *Science* 322, 1843–1845. doi: 10.1126/science.1165771
102. Marrone, P. G. (2014). "The market and potential for biopesticides," in *Biopesticides: State of the Art and Future Opportunities, ACS*

Symposium Series, eds A. D. Gross, J. R. Coats, S. O. Duke, and J. N. Seiber (Washington, DC: American Chemical Society), 245–258. doi: 10.1021/bk-2014-1172

103. Mayerhofer, G., Schwaiger-Nemirova, I., Kuhn, T., Girsch, L., and Allerberger, F. (2009). Detecting streptomycin in apples from orchards treated for fire blight. *J. Antimicrob. Chemother.* 63, 1076–1077. doi: 10.1093/jac/dkp055
104. McCarter, S. M. (1983). Survival of *Pseudomonas syringae* pv. *tomato* in Association with tomato seed, soil, host tissue, and epiphytic weed hosts in Georgia. *Phytopathology* 73, 1393–1398. doi: 10.1094/Phyto-73-1393
105. McDonald, B. A., and Linde, C. (2002). Pathogen population genetics, evolutionary potential, and durable resistance. *Annu. Rev. Phytopathol.* 40, 349–379. doi: 10.1146/annurev.phyto.40.120501.101443
106. McKenna, F., El-Tarabily, K. A., Hardy, G. E. S. J., and Dell, B. (2001). Novel in vivo use of a polyvalent *Streptomyces* phage to disinfest *Streptomyces scabies*-infected seed potatoes. *Plant Pathol.* 50, 666–675. doi: 10.1046/j.1365-3059.2001.00648.x
107. Mikiciński, A., Sobiczewski, P., Puławska, J., and Maciorowski, R. (2016). Control of fire blight (*Erwinia amylovora*) by a novel strain 49M of *Pseudomonas graminis* from the phyllosphere of apple (*Malus* spp.). *Eur. J. Plant Pathol.* 145, 265–276. doi: 10.1007/s10658-015-0837-y
108. Müller, I., Lurz, R., and Geider, K. (2012). Tasmancin and lysogenic bacteriophages induced from *Erwinia tasmaniensis* strains. *Microbiol. Res.* 167, 381–387. doi: 10.1016/j.micres.2012.01.005
109. Norelli, J. L., Jones, A. L., and Aldwinckle, H. S. (2003). Fire blight management in the twenty first century using new technologies. *Plant Dis.* 87, 756–765. doi: 10.1094/PDIS.2003.87.7.756
110. Obradovic, A., Jones, J. B., Momol, M. T., Balogh, B., and Olson, S. M. (2004). Management of tomato bacterial spot in the field by foliar applications of bacteriophages and SAR inducers. *Plant Dis.* 88, 736–740. doi: 10.1094/PDIS.2004.88.7.736
111. Okabe, N., and Goto, M. (1963). Bacteriophages of plant

pathogens. *Annu. Rev. Phytopathol.* 1, 397–418. doi: 10.1146/annurev.py.01.090163.002145

112. OmniLytics (2004). *AgriPhage Product Overview | OmniLytics [WWW Document].* Available at: http://www.ns.omnilytics.com/products/agriphage/agriphage_info/agriphage_overview.html

113. OmniLytics (2006). *OmniLytics Receives OMRI Listing for AgriPhage | OmniLytics | News [WWW Document].* Available at: http://www.ns.omnilytics.com/news/news015.html.Ordax, M., Marco-Noales, E., López, M. M., and Biosca, E. G. (2006). Survival strategy of *Erwinia amylovora* against copper: induction of the viable-but-nonculturable state. *Appl. Environ. Microbiol.* 72, 3482–3488. doi: 10.1128/AEM.72.5.3482-3488.2006

114. Parkinson, N., Bryant, R., Bew, J., and Elphinstone, J. (2011). Rapid phylogenetic identification of members of the *Pseudomonas*syringae species complex using the rpoD locus. *Plant Pathol.* 60, 338–344. doi: 10.1111/j.1365-3059.2010.02366.x

115. Parma, D. H., Snyder, M., Sobolevski, S., Nawroz, M., Brody, E., and Gold, L. (1992). The rex system of bacteriophage-lambda – tolerance and altruistic cell-death. *Genes Dev.* 6, 497–510. doi: 10.1101/gad.6.3.497

116. Pérombelon, M. C. M. (2002). Potato diseases caused by soft rot erwinias: an overview of pathogenesis. *Plant Pathol.* 51, 1–12. doi: 10.1046/j.0032-0862.2001.Shorttitle.doc.x

117. Pietrzak, U., and McPhail, D. C. (2004). Copper accumulation, distribution and fractionation in vineyard soils of Victoria, Australia. *Geoderma* 122, 151–166. doi: 10.1016/j.geoderma.2004.01.005

118. Piqué, N., Miñana-Galbis, D., Merino, S., and Tomás, J. M. (2015). Virulence factors of *Erwinia amylovora*: a review. *Int. J. Mol. Sci.* 16, 12836–12854. doi: 10.3390/ijms160612836

119. Pirhonen, M., Heino, P., Helander, I., Harju, P., and Palva, E. T. (1988). Bacteriophage T4 resistant mutants of the plant pathogen *Erwinia carotovora. Microb. Pathog.* 4, 359–367. doi: 10.1016/0882-4010(88)90063-0

120. Pirhonen, M., and Palva, E. T. (1988). Occurrence of bacteriophage T4 receptor in *Erwinia carotovora. MGG Mol. Gen. Genet.* 214,

170–172. doi: 10.1007/BF00340198

121. Preston, G. M. (2000). *Pseudomonas syringae* pv. *tomato*: the right pathogen, of the right plant, at the right time. *Mol. Plant Pathol.* 1, 263–275. doi: 10.1046/j.1364-3703.2000.00036.x

122. Prins, H., and Breukers, A., (2008). *In de puree? De gevolgen van aantasting door Erwinia voor de Pootaardappelsector in kaartgebracht.* LEI Report, Hague, The Netherlands.

123. Przybilski, R., Richter, C., Gristwood, T., Clulow, J. S., Vercoe, R. B., and Fineran, P. C. (2011). Csy4 is responsible for CRISPR RNA processing in *Pectobacterium atrosepticum*. *RNA Biol.* 8, 517–528. doi: 10.4161/rna.8.3.15190

124. Qiao, J., Qiao, X., Sun, Y., and Mindich, L. (2010). Role of host protein glutaredoxin 3 in the control of transcription during bacteriophage Phi2954 infection. *Proc. Natl. Acad. Sci. U.S.A.* 107, 6000–6004. doi: 10.1073/pnas.1000383107

125. Rao, Y. P., and Srivastava, D. N. (1973). "Application of phages in investigation of epidemiology of bacterial blight disease of rice," in *Proceedings of the Indian National Science Academy: Epidemiology, Forecasting and Control of Plant Diseases*, Vol. 37, ed. S. P. Raychandhari (New Delhi: Indian National Science Academy), 314–321.

126. Ravensdale, M., Blom, T. J., Gracia-Garza, J. A., Svircev, A. M., and Smith, R. J. (2010). Bacteriophages and the control of *Erwinia carotovora* subsp. *carotovora*. *Can. J. Plant Pathol.* 29, 121–130. doi: 10.1080/07060660709507448

127. Ray, D. K., Mueller, N. D., West, P. C., and Foley, J. A. (2013). Yield trends are insufficient to double global crop production by 2050. *PLoS ONE* 8:e66428. doi: 10.1371/journal.pone.0066428

128. Rezzonico, F., Smits, T. H. M., and Duffy, B. (2011). Diversity, evolution, and functionality of clustered regularly interspaced short palindromic repeat (CRISPR) regions in the fire blight pathogen *Erwinia amylovora*. *Appl. Environ. Microbiol.* 77, 3819–3829. doi: 10.1128/AEM.00177-11

129. Ritchie, D. F., and Dittapongpitch, V. (1991). Copper- and streptomycin-resistant strains and host differentiated races of *Xanthomonas campestris* pv. *vesicatoria* in North Carolina. *Plant*

Dis. 75, 733–736. doi: 10.1094/PD-75-0733

130. Rodrigues, C. M., Takita, M. A., Coletta-Filho, H. D., Olivato, J. C., Caserta, R., Machado, M. A., et al. (2008). Copper resistance of biofilm cells of the plant pathogen *Xylella fastidiosa*. *Appl. Microbiol. Biotechnol.* 77, 1145–1157. doi: 10.1007/s00253-007-1232-1

131. Rombouts, S., Volckaert, A., Venneman, S., Declercq, B., Vandenheuvel, D., Allonsius, C. N., et al. (2016). Characterization of novel bacteriophages for biocontrol of bacterial blight in leek caused by *Pseudomonas* syringae pv. *porri*. *Front. Microbiol.* 7:279. doi: 10.3389/fmicb.2016.00279

132. Russo, N. L., Burr, T. J., Breth, D. I., and Aldwinckle, H. S. (2008). Isolation of streptomycin-resistant isolates of *Erwinia amylovora* in New York. *Plant Dis.* 92, 714–718. doi: 10.1094/PDIS-92-5-0714

133. Ryan, R. P., Vorhölter, F.-J., Potnis, N., Jones, J. B., Van Sluys, M.-A., Bogdanove, A. J., et al. (2011). Pathogenomics of *Xanthomonas*: understanding bacterium–plant interactions. *Nat. Rev. Microbiol.* 9, 344–355. doi: 10.1038/nrmicro2558

134. Samson, R., Legendre, J. B., Christen, R., Fischer-Le Saux, M., Achouak, W., and Gardan, L. (2005). Transfer of *Pectobacterium chrysanthemi* (Burkholder et al. 1953) Brenner et al. 1973 and *Brenneria paradisiaca* to the genus *Dickeya* gen. nov. as *Dickeya chrysanthemi* comb. nov. and *Dickeya paradisiaca* comb. nov. and deli. *Int. J. Syst. Evol. Microbiol.* 55, 1415–1427. doi: 10.1099/ijs.0.02791-0

135. Sanchez Perez, A., Mejia, L., Fegan, M., and Allen, C. (2008). Diversity and distribution of *Ralstonia solanacearum* strains in Guatemala and rare occurrence of tomato fruit infection. *Plant Pathol.* 57, 320–331. doi: 10.1111/j.1365-3059.2007.01769.x

136. Sato, M. (1983). Phage induction from lysogenic strains of *Pseudomonas syringae* pathovar mori by the extract from mulberry leaves. *Ann. Phytopathol. Soc. Japan* 49, 259–261. doi: 10.3186/jjphytopath.49.259

137. Scherf, J. M., Milling, A., and Allen, C. (2010). Moderate temperature fluctuations rapidly reduce the viability of *Ralstonia solanacearum*race 3, biovar 2, in infected geranium, tomato, and

potato plants. *Appl. Environ. Microbiol.* 76, 7061–7067. doi: 10.1128/AEM.01580-10

138. Schiffenbauer, M., and Stotzky, G. (1982). Adsorption of coliphages T1 and T7 to clay minerals. *Appl. Environ. Microbiol.* 43, 590–596.
139. Schofield, D., Bull, C. T., Rubio, I., Wechter, W. P., Westwater, C., and Molineux, I. J. (2013). "Light-tagged" bacteriophage as a diagnostic tool for the detection of phytopathogens. *Bioengineered* 4, 50–54. doi: 10.4161/bioe.22159
140. Schoonejans, E., Expert, D., and Toussaint, A. (1987). Characterization and virulence properties of *Erwinia chrysanthemi* lipopolysaccharide-defective, phi EC2-resistant mutants. *J. Bacteriol.* 169, 4011–4017. doi: 10.1128/jb.169.9.4011-4017.1987
141. Semenova, E., Nagornykh, M., Pyatnitskiy, M., Artamonova, I. I., and Severinov, K. (2009). Analysis of CRISPR system function in plant pathogen *Xanthomonas oryzae*. *FEMS Microbiol. Lett.* 296, 110–116. doi: 10.1111/j.1574-6968.2009.01626.x
142. Smitley, D. R., and McCarter, S. M. (1982). Spread of *Pseudomonas syringae* pv. *tomato* and role of epiphytic populations and environmental conditions in disease development. *Plant Dis.* 66, 713–717. doi: 10.1094/PD-66-713
143. Strange, R. N., and Scott, P. R. (2005). Plant disease: a threat to global food security. *Annu. Rev. Phytopathol.* 43, 83–116. doi: 10.1146/annurev.phyto.43.113004.133839
144. Straub, T. M., Pepper, I. L., and Gerba, C. P. (1992). Persistence of viruses in desert soils amended with anaerobically digested sewage sludge. *Appl. Environ. Microbiol.* 58, 636–641.
145. Sulakvelidze, A., Alavidze, Z., and Morris, J. G. (2001). Bacteriophage therapy. *Antimicrob. Agents Chemother.* 45, 649–659. doi: 10.1128/AAC.45.3.649-659.2001
146. Suttle, C. A. (1994). The significance of viruses to mortality in aquatic microbial communities. *Microb. Ecol.* 28, 237–243. doi: 10.1007/BF00166813
147. Svircev, A. M., Lehman, S. M., Kim, W. S., Barszcz, E., Schneider, K. E., and Castle, A. J. (2006). "Control of the fire blight pathogen with bacteriophages" in *Proceedings of the 1st International Symposium on Biological Control of Bacterial Plant Diseases*,

Seeheim: Biologische Bundesanstalt für Land- und Forstwirtschaft, 259–261.

148. Sykes, I. K., Lanning, S., and Williams, S. T. (1981). The effect of pH on soil actinophage. *Microbiology* 122, 271–280. doi: 10.1099/00221287-122-2-271

149. Tanaka, H., Negishi, H., and Maeda, H. (1990). Control of tobacco bacterial wilt by an avirulent strain of *Pseudomonas solanacearum*M4S and its bacteriophage. *Jpn. J. Phytopathol.* 56, 243–246. doi: 10.3186/jjphytopath.56.243

150. Thieme, F., Koebnik, R., Bekel, T., Berger, C., Boch, J., Büttner, D., et al. (2005). Insights into genome plasticity and pathogenicity of the plant pathogenic bacterium *Xanthomonas campestris* pv. *vesicatoria* revealed by the complete genome sequence. *J. Bacteriol.* 187, 7254–7266. doi: 10.1128/JB.187.21.7254-7266.2005

151. Thomas, R. (1935). A bacteriophage in relation to Stewart's disease of corn. *Phytopathology* 25, 371–372.

152. Toth, I. K., Bertheau, Y., Hyman, L. J., Laplaze, L., López, M. M., McNicol, J., et al. (1999). Evaluation of phenotypic and molecular typing techniques for determining diversity in *Erwinia carotovora* subsp. *atroseptica*. *J. Appl. Microbiol.* 87, 770–781. doi: 10.1046/j.1365-2672.1999.00929.x

153. Toth, I. K., van der Wolf, J. M., Saddler, G., Lojkowska, E., Hélias, V., Pirhonen, M., et al. (2011). *Dickeya* species: an emerging problem for potato production in Europe. *Plant Pathol.* 60, 385–399. doi: 10.1111/j.1365-3059.2011.02427.x

154. Tsror (Lahkim), L., Erlich, O., Lebiush, S., Hazanovsky, M., Zig, U., Slawiak, M., et al. (2008). Assessment of recent outbreaks of *Dickeya*sp. (syn. *Erwinia chrysanthemi*) slow wilt in potato crops in Israel. *Eur. J. Plant Pathol.* 123, 311–320. doi: 10.1007/s10658-008-9368-0

155. Tumber, K. P., Alston, J. M., and Fuller, K. B. (2014). Pierce's disease costs California $104 million per year. *Calif. Agric.* 68, 20–29. doi: 10.3733/ca.v068n01p20

156. Twort, F. W. (1915). An investigation on the nature of ultramicroscopic viruses. *Lancet* 186, 1241–1243 doi: 10.1016/

S0140-6736(01)20383-3

157. UN (2013). *World Must Sustainably Produce 70 per Cent More Food by Mid-Century – UN report [WWW Document].* Available at: http://www.un.org/apps/news/story.asp?NewsID=46647#.VvxjOuIrLIU (accessed March 30, 2016).
158. van der Merwe, R. G., van Helden, P. D., Warren, R. M., Sampson, S. L., and Gey van Pittius, N. C. (2014). Phage-based detection of bacterial pathogens. *Analyst* 139, 2617–2626. doi: 10.1039/c4an00208c
159. Vanneste, J. L., and Eden-Green, S. (2000). "Migration of *Erwinia amylovora* in host plant tissues," in *Fire Blight: The Disease and Its Causative Agent. Erwinia Amylovora*, ed. J. L. Vanneste (New York, NY: CABI Publishing), 73.
160. Waleron, M., Waleron, K., and Lojkowska, E. (2013). Occurrence of *Pectobacterium wasabiae* in potato field samples. *Eur. J. Plant Pathol.* 137, 149–158. doi: 10.1007/s10658-013-0227-2
161. Wang, J.-F., Olivier, J., Thoquet, P., Mangin, B., Sauviac, L., and Grimsley, N. H. (2000). Resistance of tomato line Hawaii7996 to *Ralstonia solanacearum* Pss4 in Taiwan is controlled mainly by a major strain-specific locus. *Mol. Plant Microbe Interact.* 13, 6–13. doi: 10.1094/MPMI.2000.13.1.6
162. Whitman, W. B., Coleman, D. C., and Wiebe, W. J. (1998). Perspective prokaryotes: the unseen majority. *Proc. Natl. Acad. Sci. U.S.A* 95, 6578–6583. doi: 10.1073/pnas.95.12.6578
163. Wilkinson, L. (2001). *Félix d'Herelle and the Origins of Molecular Biology, Medical History.* Cambridge: Cambridge University Press.
164. Wommack, K. E., and Colwell, R. R. (2000). Virioplankton: viruses in aquatic ecosystems. *Microbiol. Mol. Biol. Rev.* 64, 69–114. doi: 10.1128/MMBR.64.1.69-114.2000
165. Woods, T. L., Israel, H. W., and Sherf, A. F. (1981). Isolation and partial characterization of a bacteriophage of *Erwinia stewartii* from the corn flea beetle, *Chaetocnema pulicaria*. *Prot. Ecol.* 3, 229–236.
166. Xin, X.-F., and He, S. Y. (2013). *Pseudomonas syringae* pv. *tomato* DC3000: a model pathogen for probing disease susceptibility and hormone signaling in plants. *Annu. Rev. Phytopathol.* 51, 473–498. doi: 10.1146/annurev-

phyto-082712-102321

167. Yamada, T. (2013). Filamentous phages of *Ralstonia solanacearum*: double-edged swords for pathogenic bacteria. *Front. Microbiol.* 4:325. doi: 10.3389/fmicb.2013.00325
168. Yamamoto, N., Fraser, D., and Mahler, H. R. (1968). Chelating agent shock of bacteriophage T5. *J. Virol.* 2, 944–950.
169. Yu, J. G., Lim, J. A., Song, Y. R., Heu, S., Kim, G. H., Koh, Y. J., et al. (2015). Isolation and characterization of bacteriophages against *Pseudomonas syringae* pv. *actinidiae* causing bacterial canker disease in kiwifruit. *J. Microbiol. Biotechnol.* 26, 385–393. doi: 10.4014/jmb.1509.09012
170. Yuliar, Nion, Y. A., and Toyota, K. (2015). Recent trends in control methods for bacterial wilt diseases caused by *Ralstonia solanacearum*. *Microbes Environ.* 30, 1–11. doi: 10.1264/jsme2.ME14144
171. Zhuang, J., and Jin, Y. (2003). Virus retention and transport as influenced by different forms of soil organic matter. *J. Environ. Qual.* 32, 816–823. doi: 10.2134/jeq2003.8160

CHAPTER

3

IMPACT OF BACTERIOPHAGE ON STRUCTURE AND FUNCTION OF BACTERIAL COMMUNITIES

Dr Pratibha Pant[1], Dr Prashant Pant[2], Dr Prerna Pandey[3]

[1]Dr. Pratibha Pant, is an Assistant Professor of Botany in Daulat Ram College, University of Delhi, New Delhi, India and has a teaching experience of more than 6 years. She is a plant virologist and biotechnologist by training and has done her PhD in Characterization of Begomoviral genomes from Papaya from Department of Botany, University of Delhi. During her PhD, she utilized a combination of molecular biology and Computational Biology tools to analyze begomoviral genomes and to her credit, a couple of novel begomoviruses have been isolated, sequenced and reported in public databases

[2]Dr. Prashant Pant, is an Assistant Professor of Botany in Dyal Singh College, University of Delhi, New Delhi, India. He has been teaching for more than 5 years now and his main areas have been Bioinformatics, Molecular Biology and Genomics at the college. His PhD was on understanding the Microbial Succession using a Metagenomic approach in a Barren mined-out site undergoing ecological restoration. For this, he deployed many molecular and computational Biology tool and

techniques. He has several publications (articles and book chapters) in International Journals and International Publications. Currently, he is also Chief Editor of Bioinformatics Review and a co-founder/Executive Chief Editor of a Science Journal in Hindi Language "Swadeshi Vigyan" for the native language speaking communities. Both the Journals are a service to the scientific community as they are fully open access.

[3]Dr Prerna Pandey, a biotechnologist with several years of wet lab research experience, worked at International Center for Genetic Engineering and Biotechnology, New Delhi. Her field of research in PhD included isolation and molecular characterization of geminiviruses, genome sequencing, gene annotation, and gene silencing using the RNA interference technology. She has also worked at Transasia Biomedicals and Advance Enzyme Technologies as a scientist. Prerna has published papers in peer-reviewed journals, and has submitted a number of annotated Geminiviral genome sequences in the GenBank, including two novel ones. She has also completed her editing and proof-reading courses from Society for the promotion of Editors and Proof-readers, UK and now works as a freelance scientific editor and writer. When Prerna is not busy with her assignments, she loves to explore historical places.

INTRODUCTION

Biological communities are driven by interactions as one of the characteristic attribute. Bacterial communities to show a tremendous number of active populations interacting with each other and with other life forms as well to bring about spatio-temporal changes above and below ground. This chapter discusses the changes brought about by the interactions between Bacterial communities and Bacteriophages in different environments namely aquatic and Terrestrial and the possible roles these interactive communities play in different spheres. We know that these microbial communities are ubiquitous, present everywhere life can survive with minimum basic life thriving requirements. These community composition is regulated by nutrient availability (internal recycling and influx from surrounding), availability of food and the agents which are parasitic or predatory to these microbes (Weinbauer et al., 2007). Among these factors Bacteriophages play a crucial role of controlling the distribution, dominance and thus composition of bacterial communities at large in almost all environments as they act as the deterministic component. They also play a role in recombination among bacterial populations by lateral gene transfer, thus affecting the structure and function (e.g. altering virulence, ability of forming biofilms) of the bacterial communities (Canchaya et al., 2003; Kristensen et al., 2010; Duerkop et al., 2012; Brussow et al., 2004; Busby et al., 2013; Wang et al., 2009; Carrolo et al., 2010). It has been evident from the studies that phages exist wherever microbes can inhabit and that makes equally (or more ubiquitous) than the former. As discussed in Chapter 1 of this book, the two life cycles of the phage systems (and the switching in between the two cycles) have been playing a role in bacterial evolution mainly by undergoing a process known as Transduction (Generalized and Specialized both). This along with the ecological organization has led to a classic state of co-evolution among phages and bacteria alike where both evolve in increasing specificity of interaction and the benefits thus derived (Koskella and Brockhurst, 2014, Pride et al., 2006; Heilmann et al., 2010; Levin et al., 2013; Lion and Gandon 2015).

BACTERIOPHAGE AND BACTERIAL INTERACTIONS IN TERRESTRIAL AND AQUATIC ENVIRONMENTS

Bacteriophages are omnipresent with their omnipresent host inhabiting all sorts of environments. Needless to say, the association is an ancient one and a complex one at that. With complex, we refer to the range of macro and microsites where this association could be found. For instance, it could be on surface waters of river and oceans, hydrothermal vents, in free living forms or inside a host as an endosymbiont or a parasite. All these infinite possibilities define the habitat of this association and that is remarkable. Over time, these sites and their factors (environment) have shaped the evolutionary development and shaped the life cycle processes, and the genetic content. It will not be wrong to infer that phages and their host (eubacteria and archaebacteria) have co-existed since their origins and they have been responsible for each other's development, only now the simplicity of the interaction has tweaked to a much complex version due to the entry of higher forms.

The interaction of the phage and bacterial cell is mainly of parasitic or predatory depending on the nature of the cycle although not in a true sense. Bacteria benefiting eventually by recombination could be a reason for saying so, as transduction eventually improves bacterial systems although at a very cost, that of mortality. Bacteriophages are responsible for a great deal of mortality and nutrient recycling among bacterial consortiums in terrestrial and aquatic environments. It is estimated that 10-50% bacterial mortality and even more in some environments is caused by phages (Fuhrman, 1999). By controlling the dominant species besides promoting horizontal transfer of genes they induce directional selection or fluctuating selection of strains in what appears as some sort of genetic drift (Ochman et al., 2000; Canchaya et al., 2003 and Skippington and Ragan 2011; Bohannan and Lenski, 2000; Williams, 2013, Gomez et al., 2014). Apart from acting as a regulator, they also contribute to the genetic diversity of the bacterial community by serving as vectors for transduction and thus are a source of variation, recombination and evolution in bacterial systems. Mostly, genes conferring antibiotic resistance genes, DNA methylases,, metabolic advantage conferring genes have been the transferred across bacterial groups (e.g., Chen and Novick, 2009; Modi et al., 2013; Bearson et al., 2014; Volkova et al., 2014).

This tremendous diversity of the host strain and that of host susceptibility has been considered as an evolutionary response to phage infections, due to the viral transduced genomic contents (Kudva et al.,2002, Ohnishi et al., 2001; Weitz et al., 2005). Several workers have reported influence of phages on Marine bacterial communities (Winter et al., 2004; Sandaa and Larsen, 2006). Microorganisms are an integral and very crucial component behind the biogeochemical and energy cycles in the oceans. They form more than 90% of the living biomass in the sea, out of this almost 20% is recycled everyday by phages (Suttle, 1994). According to Bergh et al., 1989 there are millions of virus-like particles in every milliliter of ocean water. This has been further strengthened by the fact that a large of auto- and heterotrophic planktons owe their mortality to viruses (Proctor and Fuhrman, 1990, Suttle et al., 1990). Thus, it seems that viruses define the structure and distribution of microbial communities and are a significant link in almost all mineral nutrient and energy cycle. Recently, more evidences have been collected to show the key roles played by the phage particles in proper functioning of aquatic ecosystems. Phages with a lytic phase directly affect abundance of a particular strain or group of bacterial species, most of these phages attacked the dominant species thus effectively "killing the winner strategy" in most of the communities and reorienting the communities with a different species as the dominant one unless its phage arrives (Wommack and Colwell, 2000; Weinbauer, 2004; Weinbauer and Rassoulzadegan, 2004; Suttle, 2007).

The composition and production of viral communities changes along with the host organism they infect, seasonally and temporally in the marine environment (Winter et al., 2004; Short and Suttle, 2003; Sandaa and Larsen, 2006). The viruses because of their extremely small size represents only 5% of the total prokaryotic biomass, although they are approximately 15-fold more than their bacterial and archaeal host cells in number. The rate of viral multiplication increases with increases in the growth rate of host cells, which act as a feedback mechanism to prevent the dominance of fast growing species or strains. Thus, all the communities and microbial consortia are effectively a playground of phages which determine the nature of biogeochemical cycles, trends in community dynamics and genetic diversity and so influence speciation (Weinbauer and Rassoulzadegan, 2004; Suttle, 2005; Sime-Ngando and Colombet, 2009; Fuhrman, 1999; Jardillier et al., 2005; Suttle, 2007). It has been estimated that every day, the cell lysis resulting from viral infection is responsible for loss of about 50% of microbial production and approximately 6-26% of the total carbon fixed by photosynthesis in the marine ecosystems (Fuhrman, 1999; Wilhelm and Suttle 1999). The cell lysis that occurs in sea as a result of phage activity, converts living organic matter into dead and dissolved organic matter, thus altering the pathways of carbon cycling (Suttle, 2005).

Thus, the activity of viruses creates a built up of organic matter and organic carbon, which remains for a longer time in surface waters where most of it is converted to carbon dioxide eventually by respiration and photodegradation (Wilhelm and Suttle, 1999; Fuhrman, 1999; Suttle, 2005). The lysis of microbes releases sufficient amount of iron to support the phytoplankton (Poorvin et al., 2004). Thus, it is the quantity and quality of cell lysate released in the surrounding as a result of phage attack that bring about a change in then microbial communities by creating a pool of nutrient affecting the biogeochemical cycle.

There is a lesser understanding of soil microbial systems in general as compared to aquatic systems. This is due to the reason that the study of microbial community is difficult in soil with heterogeneous structure, with different microenvironments. Soil bacteria are not always active, present mostly in the form of encysted cells with low metabolic rates, which are not easily affected by phages. More than 90% of phages are adsorbed to the soil particles in most of the soil environments (Kimura et al., 2008). Phages are found to be more abundant in soils where bacterial hosts proliferate such as soil rich in organic matter, garden soil, and farm fields with legumes as a major crop (where rhizobiphages are in abundance) (Williamson et al., 2005; Sharma et al., 2002).

Phages occurring naturally in the terrestrial ecosystem affect the microbial community existing in the soil. The commonly occurring soil phages are tailed phages belonging to phage groups *Myoviridae*, *Siphoviridae* and *Podoviridae*. The bacteria and phages coexist in soils of all terrestrial ecosystems, suggesting their co-evolution. The exposure of bacteria to phage results in the appearance of resistant bacterial hosts. The resistance may result because of modification of cell surface receptors recognized by the bacteriophage, which may reduce the ability of bacteria to utilize certain substrate in their environment.

Studies have been conducted on soil phages infecting useful and undesirable microorganisms such as rhizobia and plant pathogens in agriculture (Hashem *et al.*, 1996; Ezuka and Kaku 2000; Keel *et al.*, 2002). The root nodule bacteria *Rhizobium* spp. is affected by rhizobiphages, resulting in reduced nodule formation and ultimately reduction in the yield of the leguminous plants. The phage -bacterial interactions in the terrestrial ecosystems are affected by various soil factors such as soil structure, nutrient availability, soil moisture, temperature and soil pH (Abebe *et al.*, 1992; Novikova *et al.*, 1993). Viral survival in the terrestrial ecosystems is greatly affected by the temperature (Yeager and O'Brien, 1979). Studies have shown that under lower temperature conditions there is an increase in survival and latent periods (Zachary 1978; Straub *et al.*, 1992), whereas higher soil temperature result in faster inactivation and disintegration of soil viruses. Soil pH affects both, host growth as well as viral survival. Acidic pH effects the adsorption of phage to a solid surface or to the host cell surface, which is important for the long survival period. Soil pH also have an effect on penetration and the length of the latent period in phage–host interactions (Sykes et al., 1981).

ROLE OF BACTERIOPHAGE ON BACTERIAL PATHOGENICITY

Ubiquitous bacteriophages are an effective source of constraint within the Bacterial world (Brüssow., 2007) although bacteria can have different protective mechanisms against phage infection, including Restriction Modification System (Tock and Dryden, 2005) and the CRISPR-Cas system (van der Oost et al., 2014). Bacteriophages have been reported to decrease as well as increase pathogenicity among microbes. It is interesting to know that presence of bacteriophages influences the pathogenic bacteria towards strong selection both positively and negatively. A number of workers have reported strong selection pressure, drift and rapid evolution in large number of bacterial communities where in phage resistant or phage tolerant strains are strongly selected (Bohannan and Lenski,1999; Buckling and Rainey, 2002; Brockhurst et al., 2003; Morgan et al., 2005). At the same time, in many cases it is the pathogenicity of the microbes that gets better over time mainly with the contribution by the phage genome which is another side of the story.

The great variety of forms, factors and mechanisms that these microbes have evolved over time to counter host defenses and cause infection are called virulence factors. Bacterial pathogenicity is highly supported by some traits such as high growth efficiency, motility, nutrient uptake channel (which also acts as phage attachment receptor) so that the strain is an effective pathogen. Motility helps in colonization of suitable niches within and outside host (Josenhans and Suerbaum, 2002; Lane et al., 2007). Growth efficiency measure the rate of growth and multiplication and extent of exploitation once infection begins (Harrison et al., 2006; Frank,1996; de Roode et al., 2005). At the same time, it has been seen that increasing Phage resistance result in loss of function or reduced ability for these traits (causing virulence) and this implies a negative correlation between Phage resistance and Pathogenicity traits (such as loss of phage attachment receptor).

It has been hypothesized that increasing phage resistance result in a trade-off for factors causing virulence leading loss of function and so, phages can be potentially used for lowering bacterial pathogenicity in nature (Lenski,1988; Brockhurst et al., 2004; Heierson et al., 1986, Paruchuri and Harshey,1987; Bohannan and Lenski, 1999). To put the emphasis on the entire interaction, one can say that the recognition of cell surface receptor by bacteriophage determines bacteriophage specificity and host range (Hyman and Abedon, 2010). Any change or manipulation or removal in bacteriophage receptors will confer resistance to bacteria; however, since these cell surface receptors also have several other roles such as they act as virulence factor, nutrient uptake channels etc., as mentioned previously (Boneca, 2005; Proft and Baker, 2009; Brown et al., 2013), the modifications in these receptors may result in lower fitness compared to non-resistant strains, making the resistant bacteria less virulent as compared to non-resistant strains (Heierson et al., 1986; Park et al., 2000; Laanto et al., 2012).

This has been shown to occur in *Serratia marcescens (*Flyg et al., 1980)*, Bacillus thuringiensis* (Heierson et al., 1986) and *Salmonella enterica* serovar Entereditis (Santader and Robeson, 2007). Thus, with increased resistance comes heavy losses to pathogenicity, reduced population size, decreased exploitation of the bacterial pathogen host (Harrison et al., 2006; Frank, 1996; de Roode et al., 2005; Bohannan and Lenski, 1999; Lenski ,1988; Brockhurst et al.,2004). On the other hand there are cases well published discussing the role of bacteriophages in increasing bacterial virulence (Wagner and Waldor, 2002; Brussow et al., 2004). This is well known that phages integrate into host genome and cause genetic manipulation at various levels. Recently, with Metagenomics approach of handling community samples and with High through put Next Generation Sequencing, a vast amount of sequence data is available that provides deep insights into the genetic exchange that has been going on between the phages and the microbes. The sequence analyses from different microbiomes have revealed that a number of genes which confer pathogenicity, increased growth efficiency, and toxin production (e.g. Exotoxin genes) are of Phage origin. Such transfers of genes have been reported from Marine, Animal gut system, and sewage treatment plants, sites undergoing restoration.

With these transfers occurring, the phage during Lysogeny can transform a non-virulent strain to a virulent one creating a strain which adds to the pathogenic flora of an environment. In this manner. Phages acts as a gene pool for various genes associated with improving microbial success in different environments.

To conclude, the lysogenic phages affect microbial host fitness by different mechanisms such as:

By Modifying Genome Composition

During their "silent stay" in the genome of bacteria as a prophage, they are not silent in fact, rather are regulating and controlling the genome by several repressor proteins modifying several genes.

Gain of Function

Phages have contributed towards increasing virulence in host microbial systems by conferring new resistance or metabolic advantage through lysogenization. Some of the well-known examples where phages have conferred virulence are toxins of *Streptococcus pyogenes* (scarlet fever), *Staphylococcus aureus* (food poisoning), and *E. coli* (Shiga toxin) are encoded by phages, while toxins of *Corynebacterium diphtheriae* (diphtheria), *Clostridium botulinum* (botulism) are encoded by phages gamma and C1 (Barksdale and Arden, 1974; Freeman, 1951). A novel class of proteins called effector proteins add to the virulence machinery to several microbes and are phage encoded among pathovars of *E.coli*, *Shigella* spp., *Salmonella enterica* (Brüssow et al., 2004.)

Loss of Function

With insertions in to the host genome, prophages can cause frameshift mutation by breaking into a coding region causing loss of function. One such example is loss of tRNA gene locus due to Prophage insertion for which the attachment site carries DNA sequences to reconstitute the tRNA gene so that there is continuous fitness of the lysogen (Campbell,1992). Other examples are where lysogen causes disruption of genes leading to loss of function are lipases, and the β-toxin-negative phenotype by L54a and phi13 phages into the *S. aureus* genome (Coleman et al., 1991).

Protection from Lytic Phages

Some phage genes are known to increase the survival fitness of lysogen by repressing any further infection in the microbe by other virulent phages using many transcriptional repressors which provide immunity to the host against any future lytic infection. It is a kind of survival strategy for the temperate phages to prevent untimely or sudden disruption of the prophage.

Increasing Genetic Diversity at Species and Strain Level

Lateral gene transfer by lysogenic conversion and transduction (both generalized and specialized) contribute towards increased genetic diversity among bacterial species and strains serving as host to a number of allopatric and sympatric phages. This creates a highly-evolved pool of microbial community more specialized toward several pathogenic and adaptive traits. It also improves the genetic base of the population making it broader and more adaptive to changing environment. It has been observed that this improvement result in strainal differences and it has been estimated that such phage mediated gene transfer are to the tunes of 20×10^{15} gene transfer events per s globally (Desiere et al., 2001; Bushman, 2002). Some prophages carry additional, apparently not required set of genes which are called Morons, postulated to change phenotype or fitness of the lysogen thus potentially are a weapon to develop new pathogens, or pathogen adapted for new hosts. This has led to shift the age-old host-pathogen interaction concept to now include the element of phage in it to become host-pathogen-phage interaction where phages are playing a crucial role in selecting for a reformed pathogen whether weaker and more adaptive or stronger (stringent) and less adaptive or selecting for a non-pathogenic phage resistant one. The possibilities are many but the influence is worth understanding.

ROLE OF BACTERIOPHAGE IN BACTERIAL CLASSIFICATION

The relationship that a given phage will infect with few selected or a single bacteria species or a group is the basis of deploying bacteriophage in revealing bacterial systematics and taxonomic similarities. This way of classifying helps in two ways. One as a diagnostic method for ascertaining a fairly well-defined bacterium with several correlated characters (Katznelson and Sutton, 1951). Secondly, for cases where no clarity has been established using conventional methods of classifying and typing bacteria, phages show potential in resolving difference between closely related taxa as some of the phages are species specific and strain specific too. Phages which infect a specific strain of a species have been used for phage-typing. In this manner, phage typing helps in establishing possible number of OTUs in a given clinical or environmental sample. However, distinguishing very closely related bacterial strains which are otherwise difficult to distinguish by other methods has been the success of phage typing method. There are two widely used phage typing protocols one for *Salmonella typhi* and the other one for *S. aureus* (Anderson and Williams, 1956). The phages under these protocols are used to distinguish the species and strain to strain differences yet sometimes there is complete failure of lysis. The failure is not because of non-adsorption rather it is because of an existing prophage in the bacterial strain preventing further attack (Rountree, 1956; Uetakeh and Burrous, 1958).

REFERENCES

1. Abebe H.M., Sadowsky M.J., Kinkle B.K., Schmidt E.L. (1992). Lysogeny in *Bradyrhizobium japonicum* and its effect on soybean nodulation. *Appl Environ Microbiol*. 58: 3360–3366.
2. Anderson S., Williams E.O. (1956). Bacteriophage typing of enteric pathogens and staphylococci and its use in epidemiology. J *Din Path*. 9: 94.
3. Barksdale L., Arden S. B. (1974). Persisting bacteriophage infections, lysogeny, and phage conversions. *Annu Rev Microbiol*. 28:265-299.
4. Bearson B. L., Allen H. K., Brunelle B. W., Lee I. S., Casjens S. R., Stanton T. B. (2014). The agricultural antibiotic carbadox induces phage-mediated gene transfer in *Salmonella*. *Front Microbiol*. 5:52.
5. Bergh O., Børsheim K. Y., Bratbak G., Heldal M. (1989). High abundance of viruses found in aquatic environments. *Nature*. 340: 467–468.
6. Bohannan B. J. M., Lenski R. E. (1999). Effect of prey heterogeneity

on the response of a model food chain to resource enrichment. *Am Nat.* 153: 73–82.

7. Bohannan B. J. M., Lenski R. E. (2003). Linking genetic change to community evolution: insights from studies of bacteria and bacteriophage. *Ecol Lett.* 3:362–377.
8. Boneca I. G. (2005). The role of peptidoglycan in pathogenesis. *Curr Opin Microbiol.* 8: 46–53.
9. Brockhurst M. A., Morgan A. D., Rainey P. B., Buckling A. (2003). Population mixing accelerates coevolution. *Ecol Lett.* 6: 975–979.
10. Brockhurst M. A., Rainey P. B., Buckling A. (2004) The effect of spatial heterogeneity and parasites on the evolution of host diversity. *Proc R Soc Lond B.* 271: 107–111.
11. Brown S., Maria J. P. S., Walker S. (2013). Wall teichoic acids of gram-positive bacteria. *Ann Rev Microbiol.* 67, ed. S. Gottesman (Palo Alto: Annual Reviews), 313–336.
12. Brussow H., Canchaya C., Hardt W-D. (2004). Phages and the evolution of bacterial pathogens: from genomic rearrangements to lysogenic conversion. *Microbiol Mol Biol Rev.* 68: 560–602.
13. Brüssow H. (2007). Bacteria between protests and phages: from antipredation strategies to the evolution of pathogenicity. *Mol Microbiol.* 65: 583–589.
14. Buckling A., Rainey P. B. (2002). Antagonistic coevolution between a bacterium and a bacteriophage. *Proc Roy Soc Lond B.* 269: 931–936.A.
15. Busby B., Kristensen D. M., Koonin, E. V. (2013). Contribution of phage-derived genomic islands to the virulence of facultative bacterial pathogens. *Environ Microbiol.* 15: 307–312.
16. **Bushman F.** (2002). Lateral DNA transfer: mechanisms and consequences. Cold Spring Harbor Laboratory Press, Cold Spring Harbor, N.Y
17. **Campbell A.** (1992). Chromosomal insertion sites for phages and plasmids. *J Bacteriol.* 174**:**7495-7499.
18. Canchaya C., Fournous G., Chibani-Chennoufi S., Dillmann M-L.,

Brussow H. (2003). Phage as agents of lateral gene transfer. *Curr Opin Microbiol.* 6: 417–424.

19. Carrolo M., Frias M. J., Pinto F. R., Melo-Cristino J., Ramirez M. (2010). Prophage spontaneous activation promotes DNA release enhancing biofilm formation in Streptococcus pneumoniae. *PLoS One* 5: e15678.
20. Chen J., Novick R. P. (2009). Phage-mediated intergeneric transfer of toxin genes. *Science*. 323:139–141.
21. Coleman D., Knights J., Russell R., Shanley D., Birkbeck T. H., Dougan G., Charles I. (1991). Insertional inactivation of the *Staphylococcus aureus* beta-toxin by bacteriophage phi 13 occurs by site- and orientation-specific integration of the phi 13 genome. *Mol Microbiol*. 5:933-939.
22. de Roode J. C., Pansini R., Cheesman S. J., Helinski M. E. H., Hujben S. (2005). Virulence and competitive ability in genetically diverse malaria infections. *Proc Natl Acad Sci USA*. 102: 7624–7628.
23. Desiere F., McShan W. M., van Sinderen D., Ferretti J. J., Brüssow H. (2001). Comparative genomics reveals close genetic relationships between phages from dairy bacteria and pathogenic streptococci: evolutionary implications for prophage-host interactions. *Virology*. 288:325–341.
24. Duerkop B. A., Clements C. V., Rollins D., Rodrigues J. L. M., Hooper L. V. (2012). A composite bacteriophage alters colonization by an intestinal commensal bacterium. *Proc Natl Acad Sci USA*. 109: 17621–17626.
25. Ezuka A., Kaku H. (2000). A historical review of bacterial blight of rice. *Bull Natl Inst Agrobiol Resour. (Japan)*. 15:1–207.
26. Flyg C., Kenne K., Boman H. G. (1980). Insect pathogenic properties of *Serratia marcescens*: phageresistant mutants with a decreased resistance to *Cecropia* immunity and a decreased virulence to *drosophila*. *J Gen Microbiol.* 120: 173–181.
27. Frank S.A. (1996). Models of parasite virulence. *Q Rev Biol.* 71: 37–78.

28. **Freeman V. J.** (1951). Studies on the virulence of bacteriophage infected strains of *Corynebacterium diphtheriae*. *J Bacteriol.* 61:675-688.

29. Fuhrman J. A. (1999). Marine viruses and their biogeochemical and ecological effects. *Nature.* 399: 541-548.

30. Gomez P., Ashby B., Buckling A. (2014). Population mixing promotes arms race host-parasite coevolution. *Proc R Soc B Biol Sci.* 282:20142297–7.

31. Harrison E. F., Browning L., Vos M., Buckling A. (2006). Cooperation and virulence in acute *Pseudomonas aeruginosa* infections. *BMC Biology.* 4: 21.

32. Hashem F. M., Kuykendall L. D., Udell S. E., Thomas P. M. (1996). Phage susceptibility and plasmid profile analysis of *Sinorhizobium fredii*. *Plant Soil.* 186: 127–134.

33. Heierson A., Sidén I., Kivaisi A., Boman H. (1986). Bacteriophage-resistant mutants of *Bacillus thuringiensis* with decreased virulence in pupae of *Hyalophora cecropia*. *J Bacteriol.* 167: 18–24.

34. Heilmann S., Sneppen K., Krishna S. (2010). Sustainability of virulence in a phagebacterial ecosystem. *J Virol.* 84:3016–3022.

35. Hyman P., Abedon S. T. (2010). Bacteriophage host range and bacterial resistance, in *Advances in Applied Microbiology*, Vol. 70, eds A. I. Laskin, S. Sariaslani, and G. M. Gadd (San Diego: Elsevier Academic Press Inc), 217–248.

36. Jardillier L., Bettarel Y., Richardot M. (2005). Effects of viruses and predators on prokaryotic community composition. *Microb Ecol.* 50: 557–569

37. Jardillier L., Boucher D., Personnic S., Jacquet S., Thénot A., Sargos D., Amblard C., Debroas D. (2005). Relative importance of nutrients and mortality factors on prokaryotic community composition in two lakes of different trophic status: microcosm experiments. *FEMS Microbiol Ecol.* 53: 429-443.

38. Josenhans C., Suerbaum S. (2002). The role of motility as a virulence factor in bacteria. *Int J Med Microbiol.* 291: 605–614.

39. Katznelson N., Sutton D. (1951). A rapid plaque count method for

the detection of bacteria as applied to the demonstration of internally borne bacterial infections of seed. *J Bact.* 61: 689.

40. Keel C., Ucurum Z., Michaux P., Adrian M., Haas D. (2002). Deleterious impact of a virulent bacteriophage on survival and biocontrol activity of *Pseudomonas fluorescens* strain CHA0 in natural soil. *Mol Plant Microbe In.* 15: 567–576.
41. Kimura M., Jia Z.-J., Nakayama N., Asakawa S. (2008). Ecology of viruses in soils: Past, present and future perspectives. *Soil Sci Plant Nutri.* 54:1-32.
42. Koskella B., Brockhurst M. A. (2014). Bacteria–phage coevolution as a driver of ecological and evolutionary processes in microbial communities. *FEMS Microbiol Rev.* 38:916–931.
43. Koskella B., Parr N. (2015). The evolution of bacterial resistance against bacteriophages in the horse chestnut phyllosphere is general across both space and time. *Philos Trans R Soc B Biol Sci.* 370:20140297.
44. Kristensen D. M., Mushegian A. R., Dolja V. V., Koonin E. V. (2010). New dimensions of the virus world discovered through metagenomics. *Trends Microbiol.* 18: 11–19.
45. Kudva I. T., Evans P. S., Perna N. T., Barrett T. J., Ausubel F. M., Blattner F. R., Calderwood S. B. (2002). Strains of Escherichia coli O157:H7 differ primarily by insertions or deletions, not single-nucleotide polymorphisms. *J Bacteriol.* 184:1873–1879.
46. Laanto E., Bamford J. K. H., Laakso J., Sundberg L.-R. (2012). Phage-driven loss of virulence in a fish pathogenic bacterium. *PLoS ONE.* 7:e53157.
47. Lane M. C., Alteri C. J., Smith S., Mobley H. L. T. (2007) Expression of flagella is coincident with uropathogenic *Escherichia coli* ascension to the upper urinary tract. *Proc Natl Acad Sci USA.* 16: 16669–16674.
48. Lenski R. E. (1988) Dynamics of interactions between bacteria and virulent bacteriophage. *Adv Microb Ecol.* 10: 1–44.
49. Levin B. R., Moineau S., Bushman M., Barrangou R. (2013). The population and evolutionary dynamics of phage and bacteria with CRISPR–mediated immunity. *PLoS Genet.* 9: e1003312.

50. Lion S., Gandon S. (2015). Evolution of spatially structured host-parasite interactions. *J Evol Biol.* 28:10–28.
51. Modi S. R., Lee H. H., Spina C. S., Collins J. J. (2013). Antibiotic treatment expands the resistance reservoir and ecological network of the phage metagenome. *Nature.* 499:219–222.
52. Morgan A. D., Gandon S., Buckling A. (2005) The effect of migration on local adaptation in a coevolving host-parasite system. *Nature.* 437: 253–256.
53. Novikova N. I., Pavlova E. A., Limeshchenko E. V. (1993). Phage sensitivity and host range of Rhizobium strains isolated from root nodules of temperate legumes. *Plant Soil.* 151: 45–53.
54. Ochman H., Lawrence J. G., Groisman E. A. (2000). Lateral gene transfer and the nature of bacterial innovation. *Nature.* 405:299–304.
55. Ohnishi M., Kurokawa K., Hayashi T. (2001). Diversification of Escherichia coli genomes: are bacteriophages the major contributors? *Trends Microbiol.* 9:481– 485.
56. Park S. C., Shimamura I., Fukunaga M., Mori K. I., Nakai T. (2000). Isolation of bacteriophages specific to a fish pathogen, *Pseudomonas plecoglossicida*, as a candidate for disease control. *Appl. Environ. Microbiol.* 66: 1416–1422.
57. Paruchuri D. K., Harshey R. M. (1987). Flagellar variation in *Serratia marcescens* is associated with color variation. *J Bacteriol.* 169: 61–65.
58. Paul J. H., Frischer M. E., Thurmond J. M. (1991). Gene-transfer in marine water column and sediment microcosms by natural plasmid transformation. *Appl Environ Microb.* 57: 1509–1515
59. Poorvin L., Rinta-Kanto J. M., Hutchins D. A., Wilhelm, S. W. (2004). Viral release of iron and its bioavailability to marine plankton. *Limnol Oceanogr.* 5: doi: 10.4319/lo.2004.49.5.1734.
60. Pride D., Wassenaar T., Ghose C., Blaser M. (2006). Evidence of host-virus co-evolution in tetranucleotide usage patterns of bacteriophages and eukaryotic viruses. *BMC Genomics.* 7:8.
61. Proctor L. M., Fuhrman J. A. (1990). Viral mortality of marine bacteria and cyanobacteria. *Nature.* 343: 60–62.

62. Proft T., Baker E. N. (2009). Pili in Gram-negative and Gram-positive bacteria - structure, assembly and their role in disease. *Cell Mol Life Sci.* 66: 613–635.
63. Rountreep M. (1956). Variations in a related series of staphylococcal bacteriophages. *J Gen Microbiol.* 15: 266.
64. Sandaa R. A., Larsen A. (2006). Seasonal variations in virus-host populations in Norwegian coastal waters: focusing on the cyanophage community infecting marine *Synechococcus* spp. *Appl Environ Microbiol.* 72: 4610–4618.
65. Santader J., Robeson J. (2007). Phage-resistance of *Salmonella enterica* serovar Enteritidis and pathogenesis in *Caenorhabditis elegans* is mediated by the lipopolysaccharide. *Electron J Biotechnol.* 10: 627–632.
66. Sharma R. S., Mohmmed A., Babu C. R. (2002). Diversity among rhizobiophages from rhizospheres of legumes inhabiting three ecogeographical regions of India. *Soil Biol Biochem.* 34: 965–973.
67. Short, S. M. and Suttle, C. A., (2003). Temporal dynamics of natural communities of marine algal viruses and eukaryotes. *Aquat. Microb. Ecol.* 32, 107–119.
68. Skippington E., Ragan M.A. (2011). Lateral genetic transfer and the construction of genetic exchange communities. *FEMS Microbiol Rev.*35:707–735.
69. Straub T. M., Pepper I. L., Gerba C. P. (1992). Persistence of virusesin desert soils amended with anaerobically digestedsewage sludge. *Appl Environ Microbiol.* 58: 636–641.
70. Suttle C. A. (2005). Viruses in the sea. *Nature.* 437: 356–361.
71. Suttle C. A. (2007). Marine viruses-major players in the global ecosystem. *Nat Rev Microbiol.* 5: 801-812.
72. Suttle C. A., Chan A. M., Cottrell M. T. (1990). Infection of phytoplankton by viruses and reduction of primary productivity. *Nature.* 347: 467–469.
73. Suttle C.A. (1994). The significance of viruses to mortality in aquatic microbial communities. *Microb Ecol.* 28: 237–243.
74. Sykes I. K., Lanning S., Williams S. T. (1981). The effect of pH on soil actinophage. *J Gen Microbiol.* 122: 271–280.

75. Tock M. R., Dryden D. T. F. (2005). The biology of restriction and anti-restriction. *Curr Opin. Microbiol.* 8: 466–472.

76. Uetakeh L. E., Burroujs W. (1958). Conversion of somatic antigens in *Salmonella* by phage infection leading to lysis or lysogeny. *Virology.* 5: 68.

77. van der Oost J., Westra E. R., Jackson R. N., Wiedenheft B. (2014). Unravelling the structural and mechanistic basis of CRISPR-Cas systems. *Nat Rev Microbiol.* 12: 479–492.

78. Volkova V. V., Lu Z., Besser T., Grohn Y. T. (2014). Modeling infection dynamics of bacteriophages in enteric *Escherichia coli*: estimating the contribution of transduction to antimicrobial gene spread. *Appl Environ Microbiol.* 80(14):4350-4362.

79. Wagner P. L., Waldor M. K. (2002). Bacteriophage control of bacterial virulence. *Infect Immun.* 70: 3985–3993.

80. Wang X., Kim Y., Wood T. K. (2009). Control and benefits of CP4-57 prophage excision in Escherichia coli biofilms. *ISME J.* 3(10):1164-79. doi: 10.1038/ismej.2009.59.

81. Weinbauer M. G, Rassoulzadegan F. (2004). Are viruses driving microbial diversification and diversity? *Environ Microbiol.* 6: 1-11.

82. Weinbauer M. G. (2004) Ecology of prokaryotic viruses. *FEMS Microbiol Rev.* 28: 127–181.

83. Weinbauer M. G., Agis M., Bonilla-Findji O., Malits A., Winter C. *Bacteriophage in the Environment in Bacteriophage: Genetics and Molecular Biology,* edited by Stephen Mc Grath, Douwe van Sinderen .Caister academic press Norfolk , UK, 2007 Page: 61-92

84. Weitz J. S., Hartman H., Levin S. A. (2005). Coevolutionary arms races between bacteria and bacteriophage. *Proc Natl Acad Sci U S A.* 102:9535– 9540.

85. Williams H. T. (2013). Phage-induced diversification improves host evolvability. *BMC Evol Biol.* 13:17.

86. Williamson K. E., Radosevich M., Wommack K. E. (2005). Abundance and Diversity of Viruses in Six Delaware Soils. *Appl Environ Microbiol.* 71:3119-3125.

87. Winter C., Herndl G. J., Weinbauer M. G. (2004). Diel cycles in viral infection of bacterioplankton in the North Sea. *Aquat. Microb.*

Ecol. 35, 207–216.

88. Wommack K. E., Colwell R. R. (2000). Virioplankton: viruses in aquatic ecosystems. *Microbiol Mol Biol Rev*. 64: 69-114.
89. Yeager J. G., O'Brien R. T. (1979). Enterovirus inactivation in soil. *Appl Environ Microbiol*. 38: 694–701.
90. Zachary A. (1978). An ecological study of bacteriophages of Vibrio natriegens. *Can J Microbiol*. 24: 321–324.

CHAPTER

4

BACTERIOPHAGES AFFECT EVOLUTION OF BACTERIAL COMMUNITIES IN SPATIALLY DISTRIBUTED HABITATS: A SIMULATION STUDY

Alexandra Igorevna Klimenko[1,2], Yury Georgievich Matushkin[1], Nikolay Alexandrovich Kolchanov[1,2] and Sergey Alexandrovich Lashin[1,2]

[1]Institute of Cytology and Genetics SB RAS, Lavrentiev Avenue 10, Novosibirsk 630090, Russia

[2]Novosibirsk State University, Pirogova st. 2, Novosibirsk 630090, Russia

ABSTRACT

Background

Bacteriophages are known to be one of the driving forces of bacterial evolution. Besides promoting horizontal transfer of genes between cells, they may induce directional selection of cells (for instance, according to more or less resistance to phage infection). Switching between lysogenic and lytic pathways results in various types of (co)evolution in host-phage systems. Spatial (more generally, ecological) organization of the living environment is another factor affecting evolution. In this study, we have simulated and analyzed a series of computer models of microbial communities evolving in spatially distributed environments under the pressure of phage infection.

Results

We modeled evolving microbial communities living in spatially distributed flowing environments. Non-specific nutrient supplied in the only spatial direction, resulting in its non-uniform distribution in environment. We varied the time and the location of initial phage infestation of cells as well as switched chemotaxis on and off. Simulations were performed with the Haploid evolutionary constructor software (http://evol-constructor.bionet.nsc.ru/).

Conclusion

Simulations have shown that the spatial location of initial phage invasion may lead to different evolutionary scenarios. Phage infection decreases the speciation rate by more than one order as far as intensified selection blocks the origin of novel viable populations/species, which could carve out potential ecological niches. The dependence of speciation rate on the invasion node location varied on the time of invasion. Speciation rate was found to be lower when the phage invaded fully formed community of sedentary cells (at middle and late times) at the species-rich regions. This is especially noticeable in the case of late-time invasion.

Our simulation study has shown that phage infection affects evolution of microbial community slowing down speciation and stabilizing the system as a whole. This influencing varied in its efficiency depending on spatially-ecological factors as well as community state at the moment of phage invasion.

Keywords: Microbial community, Bacteria, Archaea, Bacteriophage, Phage, Ecological simulation, Evolution, Evolutionary modeling, Prokaryotes

BACKGROUND

Bacteriophages are known to be one of the driving forces of bacterial evolution [1]–[3]. It is generally thought that phages are responsible for about 10–50 % of the total bacterial mortality in surface waters, and 50–100 % in environments that are unfriendly to protists, such as low-oxygen lake waters [4]. Besides promoting horizontal transfer of genes between cells [5]–[7], they may induce either directional selection of cells [8], for instance, according to more or less resistance to phage infection [9]–[12], or fluctuating selection [10], [13]. Switching between lysogenic and lytic pathways results in various types of (co)evolution in bacterial cell-phage systems [14], [15]. Spatial (more generally, ecological) organization of the living environment is another factor affecting evolution [16]–[18]. A number of studies, both experimental and theoretical have recently been published on various aspects of phage-bacteria evolution.

The population mixing increasing host exposure to phages via selection for greater resistance and infectivity ranges was proved to promote arms-race dynamics [13]. It explains the variation in coevolutionary dynamics between different host–parasite systems, and more specifically the observed discrepancies between laboratory and field bacteria–virus coevolutionary studies [13]. Using a spatially explicit, individual-based model, it was shown that less infective pathogens may have an advantage in spatially structured populations, even when well-mixed models predict that they will not [19].

In a time-shift experiment with both sympatric and allopatric phages from either contemporary or earlier points in the season, it was demonstrated that bacterial resistance is higher against phages from the past, regardless of spatial sympatry or how much earlier in the season phages were collected [20]. It was also shown that future bacterial hosts are more resistant to both sympatric and allopatric phages than contemporary bacterial hosts. Nutrients availability was both theoretically and experimentally shown to affect the relative extent of escalation of resistance and infectivity (arms race dynamic) and fluctuating selection (fluctuating selection dynamic) in experimentally coevolving populations of bacteria and viruses [21]. Scanlan and colleagues have shown that in addition to affecting genome-wide evolution in loci not obviously linked to parasite resistance, coevolution can also constrain the acquisition of mutations beneficial for growth in the abiotic environment [22]. The conditions for achieving coexistence on the edge between two habitats, one of which is a bacterial refuge with conditions hostile to phage whereas the other is phage friendly were theoretically studied in [23]. They analyzed how bacterial density-dependent, or quorum-sensing, mechanisms such as the formation of biofilm can produce such refuges and edges in a self-organized manner.

A multiscale model of dynamic coevolution between hosts and viruses in an ecological context incorporating CRISPR immunity principles was presented in [24]. Hosts and viruses were shown to coevolve to form highly diverse communities. They observed evolutionary dynamics consistent with both incomplete selective sweeps of novel strains (as single strains and coalitions) and the recurrence of previously rare strains. Coalitions of multiple dominant host strains were predicted to arise because host strains can express nearly identical immune phenotypes mediated by CRISPR defense albeit with different genotypes [24]. The evolution of generalism in well-mixed populations was found to be highly sensitive to the severity of associated fitness costs, but the constraining effect of costs on the evolution of generalism is lessened in spatially structured populations [25]. The contrasting outcomes between the two environments was explained by different scales of competition (i.e., global vs. local). They suggested that local interactions may have important effects on the evolution of generalism in host-parasite interactions, particularly in the presence of high fitness costs [25].

In our previous study [26] we modeled the opposite trends of genome amplification/simplification occurred in microbial communities via gene horizontal transfer/gene loss. It was demonstrated that species with reduced genomes tend to replace genetically and metabolically rich species under highly favorable environmental conditions. Under unfavorable conditions, the opposite tendency was observed. It was also shown that phages invasion into the system radically changed the current evolutionary trends.

A wide range of mathematical and computational techniques is used for modeling and simulation of phage-bacterium systems. There are both classical ODE/PDE approaches (see examples in [11], [27]–[31]) and modern agent-based and/or multiscale approaches (see examples in [19], [23], [25], [32], [33]).

In this study, we have built and simulated a series of computer models of microbial communities evolving in spatially distributed environments under the pressure of phage infection. Communities inhabited spatially distributed flowing environments. Non-specific nutrient supplied in the only spatial direction, resulting in its non-uniform distribution in environment. We varied the time and the location of initial phage invasion as well as switched chemotaxis on and off and observed that these factors may lead to different evolutionary scenarios.

METHODS

Model Overview

We have used the Haploid evolutionary constructor (HEC) [34], [35] to build the model and to perform the simulations. The HEC models microbial community consisting of different microbial species which we call populations. The hierarchical scheme of HEC models is shown in Fig. 1. Each population consists of cells sharing the same metabolic specificity i.e. nutrients consumed and products secreted. We assume that metabolic pathways of substrates utilization and products synthesis controlled by corresponding gene networks are the same for all cells of a population. Therefore, we consider genes just as numerical parameters of these gene network models. The parameters may be responsible for efficiency of either utilization or synthesis of metabolites. Note that the HEC architecture provides various mathematical formalisms to be used to describe a gene network model mathematically (see details in [35]). In this case, mutations just change the numerical value of a certain parameter in one or several cells (which is interpreted as an origin of novel allelic variant) resulting in genetic polymorphism in a population. On the other hand, selection of more fit allelic combinations leads to the extinction of weaker alleles decreasing the polymorphism. Horizontal transfer of genes between cells of different populations is modeled in HEC via rearrangement of recipient cell's gene network by embedding additional nodes into this network (Fig. 2). Loss of genes is modeled in the same manner, nodes, of course, are deleted. We suppose in our model, that such events of horizontal gene transfer and gene loss are associated with the origin of novel species as they significantly change the metabolic properties of cells. Technically, we use in HEC the super-individual concept [36], [37].

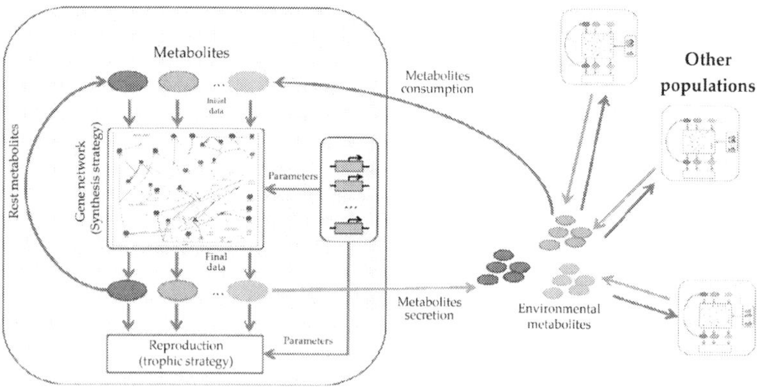

Figure 1: The hierarchical scheme of the HEC model (according to [35], [38]).

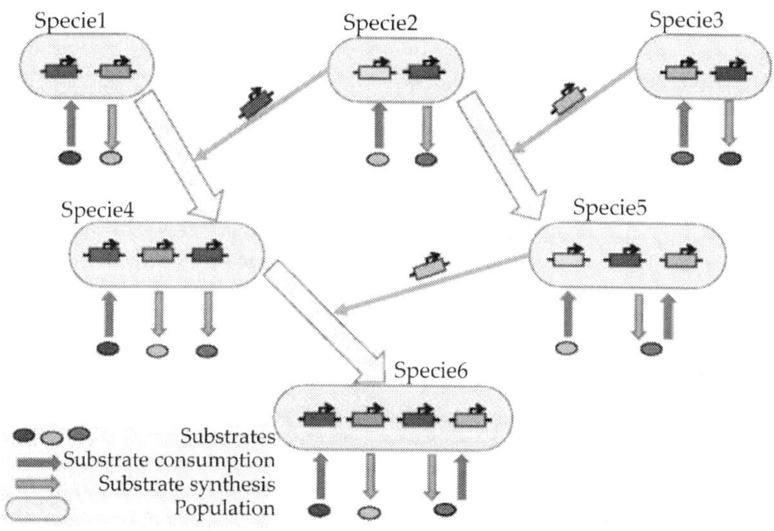

Figure 2: Simulation of horizontal gene transfer in the HEC (according to [40]).

Spatially Distributed Habitats

We describe the spatial organization of environment by a grid of so-called point environments or nodes – small well-mixed volumes containing cells, phage particles and substances. These volumes exchange their contents affected by diffusion and flow as well as during the bacterial chemotaxis activity (Fig. 3, detailed description in [38]). Flow is a directed force mixing contents of the spatial system and drawing particles from its beginning to the end. On the contrary diffusion is considered as undirected dissemination of the point environment's contents. Chemotaxis is an ability of bacterial cells to move towards the attractants and away from repellents, i.e. a capacity to move to a habitat with more chemically beneficial conditions. Such a grid of multiple well-mixed volumes is capable to simulate heterogeneous distributions of cells, substances and phage particles.

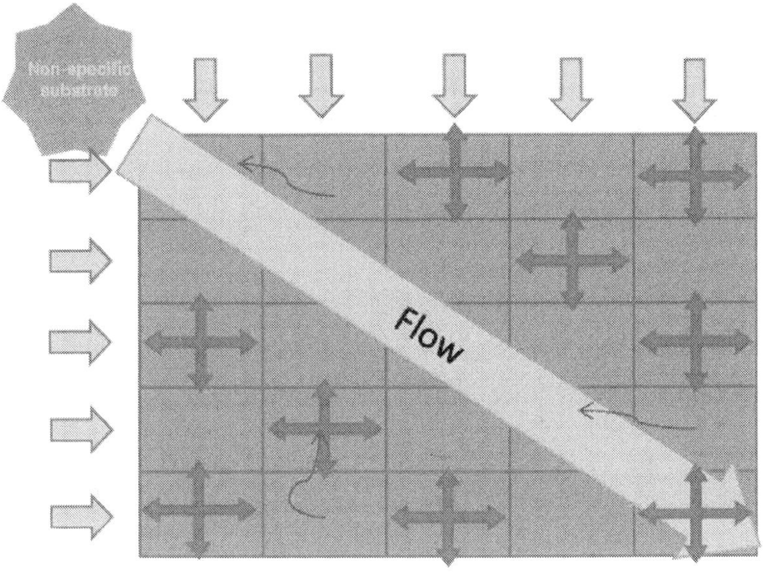

Figure 3: Spatial organization of the habitat. Mesh of 5x5 nodes. The flow determines the gradient of non-specific substrates. Green arrows depict the directions of diffusion. Red arrows show an example of chemotactic behavior of cells.

We considered 2D spatially distributed flowing habitat represented as a 5x5 grid (Fig. 3). In-flow supplies the habitat with non-specific nutrients we call non-specific substrates through the upper and leftmost nodes. The flow and diffusion then spread nutrients over all other nodes of the habitat including not only non-specific substrates, but also metabolites synthesized and secreted by cells (specific substrates). Cells themselves are also passively transported by the flow and diffusion, but additionally may move via chemotaxis (details in [38]). Thus, the (1,1) node and its neighbors may be considered as nutrient-rich (in terms of non-specific substrates), while the (5,5) node and its neighbors are nutrient-poor (in the same terms). Here and after let us call the (1,1) node as the "top", the (3,3) – "middle", and the (5,5) – "bottom" of the habitat. It should be noted that we have tested the 10×10 and 20×20 habitats and found no principal difference with the 5×5 case.

Modeling Bacteriophage Infection

In the HEC, bacteriophages (phages) are described using special type of populations – phage populations, which are capable to infect microbial cells. In this particular case the phage was able to infect those cells which are capable to utilize S1 substrate. The infection process includes the following stages:

- infestation of healthy cells by phage invasion (from environment into some cells of a population);
- phages reproduction inside the infected cells;
- phage burst after cell lysis.

Infected cells may develop then according to either lytic or lysogenic pathway. In the first case, the infected cells die bursting new phages into the habitat. In the second case, on the contrary, no cell die, phage genes are integrated into cell genome and the cells become prophages. The choice of lysogenic or lytic scenarios depends on environmental conditions and cells well-being at the moment of infestation (it is in accordance with biologically known facts [39]). Under favorable conditions, in the case of a positive population dynamics ensured by high environmental substrate concentrations and/or better genetic adaptation compared to other populations, the infection process occurs along the lytic pathway. Otherwise, under unfavorable conditions, infection takes the lysogenic pathway. Later, if the conditions are improved, a part of the population may switch to the lytic pathway, which is followed by the death of this part of the population and generation of phages. The detailed description of the modeling technique for phages in the HEC is presented in [26]. In this study, we considered invasion of phages varying both the moment of invasion and its localization.

RESULTS AND DISCUSSION

Using the approach described above, we have developed a model of spatially distributed microbial community under bacteriophage attack. We started simulations with the simple symbiotic community of three different populations (Fig. 4) uniformly inhabited spatially distributed 2D environment (Fig. 3).

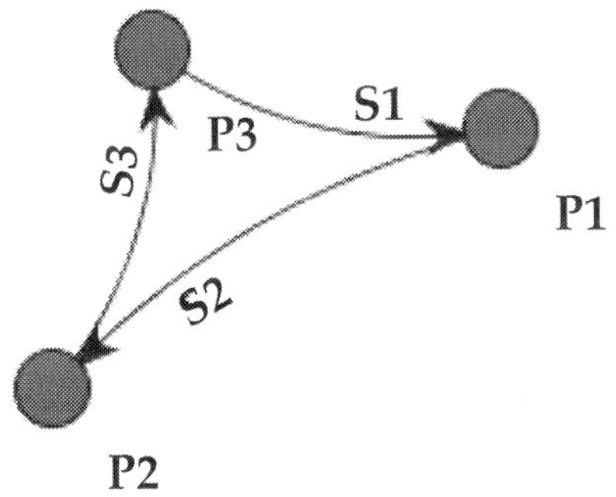

Figure 4: Trophic graph of the initial community. N1 – non-specific substrate consumed by all populations (P1, P2, P3) of the community. S1, S2, S3 – specific substrates synthesized by corresponding cells. P1→S2P2 means that cells of P1 population produce S2 substrate, which is consumed by cells of P2 population.

In the course of evolution the processes of horizontal gene transfer and gene loss stochastically simulated. It was associated with the origin of novel species (see details in the previous chapter). Initial environmental conditions promoted the origin of species possessing more and more complex genomes built combinatorically from relatively simple genomes of initial cells, as it was previously studied in [40]. We varied the time of initial phage invasion as well as its localization. The main questions we wanted to discover were the following:

- how does the phage invasion affect the ecological-evolutionary trends taking course in the community?
- what is the role played by spatial factors in found effects?
- is the bacteriophage an impediment for microbial communities' evolution?

Varying the time of initial phage invasion, we have found that infestation in general inhibits speciation, and after a certain time the species composition and the size of the community become stationary (Fig. 5). It looks as if the phages stop the evolution of the community. It should be noted that we talk about speciation in terms of variation of gene set involved in metabolic networks and the evolution of immune mechanisms is out of scope in this simulation study. It explains why we observe decrease in a speciation rate under phage invasion conditions whereas other studies [2] putting an emphasis on the evolution of defense and counter-defense strategies provide evidences of increasing phage and bacterial evolution rate. The other possible reason of the discrepancy is because we focused primarily on interactions between bacteria and temperate phages, rather than lytic phages as it has been done in the above-mentioned case.

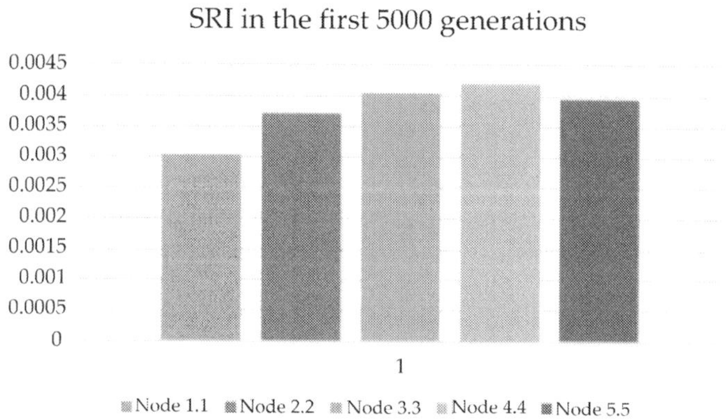

Figure 5: SRI calculated for the first 5000 generations in various nodes.

We obtained the same results when chemotaxis was switched on/off. This effect has been estimated using the Speciation Rate Index (SRI), which was calculated according to the following formula:

$$SRI = \frac{\| \{ p | p\text{-}newly\ \ emerged\ \ viable\ \ populations \} \|}{time\ \ span} \tag{1}$$

where time_span is the period speciation rate is estimated about; a population is assumed to be viable if it survived over a period of thres generations (where thres is a threshold value, in our case, 500). According to the equation 1, the higher SRI is, the more intensive the speciation is over this time period. Conversely, the lower SRI is, the fewer viable species originate in time.

Chemotaxis-Off Case

Initially, we have estimated SRI in all spatial nodes of the habitat in the absence of phage invasion (Fig. 5). Fig. 5 shows that speciation rate growth according to moving away from the in-flow (top) nodes. It could be explained by the fact that in top nodes there are several populations, which effectively utilize the non-specific substrate. These population have got dominate biomass. At the same time, novel ecological licenses associated with high concentrations of specific substrates arise in middle and top nodes due to transportation via the flow. It opens possibilities for fixation of novel species specialized on utilization of those specific substrates. That is to say, in distant, bottom nodes, higher biodiversity is associated with low biomass (Fig. 6).

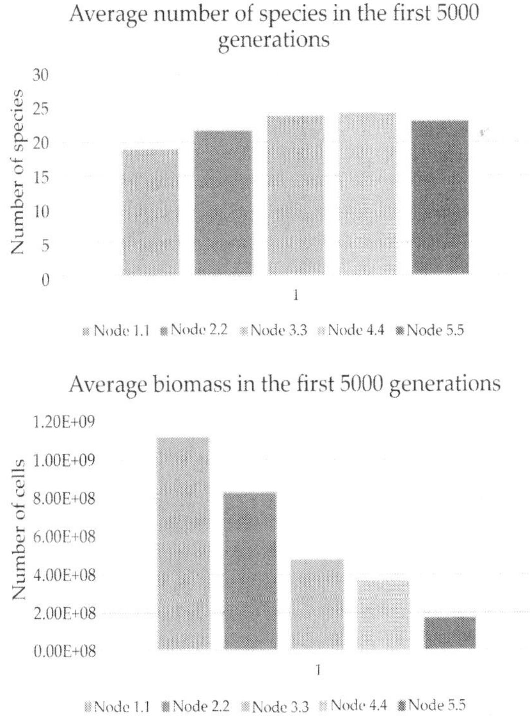

Figure 6: Average number of species (up) and biomass (down) over the first 5000 generations in various nodes.

After that, we have analyzed how the phage invasion affects speciation rate and species richness. Results obtained show the dependency of speciation rate on invasion parameters (Table 1).

Table 1: Average SRI values before and after phage invasion in various habitat locations. First 1000 generations has not been taken into account as there is a bias associated with high initial speciation

	Early-time (1st generation)		Middle-time (5000th generation		Late-time (6600th generation)	
Into node (1,1)	SRI before	SRI after	SRI before	SRI after	SRI before	SRI after
	-	0.856e-3	6.850e-3	0.520e-3	7.143e-3	0.529e-3
Into node (3,3)	SRI before	SRI after	SRI before	SRI after	SRI before	SRI after
	-	0.744e-3	7.275e-3	0.400e-3	7.746e-3	0.206e-3
Into node (5,5)	SRI before	SRI after	SRI before	SRI after	SRI before	SRI after
	-	0.578e-3	8.250e-3	0.600e-3	6.964e-3	0.412e-3

Table 1 shows that phage invasion results in decrease of speciation rate by more than one order. As is easy to see the time dependency of speciation rate varies with the distance between the node and in-flow location:

- In the case of early-time invasion, there is an inverse dependence – the farther the place of invasion, the less the speciation rate after infestation (Fig. 7).

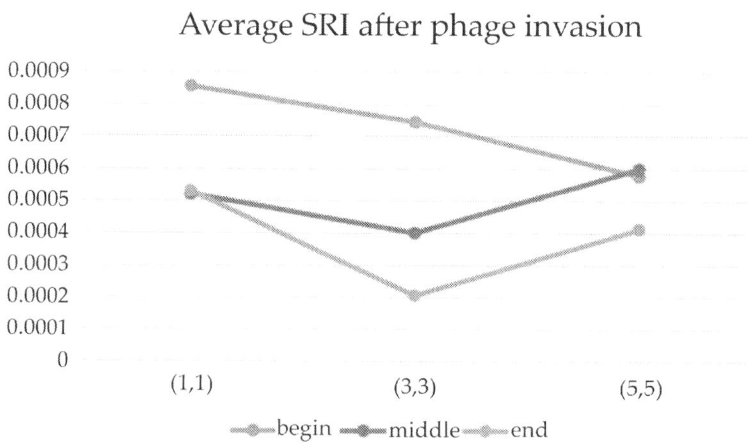

Figure 7: Dependence of SRI on time of invasion and invasion initial localization. Blue plot – early-time (1st generation), orange – middle-time (5000th generation), grey – late-time (6600th generation). X-axis – number of the node of initial invasion, Y-axis – average SRI calculated up to the end of simulation.

- In the case of middle-time invasion, the lowest speciation rate is observed when the phage invades the central node (3,3) of the habitat. It is notable that if the phage invades the most bottom node (5,5), then average SRI becomes higher compared to the most top node (1,1).
- In the case of late-time invasion, results look similar to the previous case except that average SRI is lower when the phage invaded central and bottom nodes.

As it was previously said, distant nodes are characterized by higher concentrations of the specific substrates. Therefore, these results are in accordance with earlier reported [41] suggestion that the viral effect is probably larger in eutrophic waters than in oligotrophic waters.

Table 2 shows that in the case of middle-time invasion, average SRI increases from top (1,1) nodes (non-specific substrate-rich) to bottom (5,5) nodes (non-specific substrate-poor). Additional plots show that the phage invasion leads to drastic reduction of species richness of the community. The data confirm that SRI as well as species richness are higher in the absence of phages and when there are many ecological licenses.

Table 2: Average SRI calculated over the whole simulation time with respect to the initial invasion location

	Early-time (1st generation)	Middle-time (5000th generation)	Late-time (6600th generation)
Into node (1,1)	1.3030e-3	5.6061e-3	6.5657e-3
Into node (3,3)	1.1717e-3	5.8687e-3	7.2121e-3
Into node (5,5)	1.2828e-3	6.3333e-3	6.8081e-3

Chemotaxis-on case

In cases when chemotaxis is switched on, the spatial stratification of a community in terms of biomass is more expressed. We hypothesize that this is due to the observed accumulation of specific substrates in nodes close to the bottom of the habitat. Anyway, the distributions of species richness and SRI in general look similarly to the corresponding distributions obtained for the previous case. However, the spatial differences here are more evident (Fig. 8). On the other hand, the total species number as well as SRI in this case (chemotaxis is on) are definitely lower compared to the previous case (chemotaxis is off). It is in good agreement with evolutionary biology data postulating that migration impedes speciation while isolation promotes it [42]–[44].

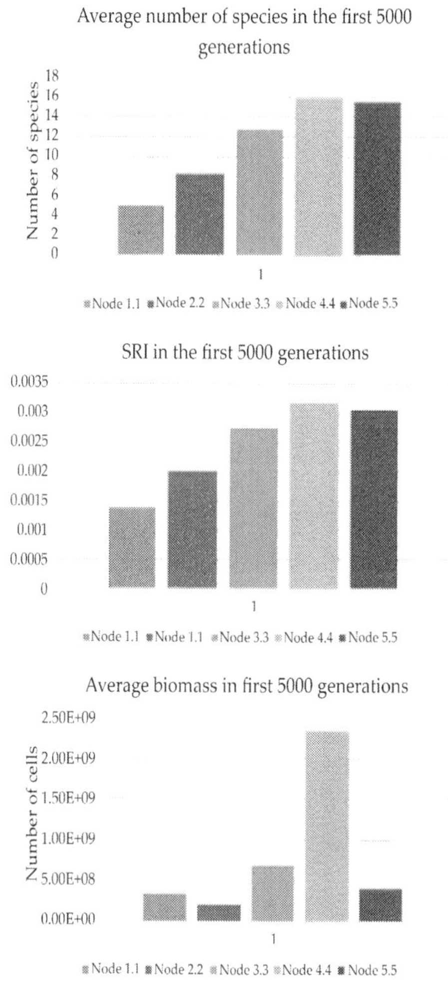

Figure 8: Distributions of average values for (a) species richness; (b) SRI; (c) total biomass in spatial nodes. Values were calculated for the first 5000 generations.

Figure 9 shows that phage invasion into the bottom node (5,5) led to the growth of speciation rate both for early-time and late-time cases. It is related to the observed fact that the phage infection could not fixate in the habitat when invaded into bottom nodes. However, in the chemotaxis-on case, if the phage invaded top nodes, the species richness observed to be higher in bottom nodes.

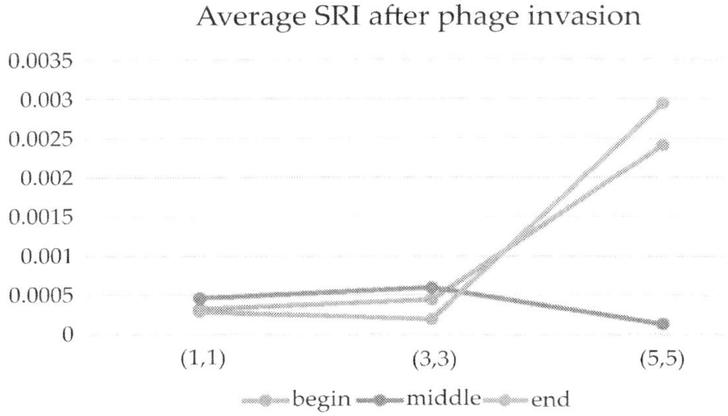

Figure 9: Dependence of SRI on time of invasion and invasion initial localization (chemotaxis is on). Blue plot – early-time (1st generation), orange – middle-time (5000th generation) , grey – late-time (6600th generation). X-axis – number of the node of initial invasion, Y-axis – average SRI calculated up to the end of simulation.

CONCLUSIONS

In this simulation study, we have shown that bacteriophages may act as constraining factors of microbial community evolution. Phage infection decreases the speciation rate by more than one order as far as intensified selection blocks the origin of novel viable populations/species, which could carve out potential ecological niches. At the same time, phages act as a stabilizing factor suspending superfluous speciation and encouraging stationary state of the system (in terms of species number).

It has also been shown that the dependence of speciation rate on the invasion node location varied on the time of invasion. Speciation rate is found to be lower when the phage invaded fully formed community (middle and late times) at the species-rich nodes (central node (3,3)). This is especially noticeable in the case of late-time invasion.

Those dependencies differ in the case of chemotaxis switched on. In both early-time and late-time cases, the speciation rate remains relatively high when phage invaded bottom nodes. This phenomenon requires additional investigations.

As a result, our simulation study shows that phage infection affects evolution of microbial community slowing down speciation and stabilizing the system as a whole. This influence varied in its efficiency depending on spatially-ecological factors as well as community state at the moment of phage invasion.

REFERENCES

1. Hendrix RW. Bacteriophage Evolution and the Role of Phages in Host Evolution. In: Waldor MK, et al. Phages: their role in bacterial pathogenesis and biotechnology. American Society of Microbiology; 2005:55–65.
2. Koskella B, Brockhurst MA: Bacteria–phage coevolution as a driver of ecological and evolutionary processes in microbial communities. FEMS Microbiol Rev. 2014, 38: 916-31. 10.1111/1574-6976.12072.
3. Pride D, Wassenaar T, Ghose C, Blaser M: Evidence of host-virus co-evolution in tetranucleotide usage patterns of bacteriophages and eukaryotic viruses. BMC Genomics. 2006, 7: 8-10.1186/1471-2164-7-8.
4. Fuhrman JA: Marine viruses and their biogeochemical and ecological effects. Nature. 1999, 399: 541-8. 10.1038/21119.
5. Ochman H, Lawrence JG, Groisman EA: Lateral gene transfer and the nature of bacterial innovation. Nature. 2000, 405: 299-304. 10.1038/35012500.
6. Canchaya C, Fournous G, Chibani-Chennoufi S, Dillmann M-L, Brüssow H: Phage as agents of lateral gene transfer. Curr Opin Microbiol. 2003, 6: 417-24. 10.1016/S1369-5274(03)00086-9.
7. Skippington E, Ragan MA: Lateral genetic transfer and the construction of genetic exchange communities. FEMS Microbiol Rev. 2011, 35: 707-35. 10.1111/j.1574-6976.2010.00261.x.
8. Bohannan BJM, Lenski RE: Linking genetic change to community evolution: insights from studies of bacteria and bacteriophage. Ecol

Lett. 2000, 3: 362-77. 10.1046/j.1461-0248.2000.00161.x.

9. Desiere F, McShan WM, van Sinderen D, Ferretti JJ, Brüssow H: Comparative genomics reveals close genetic relationships between phages from dairy bacteria and pathogenic streptococci: evolutionary implications for prophage-host interactions. Virology. 2001, 288: 325-41. 10.1006/viro.2001.1085.
10. Williams HT: Phage-induced diversification improves host evolvability. BMC Evol Biol. 2013, 13: 17-10.1186/1471-2148-13-17.
11. Berngruber TW, Froissart R, Choisy M, Gandon S: Evolution of virulence in emerging epidemics. PLoS Pathog. 2013, 9: 10.1371/journal.ppat.1003209. Article ID e1003209
12. Roberts PD, Portfors CV: Design principles of sensory processing in cerebellum-like structures. Early stage processing of electrosensory and auditory objects. Biol Cybern. 2008, 98: 491-507. 10.1007/s00422-008-0217-1.
13. Gomez P, Ashby B, Buckling A. Population mixing promotes arms race host-parasite coevolution. Proc R Soc B Biol Sci. 2014;282:20142297–7.
14. Heilmann S, Sneppen K, Krishna S: Sustainability of virulence in a phage-bacterial ecosystem. J Virol. 2010, 84: 3016-22. 10.1128/JVI.02326-09.
15. Levin BR, Moineau S, Bushman M, Barrangou R: The population and evolutionary dynamics of phage and bacteria with CRISPR–mediated immunity. PLoS Genet. 2013, 9: 10.1371/journal.pgen.1003312. Article ID e1003312
16. Perfeito L, Pereira MI, Campos PRA, Gordo I: The effect of spatial structure on adaptation in Escherichia coli. Biol Lett. 2008, 4: 57-9. 10.1098/rsbl.2007.0481.
17. Mitri S, Xavier JB, Foster KR: Social evolution in multispecies biofilms. Proc Natl Acad Sci U S A. 2011, 108 (Suppl): 10839-46. 10.1073/pnas.1100292108.
18. Lion S, Gandon S: Evolution of spatially structured host-parasite interactions. J Evol Biol. 2015, 28: 10-28. 10.1111/jeb.12551.
19. Roychoudhury P, Shrestha N, Wiss VR, Krone SM. Fitness

benefits of low infectivity in a spatially structured population of bacteriophages. Proc R Soc B Biol Sci. 2013;281:20132563–3.

20. Koskella B, Parr N: The evolution of bacterial resistance against bacteriophages in the horse chestnut phyllosphere is general across both space and time. Philos Trans R Soc B Biol Sci. 2015, 370: 20140297-10.1098/rstb.2014.0297.
21. Lopez Pascua L, Hall AR, Best A, Morgan AD, Boots M, Buckling A: Higher resources decrease fluctuating selection during host-parasite coevolution. Ecol Lett. 2014, 17: 1380-8. 10.1111/ele.12337.
22. Scanlan PD, Hall AR, Blackshields G, Friman V-P, Davis MR, Goldberg JB, et al: Coevolution with bacteriophages drives genome-wide host evolution and constrains the acquisition of abiotic-beneficial mutations. Mol Biol Evol. 2015, 32: 1425-35. 10.1093/molbev/msv032.
23. Heilmann S, Sneppen K, Krishna S: Coexistence of phage and bacteria on the boundary of self-organized refuges. Proc Natl Acad Sci. 2012, 109: 12828-33. 10.1073/pnas.1200771109.
24. Childs LM, Held NL, Young MJ, Whitaker RJ, Weitz JS: Multiscale model of crispr-induced coevolutionary dynamics: diversification at the interface of lamarck and darwin. Evolution (N Y). 2012, 66: 2015-29.
25. Ashby B, Gupta S, Buckling A: Spatial structure mitigates fitness costs in host-parasite coevolution. Am Nat. 2014, 183: E64-74. 10.1086/674826.
26. Lashin SA, Matushkin YG, Suslov VV, Kolchanov NA: Evolutionary trends in the prokaryotic community and prokaryotic community-phage systems. Russ J Genet. 2011, 47: 1487-95. 10.1134/S1022795411110123.
27. Lenski RE, Levin BR: Constraints on the coevolution of bacteria and virulent phage: a model, some experiments, and predictions for natural communities. Am Nat. 1985, 125: 582-602. 10.1086/284364.
28. Wang I-N, Dykhuizen DE, Slobodkin LB: The evolution of phage lysis timing. Evol Ecol. 1996, 10: 545-58. 10.1007/BF01237884.
29. Nowak MA, May RM: Superinfection and the evolution of parasite virulence. Proc R Soc B Biol Sci. 1994, 255: 81-9. 10.1098/rspb.1994.0012.

30. Abedon ST, Culler RR: Bacteriophage evolution given spatial constraint. J Theor Biol. 2007, 248: 111-9. 10.1016/j.jtbi.2007.02.014.
31. Bull JJ, Crandall C, Rodriguez A, Krone SM: Models for the directed evolution of bacterial allelopathy: bacteriophage lysins. PeerJ. 2015, 3: 10.7717/peerj.879. Article ID e879
32. Weitz JS, Hartman H, Levin SA: Coevolutionary arms races between bacteria and bacteriophage. Proc Natl Acad Sci U S A. 2005, 102: 9535-40. 10.1073/pnas.0504062102.
33. Iranzo J, Lobkovsky AE, Wolf YI, Koonin EV: Immunity, suicide or both? Ecological determinants for the combined evolution of anti-pathogen defense systems. BMC Evol Biol. 2015, 15: 43-10.1186/s12862-015-0324-2.
34. Lashin SA, Matushkin YG: Haploid evolutionary constructor: new features and further challenges. In Silico Biol. 2012, 11: 125-35.
35. Lashin SA, Klimenko AI, Mustafin ZS, Kolchanov NA, Matushkin YG: HEC 2.0: improved simulation of the evolution of prokaryotic communities. Math Biol Bioinform. 2014, 9: 585-96. 10.17537/2014.9.585.
36. Scheffer M, Baveco JM, DeAngelis DL, Rose KA, van Nes EH: Super-individuals a simple solution for modelling large populations on an individual basis. Ecol Modell. 1995, 80: 161-70. 10.1016/0304-3800(94)00055-M.
37. Hellweger FL, Bucci V: A bunch of tiny individuals—Individual-based modeling for microbes. Ecol Modell. 2009, 220: 8-22. 10.1016/j.ecolmodel.2008.09.004.
38. Klimenko AI, Matushkin YG, Kolchanov NA, Lashin SA: Modeling evolution of spatially distributed bacterial communities : a simulation with the haploid evolutionary constructor. BMC Evol Biol. 2015, 15: S3-10.1186/1471-2148-15-S1-S3.
39. Ptashne M. A genetic switch: Gene control and phage lambda. Elsevier B.V., 1986.
40. Lashin SA, Matushkin YG, Suslov VV, Kolchanov NA: Computer modeling of genome complexity variation trends in prokaryotic communities under varying habitat conditions. Ecol Modell. 2012, 224: 124-9. 10.1016/j.ecolmodel.2011.11.004.

41. Steward GF, Smith DC, Azam F: Abundance and production of bacteria and viruses in the Bering and Chukchi Seas. Mar Ecol Prog Ser. 1996, 131: 287-300. 10.3354/meps131287.

42. France SC, Hessler RR, Vrijenhoek RC: Genetic differentiation between spatially-disjunct populations of the deep-sea, hydrothermal vent-endemic amphipod Ventiella sulfuris. Mar Biol. 1992, 114: 551-9. 10.1007/BF00357252.

43. Gavrilets S, Li H, Vose MD: Patterns of parapatric speciation. Evolution (N Y). 2000, 54: 1126-34.

44. Won Y, Young CR, Lutz RA, Vrijenhoek RC: Dispersal barriers and isolation among deep-sea mussel populations (Mytilidae: Bathymodiolus) from eastern Pacific hydrothermal vents. Mol Ecol. 2002, 12: 169-84. 10.1046/j.1365-294X.2003.01726.x.

CHAPTER

5

ROLE OF BACTERIOPHAGE IN MOLECULAR BIOLOGY AND BIOTECHNOLOGY

Dr Prerna Pandey[1], Dr Pratibha Pant[2], Dr Prashant Pant[3],

[1]Dr Prerna Pandey, a biotechnologist with several years of wet lab research experience, worked at International Center for Genetic Engineering and Biotechnology, New Delhi. Her field of research in PhD included isolation and molecular characterization of geminiviruses, genome sequencing, gene annotation, and gene silencing using the RNA interference technology. She has also worked at Transasia Biomedicals and Advance Enzyme Technologies as a scientist. Prerna has published papers in peer-reviewed journals, and has submitted a number of annotated Geminiviral genome sequences in the GenBank, including two novel ones. She has also completed her editing and proof-reading courses from Society for the promotion of Editors and Proof-readers, UK and now works as a freelance scientific editor and writer. When Prerna is not busy with her assignments, she loves to explore historical places.

[2]Dr. Pratibha Pant, is an Assistant Professor of Botany in Daulat Ram College, University of Delhi, New Delhi, India and has a teaching

experience of more than 6 years. She is a plant virologist and biotechnologist by training and has done her PhD in Characterization of Begomoviral genomes from Papaya from Department of Botany, University of Delhi. During her PhD, she utilized a combination of molecular biology and Computational Biology tools to analyze begomoviral genomes and to her credit, a couple of novel begomoviruses have been isolated, sequenced and reported in public databases

[3]Dr. Prashant Pant, is an Assistant Professor of Botany in Dyal Singh College, University of Delhi, New Delhi, India. He has been teaching for more than 5 years now and his main areas have been Bioinformatics, Molecular Biology and Genomics at the college. His PhD was on understanding the Microbial Succession using a Metagenomic approach in a Barren mined-out site undergoing ecological restoration. For this, he deployed many molecular and computational Biology tool and techniques. He has several publications (articles and book chapters) in International Journals and International Publications. Currently, he is also Chief Editor of Bioinformatics Review and a co-founder/Executive Chief Editor of a Science Journal in Hindi Language "Swadeshi Vigyan" for the native language speaking communities. Both the Journals are a service to the scientific community as they are fully open access.

INTRODUCTION

Since their discovery, it has been well established that bacteriophages (also known as phages) are obligate intracellular parasites and are among the most ubiquitous species on the planet. An estimate of about 10^{32} phages (Hanlon, 2007) has been made from which approximately 10^4 to 10^8 particles per ml are present in water, while 10^9 particles per g are present in soil (Weinbauer, 2004).

The discovery of bacteriophages was made as early as 1915 by Federick Twort who observed the "glassy transformation" of micrococcal colonies (Twort, 1936). In 1917, Félix d'Hérelle coined the term "bacteriophage" or bacteria-eater after isolating them from *Shigella* cultures (D'Herelle, 2007). This led to the inception of the idea that phages may be used as therapeutic agents against bacterial diseases. Indeed, d'Hérelle correlated the existence of phages with disease clearance in patients. In 1921, Bruynoghe and Maisin had observed reduced pain and fever in a patient after local injection of staphylococcal phages in boils (Bruynoghe R, 1921). Furthermore, prophylactic capabilities of phages were reported by Félix d'Hérelle in 1921, where prior exposure of chickens to *Bacillus gallinarum* protected the chickens from avian typhosis caused by the bacterial strain (Summers, 2001). These observations gave a strong foothold for the development of phage based therapies against bacterial diseases but the interest in this field declined after the success of antibiotics. Phage biology changed its course and their application in studying molecular genetics and molecular biology was investigated. Revolutionizing role has been played by bacteriophages in bacterial evolution, and thus shaping global ecology today.

In the current chapter, we describe the various contributions made by bacteriophages and the derived functional components in the field of molecular biology and biotechnology. We further describe the generation of engineered bacteriophages and their applications. At the end, major applications like phage therapy, phage typing and phage display are described in brief.

CONTRIBUTIONS TO MOLECULAR BIOLOGY

The simple constitution of phages and ease of their use make them alluring model systems and tools to study molecular biology. T4 phages and λ phages have proved as valuable assets over the years and have contributed in making the foundation for molecular biology. Contributions from Leo Szilard, Salvador Luria, Max Delbrück and Emory Ellis helped in characterization of phages and laid the groundwork for using them as tools in molecular biology. Initial experiments demonstrating DNA as genetic material were carried out by Alfred Hershey and Martha Chase in their famous 'Waring blender experiment'.

Study of bacteria and phage biology has provided us with various valuable tools. Few of the contributions of bacteriophages which have transformed the face of molecular biology are enlisted below:

Restriction Enzymes

Restriction- modification system (or R-M system) was discovered in specific *E. coli* hosts. This system is the primary defence system observed in prokaryotes and involves the participation of host encoded restriction endonuclease and methyltransferase enzymes. Restriction endonucleases digest the phage encoded foreign DNA while methyltransferase mediated modification of particular host DNA sequences ensures recognition of self and foreign nucleic acid sequences. Interestingly, the advent of sequencing technologies have revealed that approximately 90% of the genomes harbour atleast one R-M system and approximately 80% of them harbour multiple R-M systems (http://rebase.neb.com/rebase/rebase.html).

The potential of these enzymes was recognized and has been used extensively in cloning DNA sequences. Restriction enzymes allow the recognition of specific sequences and endonucleolytic cleavage resulting in 5′ or 3′ overhangs or blunt ends.

Ligase

Gellert, Lehman, Richardson, and Hurwitz laboratories discovered DNA ligases in 1967 (Shuman, 2009). These are enzymes which can create phosphodiester bonds and participate in essential cellular processes like DNA replication and DNA repair. This enzyme has been used for several years in molecular cloning, molecular diagnostics and has paved its way into modern day SOLiD sequencing method.

Phages have served as a source of purified ligase for a long time. T4 DNA ligase is a ligase produced by T4 bacteriophage during infection of *E. coli* (Lehman, 1974). This enzyme has been purified and is characterized as a single polypeptide. Despite being a prokaryotic ligase it exhibits an ATP-dependent ligase activity. Isolation from other phages like bacteriophage T7 and South Hebron Phage (SHPh) DNA ligase (Alia' Mousa Al-Manasra, 2012) have also been performed successfully.

Cloning Vectors and Cloning Solutions

Various cloning vectors have been derived from Phage λ and they are demonstrated to provide better cloning efficiency than plasmid cloning vectors. Both insertion (e.g. λgt10 and λcharon16A) and replacement (λEMBL and λZap) vectors have been developed. M13 bacteriophage derived vectors provide advantages of having smaller size; genes cloned in these vectors can be obtained in single stranded DNA form and these can be easily used in sequencing.

Cosmids are yet other tools derived from phages. These combine the essential elements of λ system and plasmids, enabling the cloning of large DNA inserts. Similarly, phagemids; combination of plasmids and M13 bacteriophage genome have also been developed. An example of phagemid is pEMBL8, which was derived from pUC8 plasmid and the M13 genome. Moreover, for cloning of large DNA fragments artificial chromosomes from Phage P1 have also been derived.

PHAGES AND SYNTHETIC BIOLOGY

Synthetic biology aims to engineer new functions into the living systems by modification of biomolecules. Simple circuits like oscillators (Elowitz and Leibler, 2000) and switches (Gardner et al., 2000) and Boolean logic gates (Bonnet et al., 2013; Siuti et al., 2013) have been developed. These tweaks in biological systems have yielded cellular systems capable of detection of pathogenic bacteria and cancer cells, produce antimalarial agents and program cell fate. Bacteriophages have been known as molecular machineries and their parts have been applied to synthetic biology. T7 DNA dependent RNA polymerases ad bacteriophage recombinases (mainly Cre, Bxb1 and PhiC31 recombinases) have been used to generate the aforementioned genetic circuits. Genetic circuits that can record the past reactions have been engineered using phage recombinase.

Modified and improved bacteriophages have been developed using synthetic biology approaches for various applications. To effectively clear *E. coli* biofilms, DspB gene encoding disperin enzyme is introduced in an engineered T7 bacteriophage. Successful biofilm clearance was seen after introduction of this phage. Repeated infection, lytic cycles and release of disperin enzyme which breaks polysachharidc adhesion of biofilms contributed in clearance of *E. coli* biofilm in this study (Lu and Collins, 2007). Attempts to deliver antibiotic sensitizing cassettes to resistant bacteria have been made. Edgar et al. in an endeavour to combat resistance against streptomycin and fluoroquinolones (conferred by mutated *rpsL* and *gyrA* genes) in resistant cells reintroduced wild type *rpsL* and *gyrA* genes through engineered phage λ (Edgar et al., 2012).

APPLICATIONS

Both native and engineered phages find their applications in diverse fields as therapy and diagnostics.

Phage Therapy

Phage therapy was among the first envisioned application of bacteriophages. As mentioned before, both therapeutic and prophylactic have been demonstrated for phages in animal models by Félix d'Hérelle. This concept soon gained momentum and caught interest of various industries for commercialization. A surge of interest in antibiotics led to progressive decline in interest in phage therapy in many parts of the world. Antibiotic resistance has resulted in renewed interest in this field.

Encouraging results are obtained in phage therapy for treatment of human pathogens. Effective and safe treatment of ear infections caused by *Pseudomonas aeruginosa* was achieved (Wright et al., 2009). PhagoBurn clinical trial, a Phase I/II clinical trial of phage mediated treatment of burn wounds caused by *E. coli* and *P. aeruginosa* is under progress (PhagoBurn trial website).

The concept was also introduced in plants as early as 1924. Mallman and Hemstreet successfully inhibited the growth of Xanthomonas campestris pv campestris (pathogen causing rot), by using filterate from rotting cabbage (Mallmann WL, 1924). Field trials in plants have also been conducted. Effective reduction in disease incidence was reported after phage treament of Pantoea stewartii infested corn seeds (Thomas, 1935).

Advantages of Using Phages

Various advantages have been attributed to the use of bacteriophages in therapy against bacterial diseases. These are enlisted below:

- Pathogenic bacterial strains when infected with lytic bacteriophages ensure complete lysis. In contrast to antiobiotics, which may be bacteriostatic, the probability of resistance observed towards phage therapy may be lesser in longer run. Furthermore, phages may prove useful against multi-drug resistant pathogens like *Staphylococcus aureus* (Gupta and Prasad, 2011; Mann, 2008).

- Phages are relatively non-toxic as they constitute of nucleic acid and protein coat (Abedon and Thomas-Abedon, 2010; Kutter et al., 2010; Skurnik et al., 2007). Studies on phage toxicities have also shown no severe responses after using phage therapy. As a precautionary measure, phages used in therapy should be highly purified to avoid anaphylactic reaction from the bacterial components found in crude phage lysates.
- Administration of antibiotics causes side effects like disruption of normal gut flora. Superinfections by *Clostridium difficile* colitis or *Candida albicans* yeast infections are often observed after use of broad spectrum antibiotics (Carlton, 1999). Phages, due to their narrow host specificity may have minor impact on gut flora
- Mixtures of phages can be made to increase the spectrum of target bacterial population. Various formulations like solution, solid impregnation and creams can be made. Additionally route of administration of the appropriate formulation is also flexible.
- Clearance of bacterial biofilms has proved as a challenge for antibiotics. Phages have got limited success in clearing biofilms probably by removing one layer at a time.

Concerns

Further studies on phage biology and also on their application for phage therapy are imperative to give insights into the safety and success rate of this technique. Some of the concerns are as follows:

- Phage selection is an important step in development of phage therapy for a particular disease. "Obligately lytic" phages are suitable for therapy purposes. Temperate phages may convert sensitive phages resistant by a process of superinfection immunity. Moreover, low virulent phages and those with poor pharmacokinetic properties are unsuitable for therapy. Selection of phages for biological control of plant pathogens has also been challenging. Balogh in his experiments with phages to control *X. citri* (causes citrus canker) observed that phages interact differentially with their host bacteria on grapefruit leaves (Balogh, 2006), while no *in vitro* plaques were formed against bacterial spot on tomato.

- Presumptive therapy i.e. broad spectrum therapy where the causative agent for the disease is unidentified is not possible with phages where narrow spectrum of host is present.
- Emergence of phage resistance bacterial strains by mutation and natural selection pose a serious limitation to the large-scale use of phage therapy in health related and agricultural fields.

Phage Display

G. Smith in 1985 developed phage display to display polypeptides on lysogenic filamentous bacteriophages (Smith, 1985). Gene encoding the polypeptide is packaged in the single stranded DNA (ssDNA) of the filamentous phage (like f1, fd or M13) and the peptides are expressed as fusion peptides with the phage coat proteins. Majority of the peptides are displayed as a fusion with the pIII minor coat protein as compared to pVIII which is only used to display short peptides.

Phage display has been used for affinity screening of ligands, receptors, epitopes and antibody binding sites. Applications of the technique have been found in various fields including transfusion medicine, autoimmune disorders, drug discovery and cell biology. In the field of transfusion medicine, antibodies against Rh(D) (required for blood group typing), cluster of differentiation (CD) antigens (Barth et al., 2000; Loffler et al., 2000), clotting factors (Suggett et al., 2000), fetal red blood cells (Huie et al., 2001) and dendritic cells (Fitting et al., 2011). Phage display proved instrumental in studying autoimmune disorders like chronic immune thrombocytopenic purpura (AITP) (Gevorkian et al., 1998), thrombotic thrombocytopenic purpura (TTP) (Luken et al., 2006) and autoimmune diabetes (Farilla et al., 2002). Although, the use of this technique is limited in fields related to agriculture, many applications have been proposed which can open new avenues in the field. Immune system in plants has not been well studied and attempts have been made using phage display to elucidate related mechanisms. Phage libraries displaying recombinant plant proteins from cDNA of microbe challenged *Arabidopsis* were made by Rioja et al. to study the immune related proteins present in this plant species (Rioja et al., 2013). Later, phage displayed plant proteins were selected on the basis of their ability to interact with different species of *Pseudomonas*. Another attempt to apply phage display in this field is to mediate directed evolution of plants to produce protease inhibitors which can combat the continuously evolving range of proteases by the pests (Markland et al., 1996; Willats, 2002). These instances clearly reflect the plethora of applications that phage display may be used in the future.

Phage Typing

Phages display specificity of infection by interacting with specific molecules on the bacterial cell and this is exploited for detection of specific bacterial strain (Clark and March, 2006). Lawn of unknown bacterial culture is exposed to different phages and appearance of plaque (or clear zone) helps in identification of the bacterial strain. Multiple other methods have been devised to detect pathogenic bacteria like delivery of reporter genes (e.g. Green Fluorescent Protein) to the bacterial strains.

These conventional typing methods suffer from numerous limitations. These include the requirement of isolation of pure cultures of unknown bacteria, maintenance of phage stocks and the host strains for their propagation. Generation of false positive and false negative results also present a serious limitation in the application of this technique. Additionally, the bacterial host has to be viable at the time of detection for the propagation of phages. Variations of this technique with improved resolution and better detection have been developed. Amplification based phage typing methods have been recently devised. These are based on the detection of increased titre of phages in the presence of host bacteria in the sample. An example of this is the detection of *R. solanacearum* by estimation of phage titre using quantitative PCR (Kutin et al., 2009). Engineered bacteriophages have been developed to provide faster detection with higher sensitivity and specificity (Smartt et al., 2012). Phages carrying reporters (colorimetric, fluorescent or luminescent), enzyme coding genes and tags that can interact with elements like antibodies or quantum dots are being engineered (Edgar et al., 2006).

Phage typing has been used to detect human, animal and plant pathogens. Reporter phages for detection of *Pseudomonas cannabina pv. alisalensis* (causes bacterial blight of crucifers) were developed. These "*luxAB*-tagged" reporter phages were able to detect the bacteria in minutes (Schofield et al., 2013). In human pathogens, phage typing of *Mycobacteriumtuberculosis, E.coli, Pseudomonas, Salmonella, Listeria,* and *Campylobacter* species has been carried out (Barry et al., 1996).

Serious concerns have been raised for the use of phage typing due to the variation in interpretation of results obtained from independent studies. Reports from the phage typing during epidemiological outbreaks of bacterial pathogens have several times presented conflicting data. For instance, phage-typing of *Salmonella* strains after an outbreak in Denmark in two independent institutes (National Food Institute and WHO Collaborative Centre) resulted in different identification of strains (phage type U292 and phage type U276 respectively) (Baggesen et al., 2010).

BACTERIOPHAGES AND PLANT BACTERIA

Bacteriophages have been applied in agriculture majorly as a biocontrol agent for bacterial plant pathogens. Biocompatibility, specificity, ease and inexpensive production leads contribute to making phage therapy as an attractive therapeutic method for controlling bacterial diseases in plants. However, additional considerations exist in applying bacteriophages as biocontrol agents.

- Rhizosphere, or the area of soil near roots of the plant acts as an important determinant for selection of phage. pH, moisture content, soil type and organic matter can cause phage inactivation. For instance, phage adsorption varies with soil types (Goyal and Gerba, 1979). Charged particles in soil, pH and organic materials can affect phage adsorption (Goyal and Gerba, 1979; Schiffenbauer and Stotzky, 1982; Zhuang and Jin, 2003) and maintenance of phages in soil.
- Phages can be used only in combination with certain chemicals. Surfactants and chelators have been reported to inactivate the phages (Chattopadhyay et al., 2002; Yamamoto et al., 1968). Simultaneous application of copper based bacteriocides and phages have been shown to inactivate phages, but 4-7 days prior application of these bacteriocides followed by application of phages is well tolerated (Iriarte et al., 2007).
- Exposure of phages to UV light from sun is detrimental and

tremendously decreases the phage number (Balogh, 2003; Iriarte et al., 2007). Phages against *Xanthomonas campestris pv. Vesicatoria*, *Dickya solani* and *E. amylovora* phages have been reported as UV sensitive (Born et al., 2015; Czajkowski, 2014).

- Poor persistence on phyllosphere is posed as challenge in application of bacteriophages as antibacterial agents in plants.

As mentioned above, the poor persistence in rhizosphere and phyllosphere pose a severe limitation in the use of bacteriophages as biocontrol agents. Many methods have been experimented with to increase the stability of phages in these harsh environments. Avirulent species of host bacterium can help increase the phage availability (Bae et al., 2012; Iriarte et al., 2012). Moreover, avoidance of daylight and application of phages in the evening has been helpful (Balogh, 2003; Iriarte et al., 2007). Many substances, when applied along with bacteriophages, render protection from UV radiation (Born et al., 2015). Similarly, application of bacteriophage formulation to soil instead as a foliar spray may prove more effective in use of phages against bacterial pathogens (Iriarte et al., 2012).

Only limited application of bacteriophages against bacterial plant pathogens exist. Experimental evidence and field studies have been conducted to evaluate the use of phage as biocontrol agents show promise. Fire blight caused by *Erwinia amylovora* leads to significant losses of apples and pears. Boule et al used phages Ea1337-26 and Ea 2345 with *Pantoea agglomerans* to treat the disease on detached pear tree blossoms where it showed reduction by 84% and 96% respectively (Boulé, 2011). Recently, experiments on biocontrol of Pierce's disease caused by *Xylella fastidiosa* was successfully attenuated by application of phage cocktail (Sano, Salvo, Prado and Paz) to grapevines (Das et al., 2015).

Commercialization of phage was initially done by was AgriPhi, Inc. Recently, due to research carried out in green houses and fields, EPR registration was received by OmniLytics, Inc. (formerly AgriPhi, Inc.). The phages are proposed to be used for management of tomato bacterial spot.

SUMMARY

Bacteriophages owing to their simple design and flexibility in engineering have been alluring tools in molecular biology and biotechnology. Interestingly, various functional components isolated from phages have also proved instrumental in transforming the field of molecular biology. Concerns about widespread resistance in bacterial pathogens have reignited the interest in phage based therapies. Advances in synthetic biology enable us to expand the repertoire of these engineered entities which can be used for various applications in the future.

REFERENCES

1. Abedon S.T., Thomas-Abedon C. (2010). Phage therapy pharmacology. Current Pharmaceutical Biotechnology 11, 28-47.
2. Alia' Mousa Al-Manasra, F.A.-R. (2012). Cloning and expression of a new bacteriophage (SHPh) DNA ligase isolated from sewage. Journal of Genetic Engineering and Biotechnology 10, 177–184.
3. Bae, J.Y., Wu, J., Lee, H.J., Jo, E.J., Murugaiyan, S., Chung, E., and Lee, S.W. (2012). Biocontrol potential of a lytic bacteriophage PE204 against bacterial wilt of tomato. Journal of Microbiology and Biotechnology 22, 1613-1620.
4. Baggesen, D.L., Sorensen, G., Nielsen, E.M., and Wegener, H.C. (2010). Phage typing of Salmonella Typhimurium - is it still a useful tool for surveillance and outbreak investigation? Euro surveillance : European communicable disease bulletin 15, 19471.
5. Balogh, B. (2006). Characterization and use of bacteriophages associated with citrus bacterial pathogens for disease control. PhD Dissertation Gainesville.
6. Balogh, B., Jones, J. B., Momol, M. T., Olson, S. M., Obradovic, A., King, P., et al. (2003). Improved efficacy of newly formulated bacteriophages for management of bacterial spot on tomato. Plant Disease 87, 949–954.
7. Barry, M.A., Dower, W.J., and Johnston, S.A. (1996). Toward celltargeting gene therapy vectors: selection of cell-binding peptides from random peptide-presenting phage libraries. Nature Medicine 2, 299-305.

8. Barth, S., Huhn, M., Matthey, B., Tawadros, S., Schnell, R., Schinkothe, T., Diehl, V., and Engert, A. (2000). Ki-4(scFv)-ETA', a new recombinant anti-CD30 immunotoxin with highly specific cytotoxic activity against disseminated Hodgkin tumors in SCID mice. Blood 95, 3909-3914.

9. Bonnet, J., Yin, P., Ortiz, M.E., Subsoontorn, P., and Endy, D. (2013). Amplifying genetic logic gates. Science 340, 599-603.

10. Born, Y., Bosshard, L., Duffy, B., Loessner, M.J., and Fieseler, L. (2015). Protection of Erwinia amylovora bacteriophage Y2 from UV-induced damage by natural compounds. Bacteriophage 5, e1074330.

11. Boulé, J., Sholberg, P. L., Lehman, S. M., O'gorman, D. T., and Svircev, A. M. (2011). Isolation and characterization of eight bacteriophages infecting Erwinia amylovora and their potential as biological control agents in British Columbia, Canada. Canadian Journal of Plant Pathology 33.

12. Bruynoghe R, M.J. (1921). Essais de thérapeutique au moyen du bacteriophage. Comptes Rendus Des Seances De La Societe De Biologie Et De Ses Filiales. 85: 1120–1121.

13. Carlton, R.M. (1999). Phage therapy: past history and future prospects. Archivum Immunologiae et Therapiae Experimentalis. 47: 267-274.

14. Chattopadhyay, D., Chattopadhyay, S., Lyon, W.G., and Wilson, J.T. (2002). Effect of surfactants on the survival and sorption of viruses. Environmental Science & Technology. 36: 4017-4024.

15. Clark, J.R., and March, J.B. (2006). Bacteriophages and biotechnology: vaccines, gene therapy and antibacterials. Trends in Biotechnology. 24: 212-218.

16. Czajkowski, R., Ozymko, Z., and Lojkowska, E. (2014). Isolation and characterization of novel soilborne lytic bacteriophages infecting Dickeya spp. biovar 3 "D. solani." Plant Pathology. 63: 758–772.

17. D'Herelle, F. (2007). On an invisible microbe antagonistic toward dysenteric bacilli: brief note by Mr. F. D'Herelle, presented by Mr. Roux. 1917. Research in Microbiology. 158: 553-554.

18. Das, M., Bhowmick, T.S., Ahern, S.J., Young, R., and Gonzalez, C.F. (2015). Control of Pierce's Disease by Phage. PloS One. 10: e0128902.

19. Edgar, R., Friedman, N., Molshanski-Mor, S., and Qimron, U. (2012). Reversing bacterial resistance to antibiotics by phage-mediated delivery of dominant sensitive genes. Applied and Environmental Microbiology. 78: 744-751.

20. Edgar, R., McKinstry, M., Hwang, J., Oppenheim, A.B., Fekete, R.A., Giulian, G., Merril, C., Nagashima, K., and Adhya, S. (2006). High-sensitivity bacterial detection using biotin-tagged phage and quantum-dot nanocomplexes. Proceedings of the National Academy of Sciences of the United States of America. 103: 4841-4845.

21. Elowitz, M.B., and Leibler, S. (2000). A synthetic oscillatory network of transcriptional regulators. Nature. 403: 335-338.

22. Farilla, L., Tiberti, C., Luzzago, A., Yu, L., Eisenbarth, G.S., Cortese, R., Dotta, F., and Di Mario, U. (2002). Application of phage display peptide library to autoimmune diabetes: identification of IA-2/ICA512bdc dominant autoantigenic epitopes. European Journal of Immunology. 32: 1420-1427.

23. Fitting, J., Killian, D., Junghanss, C., Willenbrock, S., Murua Escobar, H., Lange, S., Nolte, I., Barth, S., and Tur, M.K. (2011). Generation of recombinant antibody fragments that target canine dendritic cells by phage display technology. Veterinary and Comparative Oncology. 9: 183-195.

24. Gardner, T.S., Cantor, C.R., and Collins, J.J. (2000). Construction of a genetic toggle switch in Escherichia coli. Nature. 403: 339-342.

25. Gevorkian, G., Manoutcharian, K., Almagro, J.C., Govezensky, T., and Dominguez, V. (1998). Identification of autoimmune thrombocytopenic purpura-related epitopes using phage-display peptide library. Clinical Immunology and Immunopathology. 86: 305-309.

26. Goyal, S.M., and Gerba, C.P. (1979). Comparative adsorption of human enteroviruses, simian rotavirus, and selected bacteriophages to soils. Applied and environmental microbiology 38, 241-247.

27. Gupta, R., and Prasad, Y. (2011). Efficacy of polyvalent bacteriophage

P-27/HP to control multidrug resistant Staphylococcus aureus associated with human infections. Current Microbiology. 62: 255-260.

28. Hanlon, G.W. (2007). Bacteriophages: an appraisal of their role in the treatment of bacterial infections. International Journal of Antimicrobial Agents. 30: 118-128.
29. Huie, M.A., Cheung, M.C., Muench, M.O., Becerril, B., Kan, Y.W., and Marks, J.D. (2001). Antibodies to human fetal erythroid cells from a nonimmune phage antibody library. Proceedings of the National Academy of Sciences of the United States of America. 98: 2682-2687.
30. Iriarte, F.B., Balogh, B., Momol, M.T., Smith, L.M., Wilson, M., and Jones, J.B. (2007). Factors affecting survival of bacteriophage on tomato leaf surfaces. Applied and Environmental Microbiology. 73: 1704-1711.
31. Iriarte, F.B., Obradovic, A., Wernsing, M.H., Jackson, L.E., Balogh, B., Hong, J.A., Momol, M.T., Jones, J.B., and Vallad, G.E. (2012). Soil-based systemic delivery and phyllosphere in vivo propagation of bacteriophages: Two possible strategies for improving bacteriophage persistence for plant disease control. Bacteriophage. 2: 215-224.
32. Kutin, R.K., Alvarez, A., and Jenkins, D.M. (2009). Detection of Ralstonia solanacearum in natural substrates using phage amplification integrated with real-time PCR assay. Journal of Microbiological Methods. 76: 241-246.
33. Kutter, E., De Vos, D., Gvasalia, G., Alavidze, Z., Gogokhia, L., Kuhl, S., and Abedon, S.T. (2010). Phage therapy in clinical practice: treatment of human infections. Current Pharmaceutical Biotechnology. 11: 69-86.
34. Lehman, I.R. (1974). DNA ligase: structure, mechanism, and function. Science. 186: 790-797.
35. Loffler, A., Kufer, P., Lutterbuse, R., Zettl, F., Daniel, P.T., Schwenkenbecher, J.M., Riethmuller, G., Dorken, B., and Bargou, R.C. (2000). A recombinant bispecific single-chain antibody, CD19 x CD3, induces rapid and high lymphoma-directed cytotoxicity by unstimulated T lymphocytes. Blood. 95: 2098-2103.

36. Lu, T.K., and Collins, J.J. (2007). Dispersing biofilms with engineered enzymatic bacteriophage. Proceedings of the National Academy of Sciences of the United States of America. 104: 11197-11202.

37. Luken, B.M., Kaijen, P.H., Turenhout, E.A., Kremer Hovinga, J.A., van Mourik, J.A., Fijnheer, R., and Voorberg, J. (2006). Multiple B-cell clones producing antibodies directed to the spacer and disintegrin/thrombospondin type-1 repeat 1 (TSP1) of ADAMTS13 in a patient with acquired thrombotic thrombocytopenic purpura. Journal of thrombosis and haemostasis. Journal of Thrombosis and Haemostasis. 4: 2355-2364.

38. Mallmann WL, H.C. (1924). Isolation of an inhibitory substance from plants. Agricultural Research. 28: 599–502.

39. Mann, N.H. (2008). The potential of phages to prevent MRSA infections. Research in Microbiology. 159: 400-405.

40. Markland, W., Ley, A.C., and Ladner, R.C. (1996). Iterative optimization of high-affinity protease inhibitors using phage display. 2. Plasma kallikrein and thrombin. Biochemistry. 35, 8058-8067.

41. Rioja, C., Van Wees, S.C., Charlton, K.A., Pieterse, C.M., Lorenzo, O., and Garcia-Sanchez, S. (2013). Wide screening of phagedisplayed libraries identifies immune targets in planta. PloS One. 8: e54654.

42. Schiffenbauer, M., and Stotzky, G. (1982). Adsorption of coliphages T1 and T7 to clay minerals. Applied and Environmental Microbiology. 43: 590-596.

43. Schofield, D., Bull, C.T., Rubio, I., Wechter, W.P., Westwater, C., and Molineux, I.J. (2013). "Light-tagged" bacteriophage as a diagnostic tool for the detection of phytopathogens. Bioengineered. 4: 50-54.

44. Shuman, S. (2009). DNA ligases: progress and prospects. The Journal of Biological Chemistry. 284: 17365-17369.

45. Siuti, P., Yazbek, J., and Lu, T.K. (2013). Synthetic circuits integrating logic and memory in living cells. Nature Biotechnology. 31: 448-452.

46. Skurnik, M., Pajunen, M., and Kiljunen, S. (2007). Biotechnological challenges of phage therapy. Biotechnology Letters. 29: 995-1003.

47. Smartt, A.E., Xu, T., Jegier, P., Carswell, J.J., Blount, S.A., Sayler,

G.S., and Ripp, S. (2012). Pathogen detection using engineered bacteriophages. Analytical and Bioanalytical Chemistry. 402: 3127 3146.

48. Smith, G.P. (1985). Filamentous fusion phage: novel expression vectors that display cloned antigens on the virion surface. Science. 228: 1315-1317.
49. Suggett, S., Kirchhofer, D., Hass, P., Lipari, T., Moran, P., Nagel, M., Judice, K., Schroeder, K., Tom, J., Lowman, H., et al. (2000). Use of phage display for the generation of human antibodies that neutralize factor IXa function. Blood Coagulation & Fibrinolysis. 11, 27-42.
50. Summers, W.C. (2001). Bacteriophage therapy. Annual Review of Microbiology. 55: 437-451.
51. Thomas, R. (1935). A bacteriophage in relation to Stewart's disease of corn. Phytopathology. 25: 371–372.
52. Twort, F.W. (1936). Further Investigations on the Nature of UltraMicroscopic Viruses and their Cultivation. The Journal of Hygiene. 36: 204-235.
53. Weinbauer, M.G. (2004). Ecology of prokaryotic viruses. FEMS Microbiology Reviews. 28: 127-181.
54. Willats, W.G. (2002). Phage display: practicalities and prospects. Plant Molecular Biology. 50: 837-854.
55. Wright, A., Hawkins, C.H., Anggard, E.E., and Harper, D.R. (2009). A controlled clinical trial of a therapeutic bacteriophage preparation in chronic otitis due to antibiotic-resistant Pseudomonas aeruginosa; a preliminary report of efficacy. Clinical Otolaryngology. 34: 349-357.
56. Yamamoto, N., Fraser, D., and Mahler, H.R. (1968). Chelating agent shock of bacteriophage T5. Journal of virology 2, 944-950. 57. Zhuang, J., and Jin, Y. (2003). Virus retention and transport as influenced by different forms of soil organic matter. Journal of Environmental Quality. 32: 816-823.

CHAPTER

6

PHAGE THERAPY: EMERGENT PROPERTY PHARMACOLOGY

Andrew J. Curtright and Stephen T. Abedon

Department of Microbiology, The Ohio State University, Mansfield, Ohio USA

Keywords: Antibiotics; Antibacterial agents; Emergent properties; Phage therapy; Pharmacology

ABSTRACT

Phage therapy is the application to bodies of bacteriophages – the viruses of bacteria. This is done to combat bacterial infections. In this article we combine three themes, phage therapy, pharmacology, and the concept of emergent properties. We do this to explore what is unique along with what is not so unique about phage therapy, as viewed from a pharmacological standpoint. At the same time, we consider the place of phage therapy, and drugs generally, within the larger scheme that is biology. We make these latter considerations predominantly in terms of emergent properties, which are system characteristics that exist, qualitatively, at 'higher' but not 'lower' levels of organization. The latter is a relevant consideration especially in exploring the pharmacology of 'living' antibacterials, as are phages, which at lower levels can be viewed as collections of numerous molecules and biochemical pathways but which at higher levels become organized into somewhat complex entities that are capable of 'self amplification'. Emergent properties are also prominently relevant to pharmacology, more generally, as they give rise to requirements for animal along with clinical testing to identify otherwise unexpected drug characteristics. Our broader goal is to stimulate a better integration of issues of phage therapy and pharmacology, predominantly in light of the substantial complexity of phages in comparison with chemical drugs and towards furthering the rational development of phagebased antibacterial strategies. Our chief conclusion is that while phages, due to their complexity, potentially possess numerous emergent properties when compared with less-complex antibacterials, in fact other aspects of phages– including their relative benignness, environmental ubiquity, and that their antibacterial properties are products of evolution – can allow for certain simplifications to pharmacological development. These simplifications are not as readily achieved by chemical drugs, such as antibiotics, which can possess negative properties that are difficult to predict.

INTRODUCTION

"In complexity there is simplicity." (35 Google hits)

Biological control, or biocontrol, is the application of specific types of organisms, to environments, to combat other types or organisms. One of its earliest forms involved the introduction of pathogenic organisms into naïve human populations, that is, biowarfare [1,2]. A less belligerent and perhaps even older approach is the application of diseases or predators into populations of nonhuman targets, such as the keeping of cats for the sake of rodent control [3]. Alternatively, biocontrol includes the application of the viruses of bacteria – known as bacteriophages or phages – into populations of unwanted bacteria, so-called phage therapy [4]. In this article we consider explicitly the use of phages to combat bacterial diseases associated with humans or animals, phage therapy *sensu stricto*, reserving the term biocontrol for all other phage uses as antibacterial agents [5].

A large number of reviews have addressed phage therapy generally as well as historically; see Abedon and Thomas-Abedon [6] along with Abedon et al. [7] for references. Here we focus less on the actual practice of phage therapy, or its history, and more on how the use of phages to combat bacterial infections can be considered in pharmacological terms. Since phage therapy, as well as biocontrol in general, is a form of applied ecology, we additionally integrate these ideas into a more general biological framework, focusing on the concept of emergent properties. This article also can be viewed as an extension of a number of recent efforts to consider the biology of phage therapy from a pharmacological perspective [5,6,8,9].

Overall, the use of phages as antibacterial agents can be contrasted with that of antibiotics. An urgency to these considerations exists because, for various reasons, antibiotics as antibacterial agents appear to possess less utility [10,11] or benignness [12,13] than once had been expected. For example, "at least 25,000 patients in Europe die per year because their bacterial infections are not treatable with available antibiotics" (p. 68) [10]. As possible solutions to this world-wide 'antibiotic crisis', phages – as semi-autonomous genetic entities which, to a first approximation, would appear to be too complex to develop as antibacterial "drugs" – might not seem promising. Conversely, in practice phages appear to be both efficacious and relatively easily developed, including within the context of human treatment [7,14]. In this article it is this paradox in particular that we seek to address. That is, how does what would appear to be an extreme in pharmacological complexity instead beget a relative pharmacological simplicity? To address this question we begin by defining the concept of emergent properties, indicating how this notion is relevant to pharmacology in general and phage therapy pharmacology in particular.

EMERGENT PROPERTIES

In drug development, emergent properties can be thought of as contributing to the high failure rate of new chemical entities in clinical trials. The behaviour(s) of most candidate drugs in humans is not predicted by their activities in the assays used in typical drug discovery efforts, explaining, in large part, the high failure rate of new drugs in clinical testing. Berg et al. [15], p. 204.

The goal of pharmacological research is the development of safe and effective pharmaceuticals that are useful in the treatment or prevention of disease. Though to a degree these efforts can be theory driven, in practice pharmacology requires substantial empirical testing, including in animals along with human patients. Animal testing as well as subsequent clinical trials are important because pharmacological data from chemical assays, *in vitro* models, or even complex systems biology approaches – given less than complete knowledge – cannot effectively capture all emergent properties associated with the systems drugs interact with [16-18]. As a consequence, whether or not a candidate drug becomes clinically useful is dependent upon factors that are not readily predictable, but instead are discovered, often at great expense, in the course of both animal and human testing. A key bottleneck in pharmacological development thus is the potential for drugs in combination with biological systems to display emergent properties.

Emergent or "nonreducible" properties [19] can be defined as higher-level aspects of systems that, qualitatively, are not directly exhibited by lower-level components of the same systems. These properties emerge because the resulting higher-level features are not apparent within lower-level aspects until those aspects interact as a more cohesive whole. Salt [20], quoting Harré [21], and considering emergent properties as an ecological concept, provides this definition: "...the property of the whole is produced by properties of the parts but is not qualitatively similar"

Ecological properties in particular emerge from an organism's interaction with its environment, and among environmental components are those aspects that enter into an organism's body from the outside world. The field of ecophysiology in fact considers explicitly how extraorganismal factors impact the organism internally, that is, physiologically. Body-drug as well as body-toxin interactions equivalently can also be viewed as inherently ecological and thereby at least potentially possessing emergent properties. Drugs, as extraorganismally sourced "environmental" components, in other words, often have substantially different properties viewed chemically within the laboratory, or even within the context of complex *in vitro or in silico* models, versus physiologically within experimental animals or human patients.

Appreciation of emergent properties typically requires an empirical component inasmuch as emergent properties, essentially by definition, are difficult to predict; this especially is experiential observation biased towards more-complete assemblages, such as approximations of whole ecosystems, versus analyses of system components as they may be studied in relative isolation. Thus, in Salt's own words (p. 145): "'An emergent property of an ecological unit is one which is wholly unpredictable from observation of the components of that unit.' The corollary is: 'An emergent property of an ecological unit is only discernable by observation of that unit itself.'" Furthermore, the idea of emergence can be contrasted with that of collective [20]. Pharmacologically, additive drug interactions might be deemed collective ('sum of the parts') whereas, for example, synergistic or antagonistic interactions between drugs – which typically represent far more complex phenomena, i.e., more, or less, than the sum of the parts – we could describe instead as emergent.

Drug-body interactions can be difficult to predict in part because of the complexity of animal bodies but also because of the variance in characteristics that exist between individuals. Thus, new understandings of body systems can 'emerge' upon animal or clinical testing. Antimicrobial drugs take this issue to an additional level. Here not only is complexity and variability inherent in the larger organism but so too the properties of target parasites, or pathogens, are complex. Thus, in addition to it being difficult to fully anticipate an antimicrobial drug's impact on the body, such as its toxicology, or the body's impact on the drug (e.g., it's rate of clearance), but so too one must study living microorganisms, rather than just their components (e.g., genes), to ascertain a drug's antimicrobial properties even in greatly simplified environments such as test tubes or Petri dishes. Yet additional complexity can emerge with antibiotic-bacterial interactions such as the now well-appreciated ability of bacterial biofilms to resist antibiotics that constituent bacteria, growing as individual cells, are quite sensitive to [22], or the more general ability of bacteria to evolve biofilm-independent mechanisms of antibacterial resistance [11,23]. Therefore, while pharmacology to a substantial extent can involve the elucidation of drug-body emergent properties, this idea can be even more applicable to consideration of the pharmacology of antimicrobial drugs.

An additional level of emergence can occur when the "drug" itself is also somewhat complex. One sees this with the use of live-attenuated vaccines (e.g., MMR) as well as whole-killed vaccines, particularly against the highly variable influenza virus. These reagents are not just interacting with living organisms – directly with ourselves and indirectly with would-be infecting pathogens – but themselves are organisms or at least their intact derivatives. Our ability to predict the biological properties of these entities such as from their chemical components, as a consequent, can be fairly limited. Vaccine development, like pharmacology in general, therefore possesses a substantial empirical component: At best we apply educated guesses to drug design, and then always brace ourselves for the emergence of unexpected properties upon animal and clinical testing. Another category of live as well as naturally attenuated "drugs" [9] are a group of viruses, phages, which can be used as antibacterial agents in the guise of what is known as phage therapy.

PHAGES

Phages are viruses that infect members of domain Bacteria, which contains all known pathogenic bacteria [24]. The majority of phages are both lytic and tailed, the former a description of how phage virions are released from infected bacteria and the latter a description of virion morphology. A tail is a capsid appendage that is responsible for virion adsorption along with certain aspects of DNA translocation into bacterial cells. Furthermore, all tailed phages have double-stranded DNA genomes. In lytic infections, mature phages are released from infected bacteria via degradation of the bacterial cell envelope. In the course of this lysis the infected bacterium is physiologically and to a large extent also structurally destroyed. See Figure 1 for summary of selected characteristics of various phage types.

The Problem of Lysogeny

Many lytic phages are also temperate. A temperate phage is able to display lysogenic as well as productive cycles. In a productive cycle, phage progeny are produced and subsequently released, such as through the above-described lytic process. In a lysogenic cycle, phages persist mostly as genetic material, termed prophages, replicating intracellularly in tandem with their bacterial hosts.

Prophages, through a process of induction, can also display productive cycles. Indeed, a majority of the infections initiated by temperate phages give rise to lytic rather than lysogenic cycles. Temperate phages nonetheless are problematic for phage therapy for a number of reasons, and consequently should be avoided for these purposes. These reasons include the display what is known as superinfection immunity, which in practical terms means that bacteria that are lysogenically infected – and consequently described as lysogens – can no longer be killed by the same phage types. It is as though the majority of an applied antibacterial were to succeed in killing the encountered bacteria but with some fraction instead providing an 'antidote' or 'vaccine' to the bactericidal activity.

In addition, many temperate phages carry genes that encode bacterial virulence factors including such infamous exotoxins as diphtheria toxin, cholera toxin, and Shiga toxin [25]. One can mostly avoid virulence factor genes by limiting phage choice for phage therapy to what can be described as "professionally lytic" phages [26]. These, defining the term narrowly, are phages that not only are "obligately lytic" (that is, unable to display temperate cycles) but which also have not recently descended from temperate phages. In phage therapy, the phages of choice tend to be professionally lytic as well as tailed. See Figure 1 for summary.

Phage Environmental Microbiology

Any description of phages would be remiss if it did not mention their ubiquity. This issue is relevant to phage therapy pharmacology in the sense that most substances that we live our lives awash in tend to be relatively benign, with phages seemingly not an exception. It has been estimated based on combinations of known phage prevalence, along with considerations of likely ratios of phages to bacterial hosts within environments, that perhaps 10^{31} or 10^{32} phage virions exist on Earth at any given moment, e.g., as discussed in Abedon [27]. This means that any time that one goes swimming in either the ocean or a less than oligotrophic lake, then a million or so phages are also bathing us for every ml surrounding our bodies [28]. A liter of water can easily contain a billion individual virions! Further, food, including especially bacteria-fermented products, e.g., [29,30], tend to be full of phages, as too can be animal bodies, in particular colons [31]. We are born into a world of phages and develop through our adulthoods experiencing a constant phage barrage. Indeed, animals have been so exposed presumably for as long as animals have been in existence, with phages an unseen and at times underappreciated but nevertheless omnipresent aspect of our world.

PHAGE THERAPY

Phage therapy is the application of phage virions to bacteriacontaining environments with the goal of bacterial control. In most instances with phage therapy "control" translates as "kill" and the bactericidal nature of obligately lytic phages can be well suited to this task. Phage therapy, strictly, is a form of biocontrol, with one organism (here, phages) employed to control another, typically pathogenic or nuisance organism (here, bacteria). We concentrate in this article on issues relevant to the phage therapy of animals, particularly as may be applied to humans [7,14]. Others have considered, by contrast, the phage-mediated biocontrol, for example, of foods [32,33], of food animals such as to reduce loads of zoonotic pathogens [34], or of plants [35].

Active versus Passive Treatments

Phage treatments, whether phage therapy or biocontrol in general, can be distinguished into two basic types, those that are dependent on phage *in situ* replication to achieve bacterial control and those that do not require such replication. The former has been dubbed an "active" treatment after the idea that it requires "active" phage replication to be effective; see Abedon and Thomas-Abedon [6] for discussion. By contrast, phage therapy that does not require active phage replication has been dubbed passive or, alternatively, inundative phage therapy.

Though seemingly a somewhat intimidating approach to antibacterial therapy, in fact inundative treatment is the phage equivalent of conventional antibiotic therapy, i.e., where enough drug is supplied via traditional approaches to dosing that bacterial infections are brought under control. In any case, unless phages are either engineered to kill bacteria without producing infectious phage particles [36] or otherwise are unable to reproduce upon infecting bacteria due to limitations in their 'spectrum of activity', then phages should still display some degree of *in situ* "self amplification" or "auto dosing" [37] in the course of their antibacterial action.

Phages as Emergent Property Pharmaceuticals

Note that this property of phage replication, in the course of their antibacterial action, can be considered to be emergent since it is a phage aspect that would not be easy to predict were we to analyze phage components, such as individual capsid proteins, solely in isolation. Indeed, numerous genes are found in the genomes particularly of larger phages for which no function has been assigned, e.g., Hendrix [38], but which in many cases presumably can contribute to phage population growth in some manner. Nevertheless, the phage property of replication and subsequent population growth is an expected consequence of phagebacterial interactions given our understanding of phage biology. Pharmacologically, therefore, phage population growth is not an emergent property in the sense of requiring animal testing to discover any more than discovery of a chemical antibiotic's basic antibacterial ability would require such testing.

By contrast, an unexpected absence of phage replication or antibacterial action, *in situ*, could be described as pharmacologically emergent, that is, a pharmacodynamic property that emerges only upon characterization of the larger system. Here this larger system would be phages plus target bacteria plus the body environment rather than phages plus target bacteria within a simpler laboratory environment. See, for example, the results of Bull et al. [39] as well as Abedon [9], along with references therein cited, for a broader discussion of what can 'go wrong' in terms of bacterial killing during phage therapy animal testing. To clarify: Phage replication when in the presence of target bacteria certainly can be viewed as a phage 'behavior' that emerges from the lower-level aspects of phage biochemistry, but that replication is less emergent in pharmacological terms because it is something that is expected to occur prior to the performance of animal or clinical testing.

Phages within the context of phage therapy thus may be considered to be emergent property pharmaceuticals, given the impressive complexity from which their population growth and even antibacterial activity in fact emerge. However, and a prominent theme within this article, the chemical complexity of phages as "drugs" – that is, the multitude of genes and gene products that together make up phages – in certain ways conspire to simplify rather than complicate their pharmacological properties. Such pharmacological simplification, as we will consider, may be most prominently observed from the perspective of phage safety.

Phages are used Therapeutically mostly "Off Label"

In principle phage therapy can be used against any bacterial disease, though in practice the diseases that are targeted tend to be ones against which alternative treatments are less available. This, particularly in Western medicine, can be against antibiotic-resistant bacteria. In the former Soviet Union, by contrast, phage therapy has been used against bacterial infections in general, owing to a tradition of phage use that developed in that country prior to widespread antibiotic availability or, for that matter, before the development of a robust understanding of phages as biological entities. In Poland phages have been employed on an "experimental" basis to treat chronically infected wounds that have not responded to conventional antibiotic treatment. What is fascinating about these efforts is that they seemingly have been quite successful, and, especially in the former Soviet republic of Georgia as well as Wroclaw, Poland, have been ongoing. Reviews by Kutter and colleagues [7,14,40] have explored these various international aspects of phage therapy use.

Though phage therapy has been employed extensively outside of North America, the technique has not been subject to a great deal of double-blind clinical trials. Phages consequently, as a class of "drugs", have been employed almost entirely "off label". Widespread double-blind clinical testing of phage therapy likely will only happen given substantial increases in the availability of resources to the field. In addition, phage therapy, as a field, appears to be hampered by a combination of (i) unfamiliarity with phage use by Western physicians – which is not to say that there is a lack of Western physicians who are enthusiastic of the idea, but instead that obtaining and then using phage therapeutics is difficult in environments where their use effectively is not even off label – (ii) questions of intellectual property rights and particularly the patenting of technologies that have been in existence for nearly 100 years; and (iii) the fact that in those countries most familiar with phage therapy the therapeutic use of phages already represents a standard of care, a situation that for ethical reasons can deter double-blind study.

Notwithstanding these considerations, it is our opinion that it would be helpful to the field to adopt a more pharmacologically rigorous perspective on phage therapy research and development. While various issues of phage therapy pharmacology have been considered elsewhere, as cited in the introduction, here we explicitly discuss their relevance or helpfulness to the idea of employing phages as antibacterial biocontrol agents and do so within a context of constraints that are imposed upon pharmacological research and development by drug emergent properties. See also Table 1 for discussion of advantages of using phages as antibacterial drugs including in comparison to chemical antibiotics, with many of these advantages discussed further in Loc-Carrillo and Abedon [37].

Table 1: Advantages Associated with Phage Use as Antibacterials[1].

Property[2]	Comment
Bactericidal	Reduced potential for bacterial development of resistance
Auto "dosing"	*In situ* activity increases numbers (though only given favorable bacterial densities)
Low inherent toxicity	Virions consist of only proteins and DNA
Low normal flora impact	Low likelihood potential for superinfection by endogenous flora, e.g., *C. difficile*
Narrow resistance evolution	Selection for resistance limited mostly to within populations of targeted bacteria
Lack of cross-resistance	Antibiotic-resistant bacteria tend to retain phage sensitivity
Rapid discovery process	Phages with large therapeutic windows are often easy to isolate
Potential for modification	Phages can be easy to molecular characterize and manipulate
Use with other agents	Versatility in formulation development and combination with other drugs
Dosing versatility	Phage formuations can take many forms and can be delivered via many routes
Biofilm clearance	Certain phages, unlike most chemical antibiotics, can be relatively good at this
Favorable pharmacokinctics	Delivery to targets or persistence *in situ* often is either good or improvable
Single-dose potential	Can provide dosing convenience; auto dosing consequence
Low-dosage potential	Of possible economic or safety utility; auto dosing consequence
Single-hit killing kinetics	Nonetheless effectively multi-hit since phages still multiply adsorb bacteria

Engineered lower toxicity	Particularly elimination of bacterial lysis, but auto dosing advantage is thus lost
Low environment impact	Narrow specture of anti-bacterial activity, lability, low inherent toxicity
Not antibiotics	Fewer societal concerns with use, such as in agriculture; avoids antibiotic allergies
Natural products	Potential appeal to natural medicinals market
Relatively low cost	As drugs, reasonable production costs
Public perception	Public perception of use of phages as antibacterials seemingly is positive

[1]Advantages are as seen relative particularly to the use of chemical antibiotics as antibacterial agents.

[2]Line break between upper and lower portions of list is indicative of what we feel are greater (upper) versus lesser (lower) advantages to phages as antibacterial drugs.

[3]Ratio of drug dosage that gives rise to toxic effects to that which gives rise to efficacy. The larger the ratio, the safer the drug and/or the easier it is to work with.

TOXICITY AND SIDE EFFECTS

Toxicity or side effects can be described as negative pharmacodynamic consequences, contrasting the directly sought after positive pharmacodynamic effects such as bacterial control. Pharmacodynamically, therefore, the key phage characteristic that contributes to the potential for phage therapy to serve as an alternative to antibiotic treatment, other than the phage ability to kill bacteria, is in fact the generally low toxicity of especially professionally lytic phages. The advantage of low-toxicity pharmaceuticals is not just their safety but also a freedom from patient monitoring to prevent toxicity as well as an ability to lengthen intervals between dosing, which together can greatly reduce costs while increasing patient convenience. In addition, low toxicity can allow in situ levels to exceed a drug's minimum effective densities, resulting in antibacterials with increased potential to control infections. As a result, it means that the phage ability to replicate in situ in the course of killing bacterial targets can be a dosing ally, allowing phage densities to increase precisely at their site of action and do so without simultaneously serving as a toxic hindrance.

Pharmacokinetics, contrasting pharmacodynamics, describes the impact of the body on drugs. Low toxicity can result in a reduction in certain pharmacokinctic concerns since higher doses, or phage self amplification, can be used to counter mechanisms of drug loss or dilution in the course of what are known, pharmacokinetically, as drug absorption and distribution. More generally, toxicity and side effects represent the most troubling consequence of pharmaceutically emergent properties and a relative absence of such aspects would presumably simplify pharmacological development. It is somewhat surprising, therefore, how phages, with a multitude of their own emergent properties, appear to lack a majority of the side effects observed with seemingly simpler, small-molecule antibacterial drugs. We thus begin our discussion of pharmacology as it applies to phages by focusing on phage safety since much of how phages are or could be used as pharmaceuticals derives from this property.

Toxicity as Well as Phage-Immune System Interactions

Drugs can be mutagenic or teratogenic, they can activate or inactivate receptors, they can be cytotoxic or mimic hormones, etc. Given appropriate phage choice, which generally means employing only professionally lytic phages for phage therapy (Figure 1), then what are presented to bodies by phages are mostly proteins and DNA that display little toxicity to body tissues. An alternative perspective is that in order for phages to display an evolved toxicity to human tissue they would have to be able to interact with human cells similarly to how they interact with target bacterial cells. Inasmuch as the processes inside a human cell are vastly different from the ones occurring inside a bacterial cell, such tissue toxicity generally is not even a potential concern nor has it demonstrated itself to be upon empirical observation; for review of phage-human- or animal-body interactions, see Merril [41].

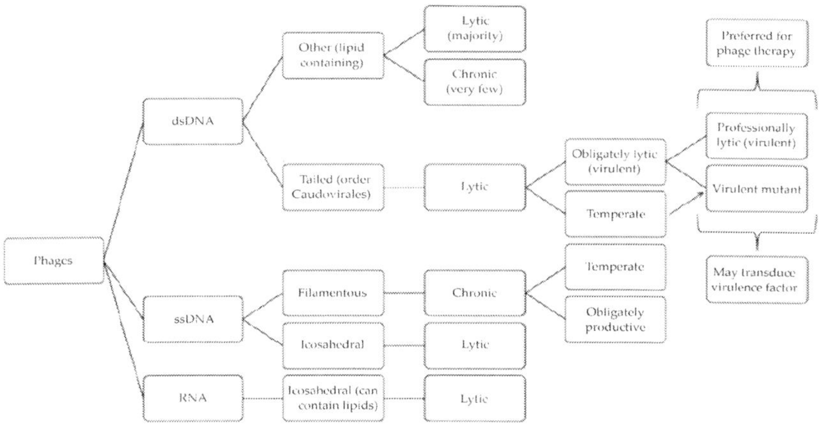

Figure 1: Types of phages. Phage therapy, as indicated, is preferably achieved employing professionaly lytic phages, that is, phages that are both obligately lytic and otherwise not substantially genetically similar to temperate phages.

The arrow between "Temperate" and "Virulent mutant" is meant to indicate descent while "Virulent mutant" as well as "Professionally lytic" are both types of "Obligately lytic" phages.

Though not necessary acting directly as toxins, protein-based drugs still can interact with host immune systems, in some cases to disastrous effect [42]. More common is simply the production of anti-drug antibodies (ADAs). ADAs are usually benign, but when protein-based drugs resemble endogenous molecules then severe complications can arise, stemming from ADA cross reactivity and subsequent autoimmune response [43]. As phages typically are not expected to be similar to endogenous molecules, and in fact are typical components of animal environments, such complications would be expected to be less likely in phage therapy application. Low likelihoods of complications stemming from phage-immune system interactions could be due, perhaps, either to bodies acquiring a tolerance to ubiquitous phages found in their environments or, instead, phages that are normally found as animal commensals evolving to minimize their potential to destructively impact their overall environment, i.e., our bodies. The latter possibility, we stress, is dependent on phage choice, that is, the use in phage therapy of professionally lytic phages and/or phages that have been shown to not carry bacterial virulence factor genes.

In practice there has been little documentation that phages used therapeutically will produce substantial side effects, or even of reductions in phage therapy effectiveness due to phage interactions with adaptive immune system [7]. At least in part this general lack of evidence of negative phage-body interactions may stem from typical phage application practices, which is topical for chronic infections or systemic use over only relatively short spans against more acute infections. That is, the impressive extent to which phages appear to safely serve as antibacterial agents upon application to bodies may not be a wholly unexpected consequence of phage-patient interaction [41].

Bacterial Dysbiosis

Besides direct interaction with animal tissues, including immune systems, a second means by which an antibacterial can negatively impact treated individuals is via the destruction of nontarget bacteria, particularly beneficial members of normal flora. The loss of these non-target bacteria can result in antibiotic-associated superinfections such as vaginal yeast infections or *Clostridium difficile*associated colitis. Generally this kind of side effect is more likely to occur when broad-spectrum rather than narrow-spectrum antibacterials are employed. Most phages, however, display narrow spectra of activity [44], a property that may be of utility even among chemical antibacterials [45]. Indeed, as Blaser [13] notes (p. 394), "We... need new, narrow-spectrum antibacterial agents to minimize collateral effects on the microbiota. ...and, importantly, better diagnostics that rapidly identify the problematic agent."

In considering the narrowness of phage spectra of activity, one may conceptualize the different steps of phage-bacterial interactions as "sieves" that generally result in prevention of non-targeted cell death. The first sieve eliminates all cells which do not carry the cognate receptor for the phage. The second eliminates all cells that are capable of resisting phage DNA uptake. Another blocks elimination of those cells that possess restriction enzymes against phage DNA or other mechanisms of active resistance by bacteria to phage infections [44,46,47], and so forth. Even when multiple phages are combined to form cocktails, the result typically is still an overall spectrum of activity that is no wider than that of the narrowest spectrum antibiotics currently employed clinically. Phages thus may be applied even prophylactically (below) with little concern of generation of bacterial dysbioses.

Drug Interactions

Many drugs are metabolized in the liver by a class of enzymes known as cytochromes. When pharmaceuticals, including antibiotics, interact with these enzymes, inhibition or activation may occur. The result is that potentially harmful drug interactions may occur including increased toxicities such as due to build up of drug densities within bodies following less effective drug removal or, alternatively, loss of drug potency due to too rapid elimination [48]. The characteristics of such drug-drug interactions are difficult to predict due to the emergent properties of the hepatic system as a whole, including drug transport and gene regulation, which are not obvious during *in vitro* toxicity or metabolism screenings. A concern with taking quinolones, for instance, is that their interaction with cytochrome P450 isoform CYP1A2 decreases its ability to metabolize other drugs, including antidepressants, and antipsychotics [49]. If patients are not carefully monitored, then harmful levels of the interacting drug can accumulate.

For phages, to at least a first approximation, such potentially harmful drug-drug interactions should not be a concern since these same liver enzymes will not play as large a role in phage elimination any more than they do in the elimination of body proteins and DNA in general. Another way of stating this is that body metabolism of phage particles should have little impact on normal mechanisms of body homeostasis. Phages thus may be more readily used in conjunction with other drugs and not just other antibacterials but other drugs that a patient may already be consuming. That is, we conjecture that phagedrug interactions do not substantially give rise to emergent properties in terms of the functioning of these other drugs, though with the caveat that this issue to our knowledge has not been systematically studied.

Release of Bacterial Toxins

Most phages, via a complex series of steps, bring anti-cell wall hydrolases into the immediate vicinity of bacterial peptidoglycan layers. The resulting bacterial lysis results in the solubilization of bacteria-associated molecules, such as endotoxin or certain exotoxins, that can be directly or indirectly toxic to body tissues. One consequence of this is that, particularly with phage systemic application, there can be a need for substantial purification of phage virions [50], i.e., as protein-based drugs in general require purification prior to use, e.g., Magalhäes et al. [51].

This issue is also a concern with regard to the lysis of bacteria as that can occur in situ, during phage therapy, which in fact is the root of occasional efforts to engineer phages that are bactericidal without being lytic [36]. Phages, though, are not unique in being lytic antibacterials since that characteristic tends to be shared with smallmolecule antibiotics such as penicillins that target cell walls. As a narrow-spectrum lytic antibacterial, however, the number of bacteria lysed by phages is fewer. Indeed, much of the complexity of phages as bacterial lysing agents can be seen as contributing to the selectivity of their lytic behavior.

This latter advantage should be relevant only to the extent that potentially lysed non-target bacteria are found in locations in which their lysis could be potentially harmful to the body, particularly within the blood such as during the treatment of septicemias. It is helpful, though, to recall that it is the bacteria being treated that ultimately are the problem, more so than the impact of antibacterials employed to eliminate them. That is, the elimination of systemically infected bacteria is not necessarily going to occur without harming the integrity of those bacteria and thereby releasing at least some bacterial toxins, though as noted phages can be engineered to reduce this damage [36].

Safety in Numbers

Phages only replicate when they are efficacious and the result, from a pharmacodynamic standpoint, is a wide breadth of possible dosing ranges during phage treatment. This range extends from low doses in anticipation of *in situ*phage self amplification – which has a low likelihood of releasing toxic bacteria-associated molecules unless bacterial targeting is successful – to high doses if rapid bacterial lysis is not a concern and/or if phage self amplification is less likely. Notwithstanding this potential for a wide breadth of dosing ranges, a standard approach towards phage therapy dosing, for both safety and economic reasons, should be the use of relatively low phage densities, e.g., such as no more than $\sim 10^8$ phages/ml, with either repeated dosing or application of greater phage densities should phage access to bacteria prove limiting. This suggestion is rather than striving under all circumstances to inundate bacteria using substantial excesses of phages, such as $>>10^8$ phages/ml.

The relatively low toxicity of phages in combination with their potential for self amplification provides substantial leeway in terms of levels of dosing, with neither "too low" phage densities (e.g., $<10^7$/ml) nor "too high" phage densities (e.g., $>10^9$/ml) necessarily resulting in poor phage pharmacodynamic effects. Less latitude exists in terms of what peak phage densities in the vicinity of target bacteria are required to achieve effective bacterial killing. Because phages display low toxicities, however, generally the problem is more one of not achieving sufficient phage densities rather than providing excessive doses. Note that pharmacodynamically sufficient peak phage densities, that is, minimally effective phage densities delivered to or generated within the immediate vicinity of target bacteria, may be assumed, as a rule of thumb, to be approximately 108 or more active phage particles per ml [6,9,52].

TYPES OF TREATMENT

Treatments with drugs can be differentiated into a number of different approaches including prophylaxis, empirical treatment, and pathogen-driven therapy. In this section we introduce these various concepts in terms of phage therapy.

Prophylaxis

The utility of prophylaxis as a treatment approach is that it can be easier to prevent an infection rather than treat an ongoing disease. Furthermore, in situ-generated bacterial lysis products should be much less abundant given the associated (or expected) lower bacterial densities. Thus, phage safety should be further enhanced in primary prophylaxis, increasing the reasonableness of such an approach and indeed potentially allowing for phage use under circumstances where application of chemical antibiotics might be avoided due to the possibility of bacteria-unrelated side effects. In addition, phages tend to not interfere with wound healing, e.g., Soothill [53].

To the extent that prophylaxis is against bacteria existing at low densities, then active treatment, that is, in situ phage self amplification, may not be possible as threshold bacterial densities required for successful phage auto dosing will not be present [6,9,52]. Prophylactic treatment also will often be empirical (below) since the pathogen to which an individual might be exposed in many instances will not be known. The use of phages prophylactically nonetheless has occurred, with success, both experimentally in animals and clinically particularly in terms of war wounds; see, for example, Abedon et al. [7] for limited review.

Empirical Therapy

Empirical therapy, as we employ the term here, is the initiation of treatment prior to full diagnosis, particularly in terms of the diagnosis of antibiotic susceptibility of an infectious agent (synonym: presumptive treatment). Although phages lack the same dysbiosis or superinfection concerns of antibiotics, their spectrum of activity generally is much narrower, making successful empirical therapy by specific phages less probable. Phage cocktails – formulations containing combination of potentially efficacious phage isolates – may help to overcome this concern, however, to the extent that a cocktail is sufficiently well formulated to cover most bacteria that would be encountered per infection type in a given geographic area.

In Georgia and the rest of the former Soviet Union phage cocktails have been continuously modified in response to changing spectra of locally prevalent bacterial pathogens [7]. Though the practitioners of this approach of ongoing formulation modification certainly would appear to have found it to be beneficial, an obvious concern is that such a strategy could be inconsistent with standard approaches to drug development or regulatory approval (such as by the U.S. Food and Drug Administration). Alternatively, continuous surveillance along with periodic formulation revision is the hallmark of influenza vaccine implementation [54].

Pathogen-Directed Therapy

In pathogen-directed therapy the infectious etiology is not just identified prior to the initiation of treatment but also is characterized in terms of its antibacterial susceptibility. Because of phage specificity, pathogen-directed therapy is the type of treatment that best suits phage therapy. Such an approach, as typically practiced such as in Poland, usually employs what can be described as the phage bank approach, where isolated bacterial pathogens are tested for susceptibility to pre-prepared phage stocks [14]. Indeed, periodically bacteria can be retested, given extended phage treatment such as of chronic infections, with the phages employed in the treatment of specific individuals revised if there is evidence of bacterial resistance [55]. It is also possible to isolate phages from environmental samples against specific bacterial strains that have been isolated from patients, though obviously this can introduce additional delays as well as costs. The development of phage cocktails can be viewed as employing just such techniques, only peformed well in advance of patient treatment.

MECHANISMS OF ACTION

Antibiotics possess a wide array of mechanisms of action. Penicillins inhibit the formation of the cell wall [56], for example, while quinolones inhibit DNA gyrase [57] and topoisomerase [58] in bacteria. By contrast, phage infections are multistep processes with phages genetical killing of bacteria generally occurring well prior to their physiologically termination via lysis [59-61]. In this section, for the sake of illustration of the diversity of phage antibacterial actions, we provide brief overviews of general modes of phage-mediated bacterial killing. See Figure 2 for illustration of the relationship of these different approaches to bacterial killing to passive versus active treatments.

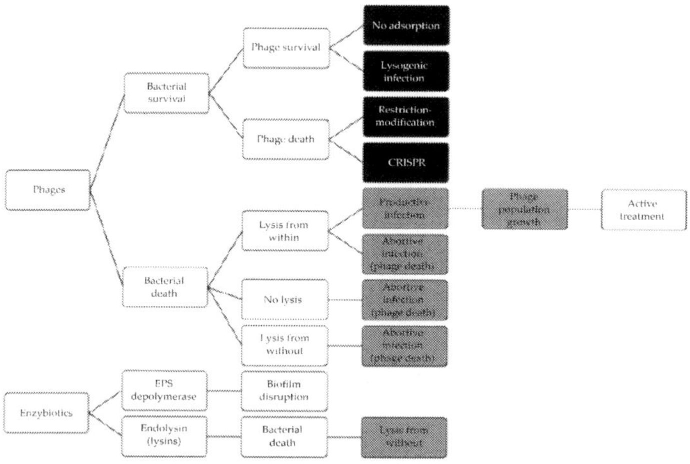

Figure 2: Phage impact on bacteria. "Phages" refers to intact phage particles whereas "Enzybiotics" are purified phage products. Treatment failure is associ-

ated with substantial bacterial survival (black outcomes with white text). Passive or inundative treatment occurs when sufficient phage numbers are provided that substantial bacterial losses are achieved without reliance on phage population growth (darker gray endpoints with white text , all but "Productive infection" and "Phage population growth", give rise to obligately passive treatments). Active treatment, indicated as a white outcome with black text, cannot substantially eradicate target bacterial populations unless sufficient phage population growth occurs in situ. Phage infections in which bacteria die and no phage progeny are produced we define as abortive. Note that CRISPR systems do not necessarily result in bacterial survival in combination with phage death but rather can also give rise to abortive infections [68].

Note to editor, the reference [62] has been temporarilty placed here to preserve numbering order since it is found in the legend to Figure 2.

Lysis From Without

Certain phages, at high multiplicities of adsorption, can display the emergent property of being able to lyse bacteria without first infecting [63]. The result can be viewed as a type of abortive infection, that is, a phage infection that results in the death of both adsorbed bacterium and adsorbing phage(s). Claims of lysis from without should always include empirical evidence that phages can lyse bacteria in a manner that (i) occurs soon after phage adsorption, (ii) does not result in phage production, and (iii) is more likely, in terms of bacterial lysis, given higher versus lower phage multiplicities of adsorption, e.g., 50 versus only 5. By contrast, and despite numerous publications which appear to suggest otherwise, simply showing that phages kill bacteria following high multiplicity adsorption does not represent evidence for lysis from without, and this is so even if phage progeny production also does not occur.

Abortive Infection

When phage infection productivity is completely blocked but bacteria nonetheless die, the result can be described as an abortive infection. These can be a consequence of expression of specific bacterial anti-phage resistance genes [44,46]. They also can result from an absence in bacteria of specific factors that are required for productive phage infection in combination with a presence of those factors necessary to support phage-mediated bacterial killing. As with lysis from without, phage therapy mediated by phages that kill bacteria solely abortively by necessity represents a form of passive rather than active treatment.

Though bacterial genome sequence determinations, including those of bacterial plasmids, in principle can allow identification of abortive infection-effecting genes, the interactions of bacterial and phage gene products can be hard to predict. Thus, whether presumptive bacterial abortive-infection genes will give rise to functional abortiveinfection systems that block the infection of specific phage types must be confirmed experimentally, as too what phages are susceptible to these systems. In practice, as with elucidation of other phageassociated emergent properties, one simply tests phage isolates against bacterial strains to determine bacterial susceptibility. Due to abortive infection systems, though, it is possible for otherwise obligately lytic phages to inhibit bacterial replication while at the same time failing to productively infect those same bacteria. The potential for phage replication in the presence of target bacteria, rather than simply killing bacteria, therefore also must be experimentally determined, such as via plaque assays or broth-culture lysis (following low-multiplicity phage application), if both active and empirical therapy are sought (Figure 2).

Lysis From Within

The more typical scenario resulting in death of a bacterium, in the course of phage therapy, occurs via a phage lytic cycle. The resulting lysis is described as a lysis from within to distinguish it from the above-noted lysis from without. Given multiple mechanisms by which phages can mortally impact infected bacteria during productive infections (above), the cell lysis event may be viewed simply as an additional layer of bacterial killing. Nonetheless, lysis from within can be an important aspect of phage therapy since bacterial killing without associated progeny release, that is, release as mediated by the lysis from within of infected bacteria, is effectively an abortive infection and therefore cannot support active phage therapy.

Enzybiotics

Certain phage molecules that possess antibacterial activity can act from without rather than from within, that is in the absence of phage presence rather than in the course of phage infection. The harnessing of purified forms of these molecules as a means of antibacterial treatment has been described as enzybiotics [64, 65]. The two most prominent enzybiotics are phage endolysins and phage extracellular polymeric substance (EPS) depolymerases.

Endolysins are the enzymes that effect lysis from within and typically, as noted, are not what actually kills bacteria though certainly are key to bacterial destruction along with phage progeny release from infected bacteria. Endolysins can, however, be used in a purified form to kill especially Gram-positive bacteria, which lack the protection afforded by the Gram-negative outer membrane. This process also can be described as lysis from without, though it is a form of lysis from without that differs from the more traditionally described phenomenon discussed above [63]. Of interest, whereas penicillin binding proteins, i.e., the targets of penicillin and related antibiotics, can be easily mutated to no longer bind penicillin while still performing their peptidoglycan synthesis function [66], the actual bonds of peptidoglycan are not as easily modified. The result is a low potential for bacterial populations to evolve endolysin resistance.

EPS depolymerases affect particularly bacterial biofilms though are not directly bactericidal [67,68]. Instead, they exert their impact on extracellular polymers associated with retaining biofilm integrity. Thus, upon lysis, enzymes can be released from phage infections that can both increase phage penetrating ability into biofilms and otherwise potentially make biofilm bacteria more susceptible to antibacterial treatment (including, potentially, to antibiotic treatment). See Figure 2 for partial comparison of enzybiotic therapy to phage therapy.

Phage Virions as Antibacterial Delivery Agents

Most targets of antibiotics are bacterial molecules that are found inside of cells. This typically means that an antibacterial must be able to pass through the cell wall and/or cell membrane in order to take effect. The first molecule contacted by phages during the infection process, however, is found on the outside of the cell and phages subsequently have an inherent ability, within the limits of their host ranges, to penetrate through bacterial cell walls and membranes. Consequently, there is no need to design or even explicitly select phages for possession of this specific ability. Rather, a phage that has been shown to replicate in the presence of a specific bacterial type too must possess a bacterial cell-envelope penetrating ability. In this section we consider this phage ability in terms of virion modularity, that is, phage possession of a delivery component (virion proteins) in combination with one or more bactericidal mechanisms (encoded by phage DNA).

Naturally Occurring Emergent-Property Antibacterial Agents

It is readily feasible to design or discover molecules that penetrate into body compartments as well as into bacteria, that interfere with bacterial metabolic processes, that are relatively lacking in side effects, that can target particular species of bacteria, or that possess reasonable stability during storage as well as following administration to patients. Identifying molecules that have all of these properties at once, however, can be quite difficult. This is why isolating natural antibacterial products has been a standard approach to antibacterial discovery since the introduction of antibiotics to clinical medicine. That is, antibiotics traditionally are natural, microorganism-produced molecules with an evolved ability to penetrate into and then either kill bacteria or reversibly inhibit their growth. Once discovered, antibacterial molecules are then chosen based upon their other desirable properties, as listed above. Numerous antibiotic features that are useful in terms of pharmacological functioning thus to a large extent are emergent properties, that is, evident only upon application to bodies. As discussed above, however, the bacteria-killing aspects of antibiotics themselves do not generally require antibiotic-body interactions to become manifest so are not similarly emergent in the larger scheme of drug pharmacology. Indeed, useful antibiotics almost by definition are chemicals that display antibacterial activities in a manner that is somewhat independent of whether they have been administered to a body environment.

Phages, too, are evolved antibacterial agents though substantially more complex than naturally occurring chemical antibacterials. Perhaps most notably, though, phages are both antibacterial delivery systems (virions) and antibacterial synthesizing agents (phage infections). As a consequence, they can be naturally benign until they are presented with target bacteria and, of course, are even capable of increasing in number once bacteria have been acquired. As a consequence, and despite their underlying complexity, professionally lytic phages tend to be highly safe antibacterials. In addition, from phage interactions with bacteria emerge the relatively straightforward phage properties of bacterial lysis and production of new, progeny phages, though the details of these infections themselves can be somewhat complex [69]. Natural selection, in part as a consequence of phage modularity, thus has resulted in substantial constraints on phage emergent properties, which, for professionally lytic phages, typically are limited to bacterial acquisition, phage progeny production, and bacterial lysis from within. These constraints exist despite or perhaps even as a consequence of the substantially greater complexity that phages display relative to naturally occurring antibiotics.

Another way of making this point is to describe professionally lytic phages as possessing relatively complex genetic programs whose primary function is the production of new phages rather than the a general interference with metabolisms, other than the metabolisms of bacteria they are directly infecting (and even then, only as such interference is useful towards the endpoint of making and then release new, progeny virions). Furthermore, when not actively infecting bacteria, the primary phage function is one of bacterial acquisition, which is accomplished in a mostly non-metabolically active manner. Phage killing power thus is focused both metabolically and spatially on target bacteria, resulting, with professionally lytic phages, in only chance interference with animals physiologically, and seemingly little likelihood of even inadvertent adverse consequences given the nonxenobiotic nature of phage molecules. Antibiotics, by contrast, exist as somewhat general disrupters of metabolic processes, resulting in the evolution of less targeted processes and thereby a greater potential for disruption of human and animal physiology.

Modularity as a Simplifying Characteristic

Phage properties can be distinguished into those that operate extracellularly versus those that operate instead intracellularly. The extracellular aspects determine formulation stability, most phage pharmacokinetic characteristics, and target bacterium acquisition properties including host range specificity. Indeed, a phage virion in principle is simply a bacterium-acquisition devise whose sole function is the delivery of the intracellular acting agents to bacterial cytoplasms. To reach bacterial targets, by contrast, chemical antibiotics must inherently cross membranes by being relatively small and hydrophobic, have targets that are found external to membranes such as cell walls, be able to pass through porins found in the Gram-negative outer membranes, or otherwise must be associated with supplied bacteria-penetrating mechanisms. Bacteria-penetrating properties, however, are not necessarily benign within the body environment to the extent that they can also give rise to eukaryotic cell penetration, or simply plasma membrane disruption. Thus, the phage potential to penetrate bacterial envelopes without first dissolving in bacterial membranes likely contributes greatly to their relative dearth of negative pharmacodynamic aspects. That is, phages as antibacterial agents may be relatively safe in part because they are large and complex rather than in spite of these properties.

Once DNA has been delivered to a bacterium's cytoplasm, the potential to encode and produce antibacterial agents are almost limitless. Indeed, not only can phages encode bacteria-toxic substances that further phage reproduction but so too phages can easily be engineered to produce antibacterial agents that have nothing to do with phage propagation, such as restriction endonucleases [70], all while preserving their inherent delivery mechanisms. By contrast, what is difficult to engineer are antibacterial activities that do not also interfere with the production of new phage virions. Fortunately, though, there often is little need to improve upon the antibacterial activity of naturally occurring phages as phages with novel properties typically may be isolated in the course of standard laboratory enrichment protocols.

In addition to this separation of extracellular and intracellular properties, phages are capable of varying their properties in response to different situations such as differences in target bacterium genotype and physiology. A result of these various phage emergent properties is that it is generally unnecessary to fully characterize every phage in terms of the mechanistic underpinnings of their antibacterial activity. A phage must be demonstrated to be safe through a combination of avoiding temperate phages and otherwise steering clear of bacterial virulence factors such as via bioinformatic analysis, but otherwise can be treated as essentially bacteria-killing black boxes which vary most importantly in terms of their individual antibacterial spectra of activity. Phages thus can be viewed as possessing assorted emergent properties that we need not fully nor, especially, mechanistically appreciate to effectively employ phages as antibacterial agents.

Modularity and the Thwarting of Bacterial Resistance Mechanisms

The intricacies of the phage infection cycle requires that many reactions successfully proceed and as a result there exist many opportunities for things to go wrong. Bacteria as a consequence have many steps available to them that they might interfere with to effect a resistance to specific phage types. Similarly, the specificity required for these various phage actions typically will limit which bacteria are susceptible to specific phage isolates. Analogous to mechanisms of antibiotic resistance, bacterial resistance to phage-mediated killing can be a consequence of both phage inactivation and modification of phage targets [44,46], though the targets which bacteria modify to prevent antibiotic-mediated killing, or factors involved in antibiotic inactivation, are typically not the same as for resisting phages. The result, minimally, is that bacterial evolution of antibiotic resistance typically is not expected to translate into cross resistance to phages.

Bacterial resistance to phages that is acquired mutationally typically involves loss or alteration of bacteria-encoded phage receptor molecules which are found on the surfaces of bacterial cells. However, and a crucial difference between phages and antibiotics as antibacterial agents, is that adsorption blocks act only on the extracellular, bacterial acquisition step of phage adsorption, which is highly variable among phages [71]. Restoration of antibacterial ability by phages consequently can require solely modification of the extracellular step, with little need to modify the intracellular infection/bactericidal program. Such adsorption affinity modification often can be attained via simple laboratory adaptation experiments [72]. Alternatively, and more conceptually similar to rational-based drug design, phages can be genetically engineered so that they display affinities for new phage receptors [36]. In actual practice the solution to the occurrence of bacterial resistance often is far simpler: One simply isolates new phages, ones which vary both in extracellular and intracellular characteristics.

To summarize, (i) phages possess narrow spectra of activity, that is, their host range [44,73]; (ii) they are affected by resistance mechanisms that are not thought to share specificities with antibioitic-resistance mechanisms; (iii) it is extremely difficult for bacteria to acquire a general resistance to phages or even resistance to a majority of possible phages, and, as a consequence, (iv) it often is relatively straightforward to isolate from the environment new, professionally lytic phage isolates that are specific for target bacterial pathogens while at the same time are unaffected by a given spectrum of bacterial anti-antibiotic as well as anti-phage resistance mechanisms. Substantial exploitation of naturally occurring anti-resistance properties found among phages in phage therapy, however, is highly dependent on a regulatory framework that allows for or even encourages the development of what potentially could be an embarrassing wealth of both safe and diverse antibacterial agents, that is, rather than focusing predominantly on a relatively few phages that have been subject to the full spectrum of regulatory inquiry. Indeed, if we simply redefined antibiotics to include phages, and then took advantage of the diverse potential for phages to overcome bacterial resistance mechanisms to both phages and antibiotics, the antibioticresistance crisis likely would be much smaller in scope [6].

PHARMACOKINETICS

Pharmacokinetics is the study of the impact of bodies on drugs where body can be defined to include not just animal cells and tissues but also associated microflora. Typically this impact is differentiated into various mechanisms that together describe barriers to the achievement of effective densities at a drug's site of action. Drugs thus are subject to absorption, distribution, metabolism, and excretion (together, ADME). These are drug movement into the blood, particularly from the gut though also by more direct routes; drug movement from the blood to other body compartments; drug chemical modification; and movement of intact drug out of the body, respectively.

Antimicrobial drugs thus are administered to patients and in some manner reach bacterial targets where they can have positive effects. All the while, however, phages are also blocked in their movement, diluted, chemically destroyed, removed from the body, or otherwise sequestered away from bacterial targets. For additional discussion of these ideas as they pertain to phage therapy, see Abedon and Thomas-Abedon [6] along with Abedon [9]. In this section we consider additional, more specific aspects of pharmacokinetics as this subject can be applied to phages. See Figure 3 for summary of pharmacokinetics in general, with associated pharmacodynamics, and as compared with those of phage therapy.

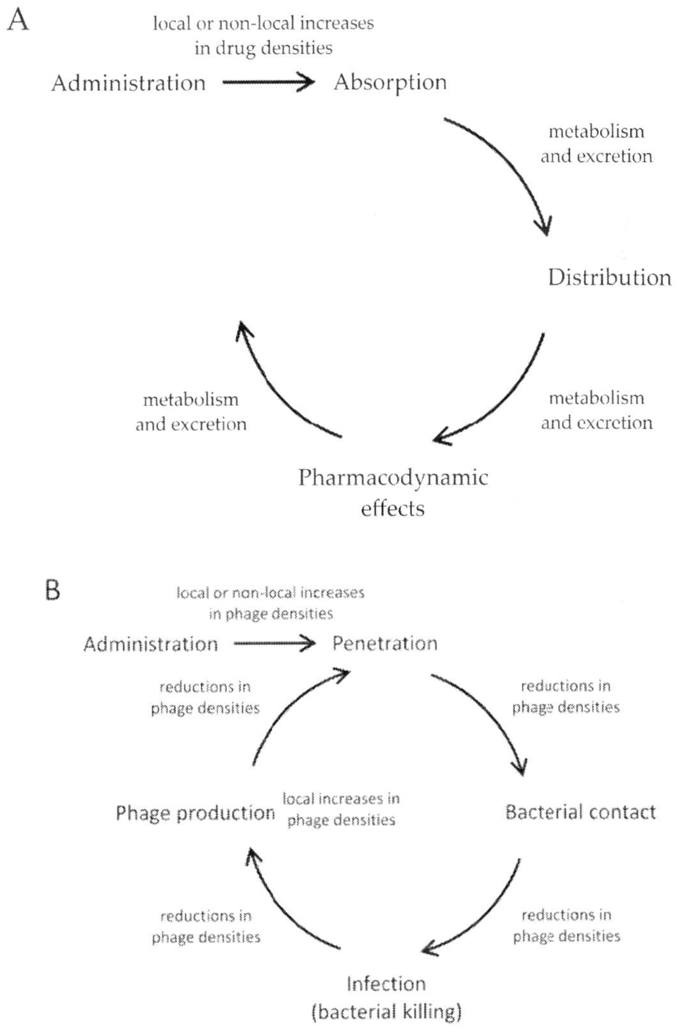

Figure 3: Pharmacokinetics basics (panel A) as compared with phage therapy basics (panel B). Phage therapy as typically practiced often differs in two basic ways from the 'life cycle' of most drugs (panel A). First, absorption, that is, uptake into the blood, is not an issue for the sake of positive pharmacodynamic

effects unless systemic treatment is sought. Second, most drugs, unlike phages, do not self amplify in situ following contact with target tissues.

Topical versus Systemic Administration

Topical phage administration, which includes application to both wounds and lungs, does not require absorption nor necessarily substantial distribution. With phages that are topically applied to bacteria, one instead can use the term penetration [9], as in phage penetration into bacterial biofilms. Such biofilm penetration by phages [67,68] may be more effective than the impact of chemical antibiotics [74]. Indeed, we can apply the phrase "active penetration" to describe the phage process [6], reflecting the possibility that phage-induced bacterial lysis, particularly of bacteria found in the outer layers of bacterial biofilms, may be aiding phage penetration. This lysis either supplies phages directly to underlying bacteria or allows nutrients to diffuse to already established phage infections.

Oral phage delivery too can involve phage penetration without accompanying absorption or distribution. High stomach acidity can inactivate phages, however, and thus "penetration" through gastric juices should always be a concern given oral phage administration. Poorly understood, though, is the issue of penetration of phages to gastrointestinal bacteria once past the stomach. In particular, the question of whether or when phages can consistently display efficient active therapy within the gastrointestinal lumen is somewhat open and efforts to reduce bacterial densities using relatively low phage densities and/or few dosings are unlikely to be effective unless active treatment is in fact possible. Thus, not only is phage inactivation in the course of passage through the stomach a concern but so too phage dilution upon mixing with food, drink, or gastric secretion as dilution can also have the effect of reducing overall phage densities [9].

Phage application to the gastrointestinal tract of animals can result in uptake into blood, a process that has been dubbed bacteriophage translocation [75]. Unfortunately, little is understood of the mechanism of this translocation, except that it is not always demonstrable [76-78]. When such absorption is anticipated following oral administration, then the application can be described as per os. By design, most antibiotics employed systemically can be delivered in this manner and since oral dosing can be highly convenient for patients, including in a non-clinical setting, it would be helpful were phages similarly deliverable. Consequently, a clear limitation to the potential of phage therapy as a general strategy of bacterial disease control is the substantial uncertainty that surrounds the phage facility to reach systemic circulation following oral delivery.

More direct approaches of administration to systemic circulation of course also exist. These, however, can be more problematic due to a combination of greater invasiveness, lower convenience, and a need for greater phage purification [50]. Contrasting per os, though, delivering phages directly to systemic circulation may be more reliably achieved. In addition, passive treatment also is possible with direct administration whereas per os phage delivery would appear to be wholly dependent on active treatment, that is, phages entering systemic circulation in low numbers that are then bolstered upon contact with target bacteria. Human treatment using these more direct approaches to systemic delivery have been used though such efforts are not nearly as common as more local applications [14]. A reasonable assumption is that the phage therapy niche as an alternative to chemical antibiotics may be biased towards local administration, probably including oral administration for gastrointestinal treatments, with phage treatment of systemic infections perhaps limited to circumstances in which antibiotic use is less permissible, for instance against antibioticresistant bacteria such as methicillin-resistant *Staphylococcus aureus* (MRSA) or vancomycin-resistant enterococci (VRE) [11]. Indeed, a number of studies have either demonstrated in humans [79,80], shown in animals [81], or proposed [82] the phage therapy or biocontrol [83] of MRSA. Phage-based treatment of MRSA may even be less expensive than treatment with chemical antibiotics [84].

Distribution

Distribution is the movement of drugs out of the blood and into other body tissues. Achievement of this movement is a function of drug molecular properties and though it probably is reasonable to speculate that the phage potential for such movement occurs for reasons that are somewhat independent of the mechanisms employed by small-molecular drugs, otherwise the means by which phages achieve such movement is poorly understood [9,85]. What is crucial to keep in mind, however, is that issues of drug distribution towards positive pharmacodynamic effects only matter to the extent that drug application is systemic and intended to reach tissues other than blood. Another key issue is that disease, including bacterial infections, can increase tissue permeability such as to phages; see Abedon [9] for discussion. Also relevant is that the pharmacokinetic property of self amplification to at least some degree can compensate for inefficiencies in phage distribution: If phages can reach infections then they can have some impact on those infections and if their ability to replicate is sufficiently robust then they may be able to achieve a substantial impact.

Metabolism

Metabolism in a pharmacokinetic sense describes the chemical modification of drugs. Though this often is a means to the end of drug elimination from the body, in some cases metabolism instead can convert a prodrug into its active form. With chemical pharmaceuticals most metabolism occurs in the liver. There cytochrome P450 family enzymes make lipophilic xenobiotics, that is, molecules not found within the body, more polar and thus more easily excreted through the renal system [86]. As phages consist mostly of proteins and DNA, they generally are not xenobiotic. In addition, in terms of phage pharmacokinetics, phage sequestration such as due to immune system action need not result in subsequent chemical modification to have phage levels reduced through metabolism so long as phages otherwise have been permanently separated from their bacterial targets.

Phages must be resupplied on an ongoing basis to the vicinity of target bacteria for phage therapy to succeed in the face of elimination from tissues. If effective phage densities are, for example, 10^8/ml, and phages are inactivated at a rate of, say, 1% per minute, then 106 new phages must be added to the system per minute just to sustain phage densities at effective levels. With active treatment not only must phages be sustained at effective levels but so too must those relatively high phage densities be attained in the first place, all of which must occur in the face of ongoing phage losses. Nonetheless, if phages are easily delivered, or otherwise can replicate sufficiently robustly *in situ*, then inactivation by immune system action need not be a substantial issue.

Note that phage self amplification *in situ* too may be described as an action of metabolism [6,9] though of course not one associated with the action of liver enzymes.

Excretion

Phage elimination from bodies is mostly due to immune system-mediated sequestration or inactivation, as opposed to excretion, which is elimination of a drug in the same form as it was administered. Phages do have some access to urine, which may be useful for treatment of urinary tract infections following systemic phage administration [9]. Other than phage delivery to such sites, excretion is not terribly relevant to phage therapy and particularly less so given topical rather than systemic application.

Conclusion

The impact of a standard chemical drug is difficult to predict due to the emergent properties that are associated with the complexity of the organisms they are treating, such as ourselves. With small-molecule antimicrobial drugs, that complexity is further increased owing to the multifaceted nature of the parasitic organisms being targeted. This means that negative impacts on target pathogens (i.e., positive pharmacodynamic effects) along with negative impacts on the body itself can be difficult to predict. When phages are properly chosen, by contrast, their negative impact on bodies can be both minimal and somewhat predictable, and this predictability can occur in spite of a lack of specific knowledge of the biochemical processes that phages encode. The result can be a much simpler path towards antibacterial development than one typically sees with chemical antibacterials, where toxicity testing often serves as a major road block [15].

The pharmacologically emergent properties of phages thus are mostly limited to their negative impact on target bacteria, that is, especially professionally lytic phages can possess biases towards pharmacologically desirable rather than undesirable effects. Phages in particular, and like antibiotics, are naturally occurring antibacterial agents whose positive pharmacodynamic effects have been honed by billions of years of organic evolution. Furthermore, to a degree the phage potential to self amplify, in situ, can obviate many pharmacokinetic issues that otherwise can negatively impact their potential to achieve minimally effective antibacterial densities in the vicinity of target bacteria. The upshot is an emergent property laden but nonetheless non-xenobiotic pharmaceutical that can be more predictable than traditional antibiotics at the 'higher levels' of antibacterial characterization, such as in terms of animal or clinical testing, even as lower-level, especially biochemical understanding of the phage impact on target bacteria can be vastly more difficult to achieve.

REFERENCES

1. Christopher GW, Cieslak TJ, Pavlin JA, Eitzen EM, Jr.(1997) Biological warfare. A historical perspective. J Am Med Assoc 278:412-417.
2. Leitenberg M(2001) Biological weapons in the twentieth century: a review and analysis. Crit Rev Microbiol 27:267-320.
3. Hokkanen HMT, Sailer RI(2008) Success in classical biological control. Crit Rev Plant Sci 3:35-72.
4. Harper DR(2006) Biological control by microorganisms. The Encyclopedia of Life Sciences. John Wiley & Sons, Chichester, pp. 1-10.
5. Abedon ST(2009) Kinetics of phage-mediated biocontrol of bacteria. Foodborne Pathog Dis 6:807-815.
6. Abedon ST, Thomas-Abedon C(2010) Phage therapy pharmacology. Curr Pharm Biotechnol 11:28-47.
7. Abedon ST, Kuhl SJ, Blasdel BG, Kutter EM(2011) Phage treatment of human infections. Bacteriophage 1:66-85.
8. Abedon ST(2009) Impact of phage properties on bacterial survival. Adams, H. T. (Ed): Contemporary Trends in Bacteriophage Research. Nova Science Publishers, Hauppauge, New York, pp. 217-235.
9. Abedon ST(2012) Bacteriophages as drugs: the pharmacology of phage therapy. Borysowski, J. Miêdzybrodzki, R. Górski, A. (Eds): Phage Therapy: Current Research and Applications. Caister Academic Press,
10. Cars O, Hedin A, Heddini A(2011) The global need for effective

antibiotics-moving towards concerted action. Drug Resist Updat 14:68-69.

11. French GL(2010) The continuing crisis in antibiotic resistance. Int J Antimicrob Agents 36 Suppl 3:S3-S7.
12. Blaser MJ, Falkow S(2009) What are the consequences of the disappearing human microbiota? Nat Rev Microbiol 7:887-894.
13. Blaser M(2011) Antibiotic overuse: Stop the killing of beneficial bacteria. Nature (London) 476:393-394.
14. Kutter E, De Vos D, Gvasalia G, Alavidze Z, Gogokhia Let al.(2010) Phage therapy in clinical practice: treatment of human infections. Curr Pharm Biotechnol 11:69-86.
15. Berg EL, Kunkel EJ, Hytopoulos E(2005) Biological complexity and drug discovery: a practical systems biology approach. Syst Biol (Stevenage) 152:201-206.
16. Butcher EC, Berg EL, Kunkel EJ(2004) Systems biology in drug discovery. Nat Biotech 22:1253-1259.
17. Trosko JE, Chang CC(2010) Factors to consider in the use of stem cells for pharmaceutic drug development and for chemical safety assessment. Toxicology 270:18-34.
18. Turner SM(2006) Stable isotopes, mass spectrometry, and molecular fluxes: applications to toxicology. J Pharmacol Toxicol Methods 53:75-85.
19. Odum EP, Barrett GW(2004) Fundamentals of Ecology.Brooks Cole
20. Salt GW(1979) A comment on the use of the term emergent properties. Am Nat 113:145-148.
21. Harré R(1972) The philosophies of science.Oxford University Press, London.
22. Cos P, Tote K, Horemans T, Maes L(2010) Biofilms: an extra hurdle for effective antimicrobial therapy. Curr Pharm Des 16:2279-2295.
23. Neu HC(1992) The crisis in antibiotic resistance. Science (New York, N Y) 257:1064-1073.
24. Gill EE, Brinkman FS(2011) The proportional lack of archaeal pathogens: Do viruses/phages hold the key? Bioessays: News And Reviews In Molecular, Cellular And Developmental Biology 33:248-254.

25. Abedon ST, LeJeune JT(2005) Why bacteriophage encode exotoxins and other virulence factors. Evol Bioinform Online 1:97-110.
26. Hyman P, Abedon ST(2008) Phage ecology of bacterial pathogenesis. Abedon, S. T. (Ed): Bacteriophage Ecology. Cambridge University Press, Cambridge, UK, pp. 353-385.
27. Abedon ST(2008) Phages, ecology, evolution. Abedon, S. T. (Ed): Bacteriophage Ecology. Cambridge University Press, Cambridge, UK, pp. 1-28.
28. Wommack KE, Colwell RR(2000) Virioplankton: viruses in aquatic ecosystems. Microbiol Mol Biol Rev 64:69-114.
29. Kilic AO, Pavlova SI, Ma WG, Tao L(1996) Analysis of Lactobacillus phages and bacteriocins in American dairy products and characterization of a phage isolated from yogurt. Appl Environ Microbiol 62:2111-2116.
30. Ackermann H-W(1997) Bacteriophage ecology. Martins, M. T. Sato, M. I. Z. Tiedje, J. M. Hagler, L. C. N. Döbereiner, J. et al. (Eds): Progress in Microbial Ecology (Proceedings of the Seventh International Symposium on Microbial Ecology). Brazilian Society for Microbiology/International Committee on Microbial Ecology, São Paulo, Brazil, pp. 335-339.
31. Letarov A, Kulikov E(2009) The bacteriophages in human- and animal body-associated microbial communities. J Appl Microbiol 107:1-13.
32. Goodridge LD(2008) Phages, bacteria, and food. Abedon, S. T. (Ed): Bacteriophage Ecology. Cambridge University Press, Cambridge, UK, pp. 302-331.
33. Hagens S, Loessner MJ(2010) Bacteriophage for biocontrol of foodborne pathogens: calculations and considerations. Curr Pharm Biotechnol 11:58-68.
34. Johnson RP, Gyles CL, Huff WE, Ojha S, Huff GRet al.(2008) Bacteriophages for prophylaxis and therapy in cattle, poultry and pigs. An Hlth Res Rev 9:201-215.
35. Balogh B, Jones JB, Iriarte FB, Momol MT(2010) Phage therapy for plant disease control. Curr Pharm Biotechnol 11:48-57.
36. Goodridge LD(2010) Designing phage therapeutics. Curr Pharm

Biotechnol 11:15-27.

37. Loc-Carrillo C, Abedon ST(2011) Pros and cons of phage therapy. Bacteriophage 1:111-114.
38. Hendrix RW(2009) Jumbo bacteriophages. Curr Top Microbiol Immunol 328:229-240.
39. Bull JJ, Vimr ER, Molineux IJ(2010) A tale of tails: Sialidase is key to success in a model of phage therapy against K1-capsulated Escherichia coli. Virology 398:79-86.
40. Sulakvelidze A, Kutter E(2005) Bacteriophage therapy in humans. Kutter, E. Sulakvelidze, A. (Eds): Bacteriophages: Biology and Application. CRC Press, Boca Raton, Florida, 381-436.
41. Merril CR(2008) Interaction of bacteriophages with animals. Abedon, S. T. (Ed): Bacteriophage Ecology. Cambridge University Press, Cambridge, UK, 332-352.
42. Stebbings R, Poole S, Thorpe R(2009) Safety of biologics, lessons learnt from TGN1412. Curr Opin Biotechnol 20:673-677.
43. Shankar G, Pendley C, Stein KE(2007) A risk-based bioanalytical strategy for the assessment of antibody immune responses against biological drugs. Nat Biotech 25:555-561.
44. Hyman P, Abedon ST(2010) Bacteriophage host range and bacterial resistance. Adv Appl Microbiol 70:217-248.
45. Fischbach MA, Walsh CT(2009) Antibiotics for emerging pathogens. Science (New York, N Y) 325:1089-1093.
46. Labrie SJ, Samson JE, Moineau S(2010) Bacteriophage resistance mechanisms. Nat Rev Microbiol 8:317-327.
47. Stern A, Sorek R(2011) The phage-host arms race: shaping the evolution of microbes. Bioessays: News And Reviews In Molecular, Cellular And Developmental Biology 33:43-51.
48. Pelkonen O, Turpeinen M, Hakkola J, Honkakoski P, Hukkanen Jet al.(2008) Inhibition and induction of human cytochrome P450 enzymes: current status. Arch Toxicol 82:667-715.
49. Fuhr U, Anders EM, Mahr G, Sorgel F, Staib AH(1992) Inhibitory potency of quinolone antibacterial agents against cytochrome P450IA2 activity in vivo and in vitro. Antimicrob Agents Chemother 36:942-948.

50. Gill JJ, Hyman P(2010) Phage choice, isolation and preparation for phage therapy. Curr Pharm Biotechnol 11:2-14.
51. Magalhaes PO, Lopes AM, Mazzola PG, Rangel-Yagui C, Penna TCet al.(2007) Methods of endotoxin removal from biological preparations: a review. J Pharm Pharm Sci 10:388-404.
52. Abedon ST(2011) Phage therapy pharmacology: calculating phage dosing. Adv Appl Microbiol
53. Soothill JS(1994) Bacteriophage prevents destruction of skin grafts by Pseudomonas aeruginosa. Burns : journal of the International Society for Burn Injuries 20:209-211.
54. Gerdil C(2003) The annual production cycle for influenza vaccine. Vaccine 21:1776-1779.
55. Slopek S, Weber-Dabrowska B, Dabrowski M, Kucharewicz-Krukowska A(1987) Results of bacteriophage treatment of suppurative bacterial infections in the years 1981-1986. Arch Immunol Ther Exp 35:569-583.
56. Waxman DJ, Strominger JL(1983) Penicillin-binding proteins and the mechanism of action of beta-lactam antibiotics. Annu Rev Biochem 52:825-869.
57. Yoshida H, Nakamura M, Bogaki M, Ito H, Kojima Tet al.(1993) Mechanism of action of quinolones against Escherichia coli DNA gyrase. Antimicrob Agents Chemother 37:839-845.
58. Levine C, Hiasa H, Marians KJ(1998) DNA gyrase and topoisomerase IV: biochemical activities, physiological roles during chromosome replication, and drug sensitivities. Biochim Biophys Acta 1400:29-43.
59. Kutter E, White T, Kashlev M, Uzan M, McKinney J et al.(1994) Effects on host genome structure and expression. Karam, J. D. (Ed): Molecular Biology of Bacteriophage T4. ASM Press, Washington, DC, pp. 357-368.
60. Sergueev K, Yu D, Austin S, Court D(2001) Cell toxicity caused by products of the p(L) operon of bacteriophage lambda. Gene 272:227-235.
61. Rees PJ, Fry BA(1981) The morphology of staphylococcal bacteriophage K and DNA metabolism in infected Staphylococcus

aureus. J Gen Virol 53:293-307.

62. Deveau H, Barrangou R, Garneau JE, Labonté J, Fremaux Cet al.(2008) Phage response to CRISPR-encoded resistance in Streptococcus thermophilus. J Bacteriol 190:1390-1400.
63. Abedon ST(2011) Lysis from without. Bacteriophage 1:46-49.
64. Hermoso JA, Garcia JL, Garcia P(2007) Taking aim on bacterial pathogens: from phage therapy to enzybiotics. Curr Opin Mirobiol 10:461-472.
65. Veiga-Crespo P, Ageitos JM, Poza M, Villa TG(2007) Enzybiotics: a look at the future, recalling the past. J Pharm Sci 96:1917-1924.
66. Zapun A, Contreras-Martel C, Vernet T(2008) Penicillin-binding proteins and beta-lactam resistance. FEMS Microbiol Rev 32:361-385.
67. Abedon ST(2010) Bacteriophages and Biofilms. Bailey, W. C. (Ed): Biofilms: Formation, Development and Properties. Nova Science Publishers, Hauppauge, New York, pp. 1-58.
68. Abedon ST(2011) Bacteriophages and Biofilms: Ecology, Phage Therapy, Plaques.Nova Science Publishers, Hauppauge, New York.
69. Abedon ST(2009) Phage evolution and ecology. Adv Appl Microbiol 67:1-45.
70. Hagens S, Habel A, von Ahsen U, von Gabain A, Bläsi U(2004) Therapy of experimental Pseudomonas infections with a nonreplicating genetically modified phage. Antimicrob Agents Chemother 48:3817-3822.
71. Randall LL, Philipson L(1980) Virus Receptors part 1 Bacterial Viruses.Chapman and Hall, London and New York.
72. Henning U, Hashemolhosseini S(1994) Receptor recognition by T-even type coliphages. Karam, J. D. (Ed): The Molecular Biology of Bacteriophage T4. ASM Press, Washington, DC 291-298.
73. Chan BK, Abedon ST(2011) Bacteriophage adaptation, with particular attention to issues of phage host range. Quiberoni, A. Reinheimer, J. (Eds): Bacteriophages in Dairy Processing. Nova Science Publishers, Hauppauge, New York
74. Gilbert P, Rickard AH, McBain AJ(2011) Biofilms and antimicrobial resistance. Fraise, A. Lambert, P. A. Maillard, J.-Y. (Eds):

Principles and Practice of Disinfection, Preservation & Sterilization (Russell, Hugo & Ayliffe's Principles and Practice of Disinfection, Preservation & Sterilization) (4). Wiley-Blackwell, ?, 128-138?

75. Górski A, Wazna E, Dabrowska B-W, Switala-Jelén K, Miedzybrodzki R(2006) Bacteriophage translocation. FEMS Immunol Med Microbiol 46:313-319.

76. Krylov VN(2001) Phage therapy in terms of bacteriophage genetics: Hopes, prospects, safety, limitations. Rus J Genet 37: 869-87.

77. Bruttin A, Brüssow H(2005) Human volunteers receiving Escherichia coli phage T4 orally: a safety test of phage therapy. Antimicrob Agents Chemother 49:2874-2878.

78. Denou E, Bruttin A, Barretto C, Ngom-Bru C, Brüssow Het al.(2009) T4 phages against Escherichia coli diarrhea: potential and problems. Virology 388:21-30.

79. Leszczynski P, Weber-Dabrowska B, Kohutnicka M, Luczak M, Gorecki A et al.(2006) Successful eradication of methicillin-resistant Staphylococcus aureus (MRSA) intestinal carrier status in a healthcare worker--case report. Folia Microbiol (Praha) 51:236-238.

80. Leszczynski P, Weber-Dabrowska B, Kohutnicka M, Luczak M, Górecki A et al.(2008) Successful eradication of methicillin-resistant Staphylococcus aureus (MRSA) intestinal carrier status in a healthcare worker - a case report. Folia Microbiol 51:236-238.

81. Capparelli R, Parlato M, Borriello G, Salvatore P, Iannelli D(2007) Experimental phage therapy against Staphylococcus aureus in mice. Antimicrob Agents Chemother 51:2765-2773.

82. Mann NH(2008) The potential of phages to prevent MRSA infections. Res Microbiol 159:400-405.

83. O'Flaherty S, Ross RP, Meaney W, Fitzgerald GF, Elbreki MFet al.(2005) Potential of the polyvalent anti-Staphylococcus bacteriophage K for control of antibiotic-resistant staphylococci from hospitals. Appl Environ Microbiol 71:1836-1842.

84. Miedzybrodzki R, Fortuna W, Weber-Dabrowska B, Gorski A(2007) Phage therapy of staphylococcal infections (including MRSA) may be less expensive than antibiotic treatment. Postepy higieny i medycyny doswiadczalnej (Online) 61:461-465.

85. Letarov AV, Golomidova AK, Tarasyan KK(2010) Ecological basis of rational phage therapy. Acta Naturae 2:60-71.

86. Meyer UA (1996) Overview of enzymes of drug metabolism. J Pharmacokinet Biopharm 24:449-459.

CHAPTER

7

ROLE OF BACTERIOPHAGE IN CONTROL OF PLANT DISEASES

Dr Pratibha Pant[1], Dr Prashant Pant[2], Dr Prerna Pandey[3]

[1]Dr. Pratibha Pant, is an Assistant Professor of Botany in Daulat Ram College, University of Delhi, New Delhi, India and has a teaching experience of more than 6 years. She is a plant virologist and biotechnologist by training and has done her PhD in Characterization of Begomoviral genomes from Papaya from Department of Botany, University of Delhi. During her PhD, she utilized a combination of molecular biology and Computational Biology tools to analyze begomoviral genomes and to her credit, a couple of novel begomoviruses have been isolated, sequenced and reported in public databases

[2]Dr. Prashant Pant, is an Assistant Professor of Botany in Dyal Singh College, University of Delhi, New Delhi, India. He has been teaching for more than 5 years now and his main areas have been Bioinformatics, Molecular Biology and Genomics at the college. His PhD was on understanding the Microbial Succession using a Metagenomic approach in a Barren mined-out site undergoing ecological restoration. For this, he deployed many molecular and computational Biology tool and

techniques. He has several publications (articles and book chapters) in International Journals and International Publications. Currently, he is also Chief Editor of Bioinformatics Review and a co-founder/Executive Chief Editor of a Science Journal in Hindi Language "Swadeshi Vigyan" for the native language speaking communities. Both the Journals are a service to the scientific community as they are fully open access.

[3]Dr Prerna Pandey, a biotechnologist with several years of wet lab research experience, worked at International Center for Genetic Engineering and Biotechnology, New Delhi. Her field of research in PhD included isolation and molecular characterization of geminiviruses, genome sequencing, gene annotation, and gene silencing using the RNA interference technology. She has also worked at Transasia Biomedicals and Advance Enzyme Technologies as a scientist. Prerna has published papers in peer-reviewed journals, and has submitted a number of annotated Geminiviral genome sequences in the GenBank, including two novel ones. She has also completed her editing and proof-reading courses from Society for the promotion of Editors and Proof-readers, UK and now works as a freelance scientific editor and writer. When Prerna is not busy with her assignments, she loves to explore historical places.

The population of humans is predicted to reach 9.6 billion in 2050 that requires a boost in global food production by 70%. (UN, 2013). This necessitates a rise of more than 80% in the field of agriculture to provide food for the population (Ray et al, 2013).

The main issue with increase in global food production is the incidence of plant diseases. Around 10% of the crops are devastated by diseases of plants (Strange and Scott, 2005). Apart from all the pathogenic bacteria like *Agrobacterium, Erwinia* etc. there are fungi, nematodes and viruses.

BACTERIOPHAGES

Bacteriophages (phages) belong to the class of viruses, described as obligate intracellular parasites of bacteria. There are seven orders of virus: *Caudovirales* (3 families), *Herpesvirales* (3 families), *Legamenvirales* (3 families), *Mononegavirales* (3 families), *Nidovirales* (3 families),

Picornavirales (3 families) and *Tymovirales* (3 families) (Doffkay et al, 2015).

Phages have two kinds of life cycles: lytic and lysogenic. Lytic phages can infect a host bacterium and produce copies of themselves and in the process, lyse or burst the host cells. Thus, these phages can be used as biocontrol agents as they kill the host cells. Lysogenic phages integrate with the host bacterial genomes and can stay latent for a while. However, upon activation, they can be induced into the lytic cycle causing destruction of the host cells.

BACTERIOPHAGES FOR CONTROL OF PATHOGENS

Felix d'Herelle, discovered a control agent that was not visible to the naked eye for dysentery bacillus. The very first use of phages to control diseases was performed by Bruynoghe and Maisin in 1921 where phages were used on skin to kill staphylococcus bacteria. This was then extended to several bacteria: Eli Lilly, E.R. Squibb and Sons and Swan-Myers manufactured phages to act against Staphylococcus (Doffkay et al, 2015).

The main targets of phages were used for treating staphylococcal infections, cholera and bubonic plague (Buttimer et al, 2017). However, many countries shifted focus to antibiotics, while certain countries like Poland and Soviet were associated with phage therapy. These research projects were in Russian that prevented access to majority scientists. With the appearance of bacteria that are multi drug resistant, the focus against shone on phage therapy to eliminate bacteria (Chanishvili, 2012).

BACTERIOPHAGES FOR CONTROL OF PLANT PATHOGENS

The biosphere reportedly has 10^{31} phages while the numbers of prokaryotes stand at 10^{30} that pitches the phages at an advantage. These phages alter the host mechanism to replicate. The majority of these viruses belong to *Caudovirales:* that have DNA as genetic material with icosahedral heads and tails. These can be classified into three groups: *Myoviridae* that possess rigid contractile tails, *Podoviridae* have short, non-contractile tails and *Siphoviridae* display long flexible tails (Buttimer et al, 2017).

The first time they were used for controlling plant infections was a filtrate of decomposing cabbages that quelled cabbage rot disease caused by *Xanthomonas campestris pv. campestris*.

Soft rot disease was targeted by Kotila and Coons in 1925 by applying phage and treating diseases caused by by *Pectobacterium atrosepticum* on potato and *Pectobacterium carotovorum* ssp *carotovorum* on carrot.

In another study, Thomas showed that phage treated seeds could resist *Pantoea stewartii*, a pathogen (Buttimer et al, 2017).

CHOICE OF BACTERIOPHAGE

The biocontrol of plant diseases using bacteriophages entails the study of its life cycle: lytic or lysogenic. Phages that follow the lytic cycle and kill their host are called virulent phages. Certain phages integrate themselves into the host genome where it is referred to as a prophage. These prophages can be triggered by certain causes like heat, UV light and even few extracts of plants. It is reported that such prophages may provide a survival edge to the host for instance, prophages ECA41 and ECA29 of *P. atrosepticum* enhance mobility. Another issue is the potential of transduction where integrated phages may pack portions of the host genome and transfer it to other organisms that could cause gene transfer.

Certain filamentous phages of *Inovirus*, cause an infection where phages are produced as an ongoing process without killing the host cell.

IDEAL FEATURES OF A PHAGE FOR PLANT DISEASE TREATMENT

- Lysis: A phage for potential use as control of pathogens can preferably be lytic or virulent to lyse the pathogenic host cell.
- Host range: A broad host range is preferred to target all possible strains of a host.
- Rapidity: A phage that lyses its host in a short time interval is desired.
- Yield: Phage should produce enormous quantity of progeny that diffuse efficiently.

Exceptions

However, a phage was isolated that was not extremely lytic called ϕRSL1 that targeted *Ralstonia solanacearum* with surprising efficiency. A plausible explanation is its co-exsistence with its host, to maintain its number while not completely destroying the host (Fujiwara et al, 2011).

In another exception, certain phages display high activity in vitro but do not show efficiency on the field. For instance, *X. citri* pv *citri* was targeted by three phages in vitro however two did not show any lysing of citrus canker viruses when tested on the field (Balogh, 2006).

Suggestion

It is desirable to explore the receptors present on the target pathogen and study the phage recognition of the host for optimal lysis. The use of cocktails of phages that recognize different receptors of a pathogen to successfully finish off the pathogen is also suggested in literature (Born et al, 2011).

ADVANTAGES OF BACTERIOPHAGES FOR CONTROL OF PLANT DISEASES

Bacteriophages can be produced rapidly, with ease and at the site of the target. Their production process has been reported as economical. As the specificity of bacteriophages is high, they target only the pathogen in question and not other bacteria. They can be combined with other treatment strategies like chemicals or antibiotics to achieve highly efficient treatment (Doffkay et al, 2015).

Phages are a normal component of the environment to which other organisms and humans are exposed with no detectable damage. This is a unique feature against biocontrol agents that are chemicals with side effects.

Another advantage is their presence in an environment at an elevated level as long as the hosts are present after which the population is optimum. This is unlike pesticides that can cause bioaccumulation (Iriarte et al, 2012).

A phage was characterized that could not infect a Nitrogen fixing bacterium *Bradyrhizobium japonicum*, however, it eliminated other bacteria that interfered with the fixation. Thus phages can have unprecedented good effects (Basit et al, 1992).

Several pathogens of plants form biofilms that impart resistance to the pathogen from commercial pesticides. Bacteriophages have developed startegies such as depolymerases on their capsids that dissolve the biofilms.

The current development of "organic" brands that circumvent the use of chemicals in food production can rely on phage control. The use of phages to control pathogens would be included in an umbrella of biocontrol for use by organic farming strategies (Buttimer et al, 2017).

ISSUES IN THE USE OF BACTERIOPHAGES

The narrow range of host strains targeted by the phages may turn against them in limiting their effectiveness (Doffkay et al, 2015). The development of a phage control agent that can target all strains of a genus is a roadblock that necessitates the use of mixtures or cocktails of phages.

Certain phages may have an expanded host range that requires the examination of potential non-target organisms that may be either harmless or useful. For instance, a T4 phage was isolated in 1988 that could eliminate both *Pectobacterium* and enteric bacteria of humans (Pirhonen and Palva, 1988).

CASE STUDY: *ERWINIA AMYLOVORA*

Bacteriophages have been tested against many plant pathogens that are summarized as: *Agrobacterium tumefaciens*, *Erwinia amylovora*, *Pseudomonas syringae* pv. *tomato*, *Ralstonia solanacearum*, *Xanthomonas arboricola* pv. *juglandis*, *Xanthomonas arboricola*pv. *pruni*, *Xanthomonas oryzae* pv. *oryzae*, and *Xanthomonas campestris* pv. *Vesicatoria* (Doffkay et al, 2015).

Erwinia amylovora causes a wilting disease called fire blight in several plants that are rosaeous that include the economically important apple, pear and quince (Doffkay et al, 2015).

The disease has affected 40 countries and causes losses of up to $100 million a year in the USA due to the various measures. It is classified as quarantine in APPPC (Asia and Pacific Plant Protection Commission),

COSAVE (Comite Regional de Sanidad Vegetal para el Cono Sur), EPPO (Europe and Mediterranean Plant Protection Organisation) and IAPC (Inter-African Phytosanitary Council). The methods employed traditionally are the *use of antibiotics or copper* as well as the *excision of the affected parts* (Buttimer et al, 2017).

However, *the treatment by streptomycin* is not applicable in countries with strict regulations. Phages have been isolated from the order *Caudovirales* with all three of its families (*Podoviridae, Myoviridae*, and *Siphoviridae*) that can target this bacterium. When three strains of phages were applied, there was a reduction in the symptoms of the fire blight disease. Another study advocated the *use of the phages along with Pantoea agglomerans, a common epiphyte* that is harmless to facilitate high infection by the phages that can be used in the fields.

Two patents, Hu P0700600 and HuP1300407 have mentioned the *use of adjuvants that impart protection from UV to the viruses*. This has resulted in the creation of a phage-based biocontrol agent in Hungary in 2012. As the product can be sprayed, it has been applied on apple and pear with promising results.

For easy uptake by the plant, studies have revealed that *spraying the phages on the phyllosphere* can facilitate good uptake in apple (Doffkay et al, 2015).

PRODUCTS AVAILABLE IN THE MARKET

The first product that received registration from the US Environmental protection agency was the phage based product " Agriphage" of Omnilytics based in USA. It controls bacterial spot or speck of tomatoes and peppers by exerting action against *X. campestris* pv. *vesicatoria*, or *P. syringiae pv.* Tomato. The product is listed on OMRI that can find use for organic farmers.

Erwiphage is a biopesticide that acts against *Erwinia amylovora* that causes fire blight of apple trees. It was developed by Enviroinvest based at Hungary.

Biolyse is a washing solution of bacteriophages that has been developed by APS biocontrol, Scotland. The product targets *Enterobacteriacea* that causes soft rot in potatoes. There are reports as to the Tesco supermarkets using the product for their potatoes.

The use of phage based agents as biocontrol requires the various challenges offered by legislation. Another is the update of the cocktails of phages used to target as many emergent strains of viruses.

This update requires registration according to EU regulations, (1107/2009 EC) that extends the time periods causing an impediment in the use of these approaches in the UK (Buttimer et al, 2017).

POTENTIAL ROADBLOCKS FOR THE PHAGES

The phyllosphere is the part of a plant above the ground from where phages are isolated with ease. The area presents several challenges to the phages they main factor being UV radiation (Buttimer et al, 2017).

The *soil and the rhizosphere* influence the survival of phages in the soil. Soil particles can adsorb phages while low pH can inactivate them. A major factor that limits them is UV radiation along with extremes in pH, temperature, drying and washing away by rains (Doffkay et al, 2015).

Another factor is the *type of soil*. It was reported that clay loam soils support phages with low moisture and elevated temperatures as compared to sandy loam soil (Straub et al., 1992). Kaolinite and montmorillonite clays have been reported possess charged surfaces that cause the phage particles to adsorb to and sequester them (Schiffenbauer and Stotzky, 1982).

The use of certain agrochemicals especially cationic, anionic, nonionic and biological detergents has adversely affected the survival of phages (Chattopadhyay et al, 2002).

Bactericidal agents that contain copper have found to inactivate phages that necessitate the use of phage after the application of the copper pesticides (Iriarte et al, 2007).

POTENTIAL CONTENDERS: PHAGE BIOCONTROL AND TARGET PATHOGENS OF PLANTS

Pseudomonas Syringe

It has a reported 50 pathovars that differ in their target plant species. *Pseudomonas syringea* pv. *tomato* causes bacterial specks or lesions on tomato that can also infect *Arabidopsis* and *Brassica*. It affects the yield as well as the fruits. The treatment entails the use of copper and streptomycin while copper resistant species have been reported of late.

Xylella Fastidiosa

It uses insect vectors to target disease in several commercial species such as grape, citrus, almond, peach, coffee and olives. Its effect on grapes is Pierce disease that affects the grape industry up to $104.4 million a year loss.

Dickeya and *Pectobacterium*

Dickeya and Pectobacterium are called as Soft Rot *Enterobacteriacea*. The former affect 10 monocot and 11 dicot families while the latter affect eleven monocot and sixteen dicot families. The use of proper culture and sequestering of infected plants are the protection measures against these infections.

Xanthomonas Species

These species affect 400 hosts that include commercially important crops such as rice, banana, tomato, and citrus fruits. *Xathomonas campestris* pv. *vesicatoria* affects spot disease in tomato across the world causing a loss of half of the crops in the US. The control of the disease requires the use of copper or streptomycin but, there have been reports of new strains that are resistant.

The above-mentioned pathogens are responsible for a large loss of the crops especially at a time when there is global food shortage. It is recommended to examine phage based measures to keep such plant diseases under check (Buttimer et al, 2017).

CURRENT PHAGE BASED THERAPIES AND EXPERIMENTS FOR PLANT DISEASES

Bloomtime® is Pantoea agglomerans that is used in biocontrol of Erwinia amylovora in apple (Mikiciński et al, 2016) .Research has indicated the combination of the above bacterium with phage yields results comparable to that of streptomycin . The research team examined eight types of phages for the control of Erwinia (Boulé et al, 2011)

The use of phages for biocontrol has been promising. Two crops have received major attention: tomato and potato.

Potato

Potato is affected by Soft Rot *Enterobacteriacea* that has limited preventive measures. Keeping this in the picture, several studies have been carried out in the use of phages for the control of Dickeya *solani.*

A T4 related phage called LIMEstone1 was tested on seed tubers of cultivar Bintje in both in vitro and field studies. Those plants received an application of *D. Solani* showed damage of upto 40% while when the phage and bacterium were applied, the damage was reduced to only 10%. Similar results of reduction was seen in another seed tuber cultivar Kondor (Adriaenssens et al, 2012).

A recent study in 2015 showed that bacteriophages ΦPD10.3 and ΦPD23.1 caused a protection from the damage by *P. carotovorum ssp carotovorum* and *P. wasabie* in by up to 80% in potato tuber slices and 95% in whole tubers (Czajkowski et al, 2015).

Tomato

Ralstonia solanacearum causes bacterial wilt in tomatoes. A research team showed that phages can be applied to the treatment of the bacterial disease. Similar results have been observed for *Xanthomonas* that causes leaf spot. Such studies have also been applied to other plants like grapefruit, leeks, grapes and orange along with mushroom with encouraging results. (Buttimer et al, 2017)

A summary of the various phage based control measures that have been employed so far for plant pathogens are summarized below (Buttimer et al, 2017):

Disease	Host	*Pathogen*	Status
Soft rot	Potato	Pectobacterium carotovorum ssp. carotovorum, Pectobacterium wasabiae, Dickeya solan*i*	ΦPD10.3 and ΦPD23.1 phages reduced the disease manifestations.
		Dickeya solani	Phages vB_DsoM_LIMEstone1 and vB_DsoM_LIMEstone2 yielded reduced infection. Another study with phages ΦD1, ΦD2, ΦD3, ΦD4, ΦD5, ΦD7, ΦD9, ΦD10, ΦD11 reduced the disease manifestations
Bacterial wilt	Tomato	Ralstonia solanacearum	Phage ΦRSL1 treatment resulted in no expression of disease. In another study,a cocktail of phage PE204 with R. solanacearum of the rhizosphere eradicated the disease in test plants.
Fire blight	Apple, pear	Erwinia amylovora	Phages ΦEa1337-26 and ΦEa 2345 reduced infection in a study in pear. The use of an associated organism Pantoea agglomerans augmented the activity of the phage and reduced infection. Another study in apple saw a reduction of the infection in half when phage ΦEa1337-26 and P. agglomerans were used.
Common scab	Potato	Streptomyces scabies	Phage ΦAS1 treated tubers produced very few numbers of progeny with lesions.
	Radish		Phages Stsc1 and Stsc3 prevented the disease in seedlings.
Bacterial blight	Leek	Pseudomonas syringae pv. porri	Several phages like vB_PsyM_KIL1, vB_PsyM_KIL2, vB_PsyM_KIL3, and vB_PsyM_KIL3b were efficient in control of the disease.
Brown blotch disease	Mushroom	Pseudomonas tolaasii	The use of phage and pathogen completely eradicated the formation of blotches.
Soft rot	Lettuce	Pectobacterium carotovorum ssp. carotovor*um*	Phage PP1 reduced the disease in green house studies.
Asiatic citrus canker	Grapefruit	Xanthomonas axonopodis pv. citri	Treatment with phage reduced disease by 59% in greenhouse studies and also prevented the disease.
Citrus bacterial spot	Orange	Xanthomonas axonopodis pv. citrumelo	A reduction in disease were observed in nursery studies, the maximum reduction at 48%.
Bacterial spot	Tomato	Xanthomonas campestris pv. vesicatoria	A cocktail of phages reduced the disease in greenhouse studies and fields. Another study reported that treatment with phage was comparable to that of copper mancozeb.
Xanthomonas leaf blight of onion	Onion	Xanthomonas axonopodis pv. allii	The application of phage reduced the disease as comparable to copper mancozeb.

Grapevines	Pierce's Disease	Xylella fastidiosa	A cocktail of four phages called Sano, Salvo, Prado and Paz reduced the levels of the pathogen on the plants. It was also shown that the treatment caused prophylaxis.

APPLICATION/ DELIVERY SYSTEMS OF PHAGE BIOCONTROL AGENTS ON PLANTS

As previously discussed, UV radiation is a deterrent for the survival of phages in the phyllosphere. To overcome this issue the use of carriers that include harmless strains of the pathogen or other bacteria present in the surroundings.

Another approach is the application of the phages at a time when sunlight is not present. It has been reported that the presence of phage and its action was enhanced in tomato when the phages were applied in the evening.

The use of certain adjuvants or substances to protect the phages from the effects of UV has been tested. Several compounds were found to be encouraging on bacteriophage Y2 that targets *Erwinia amylovora* that included: casein, soy peptone and also purified aromatic amino acids, astaxathin and Tween 80 along with extracts of carrot, red pepper and beetroot. These compounds did not affect the activity of the phage and absorbed the UV to protect the phages (Born et al ,2015).

A further method to assist the presence of phage in the phyllosphere is the application of phage to soil. A study in 2012 examined the application of phages to soil and foliage of tomato using the phages of OmniLytics Inc. that targets *X. perforans* strain 97-2. It was observed that the phages were not recorded when applied to the foliage while those applied to the soil reached the leaves at high titres and showed good numbers (Iriarte et al, 2012).

APPLICATIONS OF NEW TECHNOLOGIES

Detection of disease: The use of quantitative PCR for the detection of R. solanacearum , following propagation of the phages. The current method for detection of the pathogen involves the study of bacterial milk that causes damage to the plant. Alternatively, if the appropriate phage is added to the suspected sample, the presence of bacterium would cause a rise in phage titre that can be detected using PCR (Reviewed by Buttimer et al, 2017).

Another interesting study involved the use of a phage that targeted a plant pathogen for bioluminescence. A phage that targets Pseudomonas cannabina pv. alisalensis, causative of blight of crucifers was tagged with luxAB system. It was reported that this bioluminescent system enabled the detection of infection of bacteria rapidly. They were called "light tagged bacteriophages" (Schofield et al., 2013).

PHAGES AND RESISTANCE

The potential of host bacteria to develop resistance to phage based systems akin to those of copper or antibiotics is a possibility. An advantage of using phages is that as they are biological agents that constantly adapt themselves unlike conventional chemicals.

It was observed that phage phi2954 against *Pseudomonas syringae* used a host protein for optimum activity. Though mutations caused development of hosts that lacked protein and were resistant to the phages, there were also mutant phages that evolved without the dependence on the protein (Qiao et al, 2010).

Another system bacterial hosts use to escape attack by phage is the altruistic abortive infection that enables the infected cells to die and stall the phage cycle. But, phages have evolved systems to target such escape systems. For instance, *P. atrosepticum*, a plant pathogen had a protein-RNA toxin-antitoxin to escape attack by phage. A study in 2012 showed that phiTE phage developed mutant forms that formed a pseudo antitoxin to prevent the above system (Blower et al ,2012)

Certain bacteria display CRISPR/Cas mechanisms to protect themselves from phages. The clustered regularly interspaced short palindromic repeat (CRISPR) along with CRISPR associated (Cas) proteins have been reported in several plant phytopathogens like, *E. amylovora* , *P. atrosepticum* and *Xantomonas oryzae*. A 2009 study found that phage can create mutated forms to evade detection by the CRISPR/ Cas mechanisms (Semenova et al, 2009).

Thus, though host bacteria may develop several strategies to evade infection by phage, these agents of biocontrol the bacteriophages can also show mutations to overcome the mechanisms.

SUMMARY

The chapter examined the potential, advantages and limitations of the use of bacteriophages as agents of controlling plant diseases. This research comes at a time when there are drug and pesticide resistant bacteria that are emerging. The chemicals that are employed in the treatment of such plant diseases also have several adverse effects on the environment and organisms including humans.

As phages are biological agents, they can be used in farming along with other phages as a cocktail or as augments to chemical or antibiotic treatments. They also display constant adaptation to overcome resistance mechanisms displayed by bacteria.

The growth of the biopesticide market is expected with the growing consciousness about chemical free food and the side effects of the chemical pesticides on the ecosystem. In this context, the use of phages as biopesticides looks promising as they have yielded encouraging results.

REFERENCES

1. Adriaenssens E. M., Van Vaerenbergh J., Vandenheuvel D., Dunon V., Ceyssens P.-J., De Proft M., et al. (2012). T4-related bacteriophage LIMEstone isolates for the control of soft rot on potato caused by 'Dickeya solani'. PLoS ONE 7:e33227 10.1371/journal.pone.0033227.
2. Balogh B. (2006). Characterization and Use of Bacteriophages Assciated with Citrus Bacterial Pathogens for Disease Control. Ph.D. Dissertation, Unversity of Florida; Gainesville, FL.
3. Basit H. A., Angle J. S., Salem S., Gewaily E. M. (1992). Phage coating of soybean seed reduces nodulation by indigenous soil bradyrhizobia. Can. J. Microbiol. 38: 1264–1269. 10.1139/m92-208.
4. Blower T. R., Evans T. J., Przybilski R., Fineran P. C., Salmond G. P. C. (2012). Viral evasion of a bacterial suicide system by RNA-based molecular mimicry enables infectious altruism. PLoS Genet.8:e1003023 10.1371/journal.pgen.1003023.
5. Born Y., Bosshard L., Duffy B., Loessner M. J., Fieseler L. (2015).

Protection of Erwinia amylovora bacteriophage Y2 from UV-induced damage by natural compounds. Bacteriophage 5:e107433 010.1080/21597081.2015.1074330.

6. Born Y., Fieseler L., Marazzi J., Lurz R., Duffy B., Loessner M. J. (2011). Novel virulent and broad-host-range Erwinia amylovora bacteriophages reveal a high degree of mosaicism and a relationship to Enterobacteriaceae phages. Appl. Environ. Microbiol. 77: 5945–5954. 10.1128/AEM.03022-10.
7. Boulé J., Sholberg P. L., Lehman S. M., O'gorman D. T., Svircev A. M. (2011). Isolation and characterization of eight bacteriophages infecting Erwinia amylovora and their potential as biological control agents in British Columbia, Canada. Can. J. Plant Pathol. 33: 308–317. 10.1080/07060661.2011.588250.
8. Buttimer C., McAuliffe O., Ross R.P., Hill C., O'Mahony J., Coffey A. (2017). Bacteriophages and Bacterial Plant Diseases. Front. Microbiol. 8:34.
9. Chanishvili N. (2012a). A literature review of the practical application of bacteriophage research. In Series: Bacteriology Research Developments, Virology Research Progress; Nova Science Publishers.
10. Chanishvili N. (2012b). Phage therapy-history from Twort and d'Herelle through Soviet experience to current approaches. Adv Virus Res. 83:3-40.
11. Chattopadhyay D., Chattopadhyay S., Lyon W. G., Wilson J. T. (2002). Effect of surfactants on the survival and sorption of viruses. Environ. Sci. Technol. 36: 4017–4024. 10.1021/es0114097.
12. Czajkowski R., Ozymko Z., de Jager V., Siwinska J., Smolarska A., Ossowicki A., et al. (2015). Genomic, proteomic and morphological characterization of two novel broad host lytic bacteriophages ΦPD10.3 and ΦPD23.1 infecting pectinolytic Pectobacterium spp. and Dickeya spp. PLoS ONE 10:e0119812 10.1371/journal.pone.0119812.
13. Fujiwara A., Fujisawa M., Hamasaki R., Kawasaki T., Fujie M., Yamada T. (2011). Biocontrol of Ralstonia solanacearum by treatment with lytic bacteriophages. Appl. Environ. Microbiol. 77: 4155–4162. 10.1128/AEM.02847-10.

14. Iriarte F .B., Balogh B., Momol M. T., Smith L. M., Wilson M., Jones J. B. (2007). Factors affecting survival of bacteriophage on tomato leaf surfaces. Appl. Environ. Microbiol. 73: 1704–1711. 10.1128/AEM.02118-06.

15. Iriarte F. B., Obradović A., Wernsing M. H., Jackson L.E., Balogh B., Hong J. A., et al. (2012). Soil-based systemic delivery and phyllosphere in vivo propagation of bacteriophages: two possible strategies for improving bacteriophage persistence for plant disease control. Bacteriophage. 2: 215–224. 10.4161/bact.23530.

16. Iriarte F. B., Obradović A., Wernsing M. H., Jackson L. E., Balogh B., Hong J. A., et al. (2012). Soil-based systemic delivery and phyllosphere in vivo propagation of bacteriophages: two possible strategies for improving bacteriophage persistence for plant disease control. Bacteriophage. 2: 215–224. 10.4161/bact.23530.

17. Mikiciński A., Sobiczewski P., Puławska J., Maciorowski R. (2016). Control of fire blight (Erwinia amylovora) by a novel strain 49M of Pseudomonas graminis from the phyllosphere of apple (Malus spp.). Eur. J. Plant Pathol. 145: 265–276. 10.1007/s10658-015-0837-y.

18. Pirhonen M., Palva E. T. (1988). Occurrence of bacteriophage T4 receptor in Erwinia carotovora.MGG Mol. Gen. Genet. 214: 170–172. 10.1007/BF00340198.

19. Qiao J., Qiao X., Sun Y., Mindich L. (2010). Role of host protein glutaredoxin 3 in the control of transcription during bacteriophage Phi2954 infection. Proc. Natl. Acad. Sci. U.S.A. 107: 6000–6004. 10.1073/pnas.1000383107.

20. Ray D. K., Mueller N. D., West P. C., Foley J. A. (2013). Yield trends are insufficient to double global crop production by 2050. PLoS ONE. 8:e66428 10.1371/journal.pone.0066428.

21. Schiffenbauer M., Stotzky G. (1982). Adsorption of coliphages T1 and T7 to clay minerals. Appl. Environ. Microbiol. 43: 590–596.

22. Schofield D., Bull C. T., Rubio I., Wechter W. P., Westwater C., Molineux I. J. (2013). "Light-tagged" bacteriophage as a diagnostic tool for the detection of phytopathogens. Bioengineered. 4: 50–54. 10.4161/bioe.22159.

23. Semenova E., Nagornykh M., Pyatnitskiy M., Artamonova I. I., Severinov K. (2009). Analysis of CRISPR system function in plant

pathogen Xanthomonas oryzae. FEMS Microbiol. Lett. 296: 110–116. 10.1111/j.1574-6968.2009.01626.x

24. Strange R. N., Scott P. R. (2005). Plant disease: a threat to global food security. Annu. Rev. Phytopathol. 43: 83–116. 10.1146/annurev.phyto.43.113004.133839.
25. Straub T. M., Pepper I. L., Gerba C. P. (1992). Persistence of viruses in desert soils amended with anaerobically digested sewage sludge. Appl. Environ. Microbiol. 58: 636–641.
26. UN (2013). World Must Sustainably Produce 70 per Cent More Food by Mid-Century – UN report. Availableat: http://www.un.org/apps/news/story.asp?NewsID=46647#.VvxjOuIrLIU.
27. Doffkay, Z., Dömötör, D., Kovács, T., Rákhely, G. (2015) Bacteriophage therapy against plant, animal and human pathogens. Acta Biol Szeged. 59(Suppl.2):291-302.

CHAPTER

8

DELIVERY OF PLANT BACTERIOPHAGES: METHODS, EXPECTATIONS AND CHALLENGES

Dr Prerna Pandey[1], Dr Pratibha Pant[2], Dr Prashant Pant[3]

[1]Dr Prerna Pandey, a biotechnologist with several years of wet lab research experience, worked at International Center for Genetic Engineering and Biotechnology, New Delhi. Her field of research in PhD included isolation and molecular characterization of geminiviruses, genome sequencing, gene annotation, and gene silencing using the RNA interference technology. She has also worked at Transasia Biomedicals and Advance Enzyme Technologies as a scientist. Prerna has published papers in peer-reviewed journals, and has submitted a number of annotated Geminiviral genome sequences in the GenBank, including two novel ones. She has also completed her editing and proof-reading courses from Society for the promotion of Editors and Proof-readers, UK and now works as a freelance scientific editor and writer. When Prerna is not busy with her assignments, she loves to explore historical places.

[2]Dr. Pratibha Pant, is an Assistant Professor of Botany in Daulat Ram College, University of Delhi, New Delhi, India and has a teaching experience of more than 6 years. She is a plant virologist and biotechnologist

by training and has done her PhD in Characterization of Begomoviral genomes from Papaya from Department of Botany, University of Delhi. During her PhD, she utilized a combination of molecular biology and Computational Biology tools to analyze begomoviral genomes and to her credit, a couple of novel begomoviruses have been isolated, sequenced and reported in public databases

[3]Dr. Prashant Pant, is an Assistant Professor of Botany in Dyal Singh College, University of Delhi, New Delhi, India. He has been teaching for more than 5 years now and his main areas have been Bioinformatics, Molecular Biology and Genomics at the college. His PhD was on understanding the Microbial Succession using a Metagenomic approach in a Barren mined-out site undergoing ecological restoration. For this, he deployed many molecular and computational Biology tool and techniques. He has several publications (articles and book chapters) in International Journals and International Publications. Currently, he is also Chief Editor of Bioinformatics Review and a co-founder/Executive Chief Editor of a Science Journal in Hindi Language "Swadeshi Vigyan" for the native language speaking communities. Both the Journals are a service to the scientific community as they are fully open access.

The shifting of various practices in agriculture along with various trade considerations has a role in the development of plant diseases (Gilligan, 2008). The pathogens that are responsible for plant diseases present several challenges in the development of appropriate measures to limit diseases that include: a vast variation in the pathogen that can multiply rapidly during optimal conditions (Jones et al, 2012).

CURRENT TREATMENT STRATEGIES

Current methods of addressing the issue include various methods followed during operations to quell infections, excision of the affected part with proper disposal measures. The use of pesticides to keep a check on vectors or antibiotics like streptomycin or tetracycline and copper metal has been reported.

It is reported that bacterial pathogens are more tough nuts to crack compared to fungi.

Though streptomycin has been used since more than 60 years successfully for pathogens like *Pseudomonas syringae,* there have been reports of resistance to the antibiotic. There is also the issue of the antibiotic resistance genes being transmitted to other microbes.

Copper has been utilised since a century for pathogens but there have been reports of resistance in species like *Xanthomonas* and *Pseudomonas*. An additional concern is the accumulation of copper in the environment upon prolonged usage. This necessitates research in the field of using biological agents (Frampton et al, 2012).

BIOLOGICAL CONTROL

The use of biological control for bacterial diseases includes:

- *Agrobacterium radiobacter* that produces bacteriocin to inhibit other species of pathogens (Kerr, 1974)
- Plant growth-promoting rhizobacteria (PGPR) (Ji et al, 2006).
- Nonpathogenic or attenuated species of pathogens (Liu, 1998)
- Saprophytic bacteria (Ji et al, 2006).

BACTERIOPHAGES AND AGRICULTURE

A few instances of the use of phages have been discussed in 4.2. The earliest instance of using phage on the field was performed By Thomas. Corn seeds that were infected by *Pantoea stewartii (causative agent of* Stewart's wilt of corn) were treated with phage that resulted in a drastic lowering of the disease (Thomas, 1935). The advantages of using phages for control of plant diseases have been discussed in 4.6.

ISOLATION OF PHAGES

The phages that would be used in experiments can be isolated from several sources like water, soil and plant parts. The collection can be done by soft agar overlay with several pathogens. The diversity of phages that are collected is enhanced if the strains of the bacterial hosts are diverse.

Research indicates the isolation of phages that produce clear plaques to avoid the temperate phages that are associated with lysogeny. It is also reported that the method facilitates the isolation of phages of the order order Caudovirales with the three main families: Myoviridae, Siphoviridae, and Podoviridae (Frampton et al, 2012).

Filamentous Phages that are produced continuously by extrusion without lysis of the host cell have also be found to be promising. It was reported that filamentous phages produced contradictory effects on *Ralstonia solanacearum*. While φRSS1 enhanced the virulence on tomato plant, another filamentous phage φRSM3 reduced the virulence of the pathogen. It was reported that filamentous phage φRSM3 along with a lytic phage can be a tool for controlling pathogens (Addy et al, 2012).

CHARACTERIZATION OF PHAGES

The host range of the isolated phages requires study in order to develop a phage mixture or cocktail that can target all the possible strains of the pathogen. The phages should be tested whether they induce lysis only in the target or even other bacteria that are not targets.

This step is followed by characterization through methods like Transmission electron microscopy or RFLP analysis of the genetic material to evaluate the genetic diversity of the phage. The use of next generation sequencing techniques can determine the genetic sequence of phages. The knowledge about sequence can aid in the development of PCR assays for tracking phages. Additional factors imparted by the DNA such as resistance or lysogeny can be determined.

In order to rule out the possibility of transduction, assays can help in identifying phages that have a high frequency of transduction. The study of phage growth pattern such as the duration, burst size, conditions can aid in evaluating the performance of the phage in the field (Frampton et al, 2012).

CHOICE OF PHAGES

Apart from the ability to lyse the bacterial host cells, prescreening of phages for additional properties is desirable. Assays are performed in vitro to evaluate the phages that are not completelty reflective of their performance in a biological context where additional factors such as moisture, UV etc. Come into play.

For instance, it was found that phages associate with the host under study: *X. citri* in different mechanisms. This interaction influenced the manner in which they could control the disease. It was found that of three phages that inhibited *X. citri*, in vitro: two of them did not display activity against the disease in green house trials (Balogh, 2006).

In another study, eight different bacteriophages were tested on *Xanthomonas campestris pv pruni* (causative agent of spot of plum and peach). They isolated the phage that displayed the largest target range (Saccardi et al, 1993).

In another study on *Erwinia amylovora*, that causes fire blight in apples and pears, the phages were selected that lysed the pathogen along with another organism: *Pantoea agglomerans* that inhabits the phyllosphere and is an antagonist of the pathogen (Svircev et al, 2005).

HOST BACTERIAL STRAINS

Another important consideration is the diversity of the bacterial hosts that dictates the combination of an appropriate phage. *Xanthomonas campestris pv vesicatoria*, that causes spot disease in tomato and pepper was thought to be a single species, only to later be shown of possessing 4 species.

Another study examined the different patterns of lysis across 26 bacteriophages on several strains of xanthomonads across the Caribbean and Central America.

TESTING AND TRIALS OF THE PHAGES FOR USE IN CONTROL OF PLANT DISEASES

Subsequent to the identification of the host range of the phages, the testing of their activity is performed. The scaling up of the phage production and application requires understanding of the phage. A list of several trials done using phages to control plant pathogens has been tabulated in 4.12.

Criteria to be checked: to evaluate a testing program, following pointers are to be kept in mind.

The use of techniques to check the association of the host organism and the phage is essential before testing the phages on the field. This can encompass methods such as enumeration or PCR based assays.

As the testing on a field can involve the association of other micro-organisms, the influence of such other organisms on the activity of the phage is of importance. Hence, this calls for a thorough understanding of the cause of the disease or the appearance of mutants.

The use of controls is essential to ensure the absence of contamination of the phage preparation with other microbes or pathogens. The nature and composition of water used in the fields along with the presence of agrochemicals also have to be taken into account. The outcome of a particular test depends on the special highlights of the plant-pathogen dynamics along with external factors that are more different in a field compared to a controlled atmosphere of greenhouses or labs.

FEW FACTORS AFFECTING THE USE OF PHAGES AS BIOCONTROL AGENTS

The application of phages requires that the viruses be present in high quantities near the target pathogen. Several factors that influence the success of using phages as agents of control include (Jones et al, 2012):

- The phage must multiply in the environment of application.
- The phage must have high titre in the phyllosphere (part of the plant above the ground) or rhizosphere (the roots).
- The factors and dynamics of the degradation of phage.
- The presence of moisture.

APPLICATION SRATEGIES

Various factors affect the application of bacteriophages for the treatment of plant diseases. Several factors affecting the success of the phage therapy have been dealt with in 4.10. To overcome such roadblocks, several strategies have been reported that are summarized below:

Time

As UV radiation is one of the main factors that affect phage action, the time interval at which phages are applied on the plant should be taken into account. It is desirable to apply the phages when they are not under the receiving end of sunlight: dawn or dusk. It was reported that when bacteriophages were applied on tomato plants immediately after sunset achieved significant reduction of spot disease as compared to morning time. The application of phages in the evening showed high viability on the leaf surfaces (Balogh et al, 2003).

It is suggested that phages should be applied at high amounts in the evening at regular intervals in order to enhance the persistence and action of the phages. An essential factor to note is the presence of free moisture for the bacteriophage to exert its effect. Hence, phages can be applied when the leaves are wet for enhanced action (Jones et al, 2012).

Protection from UV

Few strategies to overcome the detrimental effect of UV on phages when they are present on the plant surfaces have been discussed in 4.13.

Several combinations have been tested for delivery of phages to minimize the effects of UV. A research team reported that sucrose, Casecrete NH-400, pregelatinised corn flour or skim milk could increase the life span of the phages in greenhouse and field testing against *Xanthomonas campestris pv. Vesicatoria* (Balogh et al, 2003).

However, the presence of skim milk had an adverse effect on phages that target *Xanthomonas axonopodis* in another study despite an extension of their presence on leaves (Balogh et al, 2008).

Phages Along with Carriers

Several studies have suggested the use of several bacteria that are non-pathogenic as carriers for the phages. These bacteria aid in the replication of the phages. The bacterium used as carrier should not infect the plants and should facilitate infection by the phage.

A good example to illustrate this technique would be the use of phage with a carrier, *P. agglomerans* was effective against *Erwinia amylovora,* that causes fire blight disease in apples and pears. The carrier saprophyte facilitated the amplification of phage numbers that were directed against the pathogen.

The use of lytic phages and *P. agglomerans* caused a decrease in fire blight disease in apple trees in pots to a same extent as streptomycin

(Boulé et al, 2011). Care must be exercised in the selection of a non-pathogenic carrier to prevent the carrier from becoming infective if they were attenuated. Hence, the use of non-pathogenic organisms is preferred to amplify the phages.

Phages in the Environment of Interaction

The phyllosphere refers to the portion of a plant above the soil. It is critical that the phages should survive in this area for effecting action against pathogens if they are applied to this portion. Several factors that affect them include: elevated temperatures, pH extremes, washing away by rain and UV-A and UV-B radiations that damaged the viruses. Additionally, the phyllosphere may be subjected to dessication or chemical treatments (for example: pesticides or copper) (Jones et al, 2012).

Another potential area for applying phages is the soil or rhizosphere. A study in 2012 showed that phages showed good persistence in the soil and roots of tomato plants where they were applied. Though the phages were applied in the soil, they were measured in the leaves at good titres after 2 days following application (Iriarte et al, 2012).

In order to enhance the activity of the phages in the environment, several approaches have been reported:

The use of several formulations along with the phages to enhance their persistence in the environment. (few have been discussed above)

Bacteriophages multiply their population in the presence of the host bacteria that chemicals cannot achieve. This property can be exploited when phages can be applied to an area where the host bacteria are present. For instance, a bacteriophage against *P. solanacearum* that causes tobacco wilt was administered to plants along with an avirulent strain of *P. solanacearum* called MS4. It was found that the mixing of the non-pathogenic version and phage caused a reduction in disease (Tanaka et al, 1990).

In another study, phages were administered to pear plant to target fire blight disease caused by *Erwinia amylovora.* The administration of phages were such that they could target both the pathogen and an antagonistic bacterium to the pathogen called *P. agglomerans* present in the phyllosphere of the plant. The combination caused a reduction in the disease manifestations (Svircev et al, 2005).

PHAGES WITH OTHER MEASURES

The combination of phage based therapies can be combined with other therapies in practise or yet to be established in order to achieve eradication of the plant pathogen. A few examples are discussed below:

A combination of copper along with phages dictates the use of double stranded DNA phages as against lipid phages that were affected by the copper. This opens up the avenues to use copper along with phage therapy for plant diseases (Li Dennehy, 2011).

Another combination is that of phages and antibiotics. Such combinations could help in the easing out of resistance that is induced in the target organism either to antibiotic or phage alone (Borah et al, 2000). However, an impediment in this strategy is the restrictions in certain parts of the world like EU where streptomycin has been discontinued for combating plant diseases.

The combination of phage with copper-mancozeb was found to yield significant reductions of leaf blight of onion caused by *Xanthomonas* (Lang et al, 2007).

Another set of studies showed the effectiveness of phages along with other strategies on bacterial sopt disease of tomato. With a view of integration of treatments, several combinations such as "phages, biocontrol agents, including strains of PGPR, bacterial antagonists, SAR inducers (harpin, ASM) and copper hydroxide" in several trials.

It was reported that the plants that received phage treatment two times a week, showed a result comparable to copper-mancozeb treatments and also resulted in the production of fruits that could be sold in the market. A more efficient control of the disease was achieved on the leaf surfaces when phages were combined with inducers of stress associated response like ASM (Obradovic et al, 2004 and 2005).

PHAGE GENE BASED OPTIONS

With studies in genomics, phage genes have been explored for development of treatment options. Endolysins are enzymes (peptidoglycan hydrolases) that degrade cell wall of bacteria for the lysis and release of the viruses, and are coded by the phage genome. A research team explored the application of endolysins against *Clavibacter michiganensis subsp. Michiganensis* that causes disease in tomato with very little resistant plants and chemical treatments. The endolysins of two phages specific for the pathogen were investigated and was reported that there was potential of using the enzymes of phages called CMP1 and CN77 to eradicate the pathogen (Wittmann et al, 2010).

APPLICATION OF PHAGES IN THE SOIL

Certain pathogens enter a plant through openings and form cankers in the plant where they are shielded from chemicals that are applied for the eradication of the pathogen. For instance, *E. amylovora* and *P. syringae pv. Actinidiae* follow such a life cycle.

Research has indicated that phages can live in soil for a month also dependent on the type and characters of the soil. The use of soil phages for bacteria that infect tubers is desirable. A study reported that phages that targeted *Dickeya solani* could prevent soft rot disease in potato tubers (Adriaenssens et al, 2012).

OVERCOMING RESISTANCE TO PHAGES

There are several mechanisms by which host bacteria exhibit or develop resistance against phages. The mechanisms along with the solution that can be applied are described below:

Phage Adsorption Inhibition

Phages attach on to the host cells using certain receptors on the surface of the cells such as LPS, pili and flagella. Mutations in the receptors prevent infection by the virus. For instance, φAT1 phage could not infect *P. atrosepticum* that developed mutations in the flagella. Mutations in Type IV pili of *P. syringae* prevented attachment of φ6 and thus reduced the survival of the phages in the phyllosphere.

This entails the use of a cocktail of mixture of phages to quell the resistant bacteria that arise out of mutations. Such an approach was utilized in the treatment of *E. coli* infection in animals using phages.

Apart from the use of a phage cocktail, phages can be applied on the mutant bacterial hosts in order to isolate phages that mutate their receptors to enable infection. A 2010 study used mutants of *Erwinia* in order to isolate a mutant phage (Evans et al, 2010).

Internal Cell Mechanisms

There are several molecules and factors in the cytoplasm that upon mutations can prevent infection by phage and hence, development of phage resistance.

For instance, a phage φ2954 that targets *P. syringae* utilizes glutaredoxin 3 of the host cell. When the host mutated the DNA, resistance was developed but there were also mutant phages which did not need the protein for their functioning (Qiao et al, 2010).

Such studies point out that phages can be developed and isolated in responses to mutations in the cytoplasmic factors of the host.

Another mechanism that occurs in a cell in response to phage infection is abortive infection (Abi). This mechanism causes the cells infected by the virus to commit suicide to contain infection.

For instance, *P. atrosepticum* possesses a system of ToxIN that is a type III protein-RNA toxin-antitoxin (TA) system. This forms a toxin - antitoxin system against which a phage was isolated that displayed resistance to the system. The phage formed a pseudo RNA that prevented the action of the toxin-antitoxin system to overcome the resistance displayed by the bacterium using this mechanism of molecular mimicry (Blower et al, 2012).

CRISPR/Cas Resistance

Clustered regularly interspaced short palindromic repeat (CRISPR) arrays and CRISPR associated (Cas) proteins form a system in bacterial cells that enable a bacterial cell to develop resistance against phages. These arrays use viral nucleic acids to produce small RNA molecules that destroy the genetic material of the virus along with help of Cas proteins.

This mechanism has been detected in several plant pathogens such as *E. amylovora, X. oryzae, P.atrosepticum* etc.

Despite the presence of this resistance system in bacterial pathogens, there have been reports of a phage Xop411 that developed mutations in order to counter the CRISPR/Cas system of *Xa. oryzae.*

In addition, it can be seen that bacteriophages also evolve themselves to overcome these mechanisms and target the cells.

Strategy for Resistance to Phages

An interesting mechanism patent has been described to circumvent the issue of cells that develop resistance against phages (Jackson LE, 1989).

The method involves the preparation of a mix of wild type phages along with host range mutant phages (h-mutants). This aids in the targeting of bacterial cells that are resistant to the wild type phage.

This technique was tested in field trials where a mixture of four phages: wild-type and h-mutant was applied on tomato bacterial spot two times a week. This treatment caused reduction in disease and also aided in the production of fruits that were larger as compared to those obtained with the prescribed copper-mancozeb.

Testing for resistance against phage can be done on a routine basis. Balogh et al, 2003 have described a method for the same where bacterial populations in a field can be isolated and tested for plaque formation upon infection with virus. The absence of plaques is indicative of developed resistance against the virus that calls for an appropriate measure such as the use of viruses or a cocktail of phage that could target the resistant population (Balogh et al, 2003).

CASE STUDY: BACTERIOPHAGES FOR PIERCE'S DISEASE

The organism *Xylella fastidiosa subsp. fastidiosa (Xf)* causes Pierce's Disease (PD) in grapes that threatens the wine industry. Current treatment involves the excision of affected vines or the application of neonicotinoids that target the nervous system of the insect vectors. However, the use of such chemicals has raised concerns as to their side effects.

With the goal of developing phage based options that are efficient and do not have any adverse effects on the environment. Tagged as the first phage based control of PD, a research team used four lytic Xf phages and reported reduction in numbers of pathogens and also decreased the symptoms.

To prevent the incidence of phage resistant mutants, the team used a cocktail of four phages that differed in their host targets. They also reported that the phages were not harmful to other microbes, animals, plants or humans.

Such research opens up the possibilities of the application of phage based therapies for the control of plant diseases (Das et al, 2015).

SUMMARY

The growth of plant diseases across the world call for appropriate treatment measures as damage to agriculture would adversely affect the food supply. The use of phages in control of plant pathogens has shown encouraging results though there are few issues that need to be addressed.

The main issue is the stability of phages in the phyllosphere that is being researched upon. The correct cocktail of phages along with appropriate formulations to prolong the life of the phages. The use of various strategies such as carrier organisms along with the time of application can help in optimization of the use of phages as biocontrol agents. There are already few phage based biocontrol systems that were discussed in Chapter 4. With more research propelled in this field, it is possible to use these natural agents in the control of plant diseases.

REFERENCES

1. Addy H. S., Askora A., Kawasaki T., Fujie M., Yamada T. (2012). Loss of virulence of the phytopathogen Ralstonia solanacearum through infection by phiRSM filamentous phages. *Phytopathology.* 102(5):469–477.
2. Adriaenssens E. M., Van Vaerenbergh J., Vandenheuvel D., et al. (2-12). T4-related bacteriophage LIMEstone isolates for the control of soft rot on potato caused by "Dickeya solani". *PLoS ONE.* 7(3) e33227.
3. Balogh B., Canteros B. I., Stall R. E., Jones J. B. (2008). Control of citrus canker and citrus bacterial spot with bacteriophages. *Plant Dis.* 92(7):1048–1052.
4. Balogh B., Jones J. B., Momol M. T., Olson S. M., Obradovic A., King P., et al. (2003). Improved efficacy of newly formulated bacteriophages for management of bacterial spot on tomato. *Plant Dis.* 87:949–954. doi: 10.1094/PDIS.2003.87.8.949.
5. Balogh B. *Characterization and use of bacteriophages associated*

with citrus bacterial pathogens for disease control. PhD Dissertation, Gainesville, FL: University of Florida. 2006.

6. Blower T. R., Evans T. J., Przybilski R., Fineran P. C., Salmond G. P. C. (2012). Viral evasion of a bacterial suicide system by RNA-based molecular mimicry enables infectious altruism. *PLoS Genet.*8:e1003023 10.1371/journal.pgen.1003023.
7. Borah P. K., Jindal J. K., Verma J. P. (2000). Integrated management of bacterial leaf spot of mungbean with bacteriophages of Xav and chemicals. *J Mycol Plant Pathol*. 30(1):19–21.
8. Boulé J., Sholberg P. L., Lehman S. M., O'gorman D. T., Svircev A. M. (2011). Isolation and characterization of eight bacteriophages infecting Erwinia amylovora and their potential as biological control agents in British Columbia, Canada. *Can J Plant Pathol*. 33(3):308–317.
9. Das M., Bhowmick T. S., Ahern S. J., Young R., Gonzalez C. F. (2015). Control of Pierce's Disease by Phage. *PLoS ONE*. 10(6): e0128902. https://doi.org/10.1371/journal.pone.0128902
10. Evans T. J., Trauner A., Komitopoulou E., Salmond G. P. C. (2010). Exploitation of a new flagellatropic phage of Erwinia for positive selection of bacterial mutants attenuated in plant virulence: towards phage therapy. *J Appl Microbiol*. 108(2):676–685.
11. Frampton R. A., Pitman A. R., Fineran P. C. (2012). Advances in Bacteriophage-Mediated Control of Plant Pathogens. *Int J Microbiol*. 2012:326452. doi:10.1155/2012/326452.
12. Gilligan C. A. (2008). Sustainable agriculture and plant diseases: an epidemiological perspective. *Phil Trans R Soc B*. 363(1492):741–759.
13. Iriarte F. B., Obradovic A., Wernsing M. H., Jackson L. E., Balogh B., Hong J. A, et al. (2012). Soil-based systemic delivery and phyllosphere in vivo propagation of bacteriophages: Two possible strategies for improving bacteriophage persistence for plant disease control. *Bacteriophage*. 2(4):215-224.
14. Jackson L. E. Bacteriophage prevention and control of harmful plant bacteria. US Patent No.4828999. 1989.
15. Ji P., Campbell H. L., Kloepper J. W., Jones J. B., Suslow T. V.,

Wilson M. (2006). Integrated biological control of bacterial speck and spot of tomato under field conditions using foliar biological control agents and plant growth-promoting rhizobacteria. *Biol Control.* 36:358–367. doi: 10.1016/j.biocontrol.2005.09.003.

16. Jones J. B., Vallad G. E., Iriarte F. B., et al. (2012). Considerations for using bacteriophages for plant disease control. *Bacteriophage.* 2(4):208-214. doi:10.4161/bact.23857.
17. Kerr A. (1974). Soil microbiological studies on *Agrobacterium radiobacter* and biological control of crown gall. *Soil Sci.* 118:168–172. doi: 10.1097/00010694-197409000-00006.
18. Lang J. M., Gent D. H., Schwartz H. F. (2007). Management of Xanthomonas leaf blight of onion with bacteriophages and a plant activator. *Plant Dis.* 91:871–878. doi: 10.1094/PDIS-91-7-0871.
19. Li J., Dennehy J. J. (2011). Differential bacteriophage mortality on exposure to copper. *Appl Environ Microbiol.* 77(19):6878–6883.
20. Liu T. Biological control with tomato bacterial spot with hrp-mutants of *Xanthomonas campestris* pv. *vesicatoria* MS Thesis, Gainesville, FL: University of Florida.1998.
21. Obradovic A., Jones J. B., Momol M. T., Balogh B., Olson S. M. (2004). Management of tomato bacterial spot in the field by foliar applications of bacteriophages and SAR inducers. *Plant Dis.* 88:736–740. doi: 10.1094/PDIS.2004.88.7.736.
22. Obradovic A., Jones J. B., Momol M. T., Olson S. M., Jackson L. E., Balogh B., et al. (2005). Integration of biological control agents and systemic acquired resistance inducers against bacterial spot on tomato. *Plant Dis.* 89:712–716. doi: 10.1094/PD-89-0712.
23. Qiao J., Qiao X., Sun Y., Mindich L. (2010). Role of host protein glutaredoxin 3 in the control of transcription during bacteriophage ϕ2954 infection. *Proc Natl Acad Sci U. S. A.* 107(13):6000–6004.
24. Saccardi A., Gambin E., Zaccardelli M., Barone G., Mazzucchi U. (1993). Xanthomonas campestris pv. pruni control trials with phage treatments on peaches in the orchard. *Phytopathol Mediterr.* 32:206–210.
25. Svircev A. M., Lehman S. M., Kim W. S., Barszcz E., Schneider K. E., Castle A. J. *Control of the fire blight pathogen with bacteriophages.* In: Zeller W, Ullrich C, eds. Proc 1st Int Symp Biol Control Bact

Plant Dis, October 23–26, 2005. Berlin, Germany: Die Deutsche Bibliothek – CIP-Einheitsaufnahme, 259–261.

26. Tanaka H., Negishi H., Maeda H. (1990). Control of tobacco bacterial wilt by an avirulent strain of Pseudomonas solanacearum M4S and its bacteriophage. *Ann Phytopathological Soc Jpn*. 56:243–246. doi: 10.3186/jjphytopath.56.243.
27. Thomas R. C. (1935). A bacteriophage in relation to Stewart's disease of corn. *Phytopothology*. 25:371–372.
28. Wittmann J., Eichenlaub R., Dreiseikelmann B. (2010). The endolysins of bacteriophages CMP1 and CN77 are specific for the lysis of Clavibacter michiganensis strains. *Microbiology*. 56(8):2366–2373.

CHAPTER

9

ADVANCES IN BACTERIOPHAGE-MEDIATED CONTROL OF PLANT PATHOGENS

Rebekah A. Frampton,[1] Andrew R. Pitman,[2] and Peter C. Fineran[1]

[1]Department of Microbiology & Immunology, University of Otago, Dunedin 9054, New Zealand

[2]New Zealand Institute for Plant & Food Research, Christchurch 8140, New Zealand

ABSTRACT

There is continuing pressure to maximise food production given a growing global human population. Bacterial pathogens that infect important agricultural plants (phytopathogens) can reduce plant growth and the subsequent crop yield. Currently, phytopathogens are controlled through management programmes, which can include the application of antibiotics and copper sprays. However, the emergence of resistant bacteria and the desire to reduce usage of toxic products that accumulate in the environment mean there is a need to develop alternative control agents. An attractive option is the use of specific bacteriophages (phages), viruses that specifically kill bacteria, providing a more targeted approach. Typically, phages that target the phytopathogen are isolated and characterised to determine that they have features required for biocontrol. In addition, suitable formulation and delivery to affected plants are necessary to ensure the phages survive in the environment and do not have a deleterious effect on the plant or target beneficial bacteria. Phages have been isolated for different phytopathogens and have been used successfully in a number of trials and commercially. In this paper, we address recent progress in phage-mediated control of plant pathogens and overcoming the challenges, including those posed by CRISPR/Cas and abortive infection resistance systems.

INTRODUCTION

In October 2011 the United Nations announced that the global human population had reached 7 billion. The world is facing not only this increase in population, but also a decrease in land availability for agriculture and a changing climate [1]. It is apparent that there is a requirement to increase food production to feed the growing population and a need to achieve this with diminished land and water resources [1]. A major threat to food production is plant diseases, which are influenced by changing agricultural practices and more global trade [2]. Recent topical examples include citrus greening of oranges caused by psyllids that transmit bacteria belonging to the genus Candidatus Liberibacter [3] and canker of kiwifruit caused by the bacterium Pseudomonas syringae pv. actinidiae [4]. Citrus greening has doubled the cost of orange production for growers in Florida, where the disease was first identified in 2005 [3]. In New Zealand, where Pseudomonas syringae pv. actinidiaewas discovered in late 2010, 40% of orchards have been infected resulting in a significant economic cost to the industry [5].

A variety of approaches are required to minimise the impact of bacterial plant diseases on the quantity, quality, and economy of food production. Conventional control measures involve the implementation of operating practices to prevent further infections, removal of infected plant tissue, and appropriate disposal to stop the transmission of the pathogen from one site to another. Other methods to control phytopathogens include chemicals such as pesticides, to control insect vectors, antibiotics (e.g., tetracycline and streptomycin), and copper. Copper has been used for over 100 years and antibiotics such as streptomycin have been used since the 1950s [6, 7]. Streptomycin has been used for many years for the control of pathogens, including Pseudomonas syringae pathovars, and resistance has been regularly reported following use [6–8]. Another concern with antibiotics is the spread of resistance genes to other bacteria, including other pathogens or nonpathogenic bacteria present in the environment [7]. Copper resistance has also been documented for plant pathogens, including Xanthomonas and Pseudomonas species [9–11]. Continual use of copper sprays can lead to toxic levels in the environment [12, 13]. Therefore, it is favourable to replace or integrate chemical control methods with less toxic biological methods.

There is mounting interest in using bacteriophages (phages) as biocontrol agents (BCAs) to target phytopathogens. Phages are viruses that specifically infect bacteria, yet have no direct negative effects on animals or plants. Infection of a bacterium by a virulent phage typically results in rapid viral replication, followed by the lysis of the bacterium and the release of numerous progeny phages. These phages can then proceed to infect neighbouring bacteria. Therefore, the numbers of phage will expand when target pathogens are encountered and the therapy will essentially be amplified in response to the bacterial infection. This is a distinct advantage over other treatments, such as antibiotics. Since the 1920s, within a decade of the first discovery of phages, their potential as therapeutic agents for use in agriculture was under investigation and provided some promising results (see [14] for review of the early literature). Recent years have seen a resurgence of interest in phage therapy for the control of phytopathogens (see [15–17] for recent reviews). In part, this renewed interest is due to the nontoxic nature of phages and their ability to infect antibiotic or heavy metal resistant bacteria. Successful phage therapy is being applied commercially to processed and packaged foods by Intralytix and Micreos Food Safety (formerly EBI Food Safety) and to agricultural crops by Omnilytics. There is also interest in the use of phage in the detection of phytopathogens. Indeed, many of the first phages against plant pathogens were isolated for diagnosis and strain typing [18] and recently genetic advances are yielding effective phage-based reporter systems [19, 20]. In this paper we will examine the use of phage as BCAs for bacterial plant diseases. First, we will address the initial isolation and laboratory characterisation of phages. Next, we will discuss the transition from in vitro analyses to bioassays and field/greenhouse trials to commercialisation and application. Finally, we will provide an analysis of phage resistance in phytopathogens and address how this can be avoided or minimised when developing phage BCAs.

INITIAL PHAGE CHARACTERISATION

Isolation and Host Range of Phages

The initial stage in developing a phage-based BCA involves the isolation of phages. Phages can be isolated relatively simply from soil, water, and plant material collected from multiple locations using the soft agar overlay technique and a range of host pathogens. Isolation methods are well established and covered in several reviews [21, 22], and many successful studies have resulted from this approach (e.g., [23–25]). To isolate diverse phages it is important to include a range of host bacterial strains that represent the diversity of the pathogens involved in the disease [17, 26]. Phages that produce clear plaques should be chosen preferentially to reduce the isolation of temperate phages as certain temperate phage can cause lysogenic conversion, a process whereby virulence genes carried by the prophage contribute to pathogenicity of the bacterial lysogen [27]. This approach increases the isolation of phages from the order Caudovirales (Myoviridae, Siphoviridae, and Podoviridae). Filamentous phage from the Inoviridaefamily produce a chronic infection that results in the continuous release of phage from growing bacterial cells. Filamentous phage could also be a BCA option and this is discussed in more detail in Section 3.5 [28].

Determining the host range of each phage enables the design of a cocktail capable of infecting all known pathogenic strains involved in the disease. Although the host range of most phages is usually narrow, bacteria isolated from the plant environment should be tested for lysis by the putative BCA phages to ensure a minimal impact of the phages on the wider microbial community and potential commensal strains.

Basic Characterisation of Phages

Once isolated, the phages need to be characterised to ensure they are appropriate BCAs. This information will allow for rational design of a phage cocktail and enable the tracking of phages during bioassays and field trials. Transmission electron microscopy and molecular methods, such as restriction pattern analysis of phage DNA, enable assessment of phage diversity. Identification of the phage receptors can assist in the rational selection of phages that target through different mechanisms to reduce the frequency of resistance (discussed in Section 5).

The development of next-generation sequencing has dramatically reduced the cost of determining the complete DNA sequence of a phage genome. The sequence also enables the design of quantitative PCR strategies for phage tracking in field trials [26] and if present, genes required for integration/lysogeny or that encode known toxins, antibiotic resistance, or virulence factors can be identified. Phage-mediated horizontal transfer of bacterial genes by specialised and generalised transduction should be avoided [27]. The possibility of specialised transduction is eliminated by removing temperate phages but assays are necessary to examine generalised transduction. These assays cannot rule out transduction but enable the identification and elimination of phages with a high transducing frequency [27]. It has been proposed that if a phage cocktail is used then any bacteria that receive additional DNA through transduction can still be killed by another phage in the mixture [17]. An understanding of the basic growth parameters of the phage will aid in the design of a phage BCA. Therefore, investigating the length of infection and burst size with one-step growth curves at temperatures and conditions likely to be encountered in the field iss important. It is also important to identify conditions under which the phage can be successfully stored [15]. Following phage characterisation the next step is to perform bioassays and/or field trials which are discussed in Section 3.

Alternative Biocontrol Phage Technologies

Through phage genomics, genes encoding other potential biocontrol options have been identified. One example is phage endolysins, which have been investigated for use against antibiotic resistant bacteria that colonize human mucosal membranes [29]. Endolysins are phage-encoded peptidoglycan hydrolases that work in concert with holins to ensure cell lysis occurs following phage maturation [30]. Generally, the C-terminus binds the bacterial cell wall and positions the enzymatically active N-terminus close to its target. Due to their high level of specificity, phages CMP1 and CN77 were originally used for detection and identification of plant pathogenic Clavibacter michiganensis strains [31, 32]. Clavibacter michiganensis subsp. michiganensis is a recalcitrant tomato pathogen because there are no resistant cultivars and chemical control agents are ineffective. The endolysins of CMP1 and CN77 responsible for degrading the bacterial cell wall have been pursued as potential antimicrobials [33]. Sequence analysis revealed that the similarity of the catalytic domains of the CMP1 and CN77 endolysins was low, suggesting they target different covalent bonds within the peptidoglycan. However, both endolysins are specific for Clavibacter michiganensis subsp. michiganensis, as observed for the phages [34]. The use of endolysins is still highly specific but avoids concerns of using a replicating BCA. Compared with the application of whole phage preparations, the development of lysine-based therapies is more technically challenging. For example, the generation of lysins requires greater molecular insight than phage therapy because sequencing, identification, cloning, characterization, and purification of the lysins must be performed.

TRIALS OF PHAGE BIOCONTROL

Once a selection of phytopathogen-specific phages is obtained and the initial characterisation has indicated their potential for biocontrol, the next step is to test their efficacy in relation to plant disease. First it is necessary to scale-up phage preparations, which requires knowledge about the characteristics and lifestyle of the phages [35]. To test phytopathogen control, a range of approaches have been taken, from laboratory-based bioassays through greenhouse and field trials. Table 1 summarises some results of phage trials that have been performed on a range of phytopathogens including A. tumefaciens, Dickeya solani, Pectobacterium carotovorum, Erwinia amylovora, Pseudomonas syringae, Ralstonia solanacearum, Streptomyces scabies, and Xanthomonasspecies. Many of these studies have given promising results. However, a number of factors can contribute to phage biocontrol trials that have failed. Firstly, to understand the reason for success or failure it is important to measure the dynamics of phage and host, for example, via quantitative PCR or traditional enumeration assays [26]. Secondly, field trials are biologically complex and the presence of other microbes, including other pathogens, can influence the effectiveness of the phages. In some cases pathogens of a different genus or species will cause very similar disease symptoms but are not killed by the phages [23]. Therefore, it is important to check which organism(s) caused the disease symptoms and if it is the targeted bacterial host, whether phage-resistant mutants account for the lack of phage killing or not. It is important to confirm that phage preparations are free of any pathogenic bacteria used in the production process, which highlights the requirement for phage-only controls in trials. The absence of a "gold standard" positive treatment, where available, in some studies makes evaluating the efficacy less informative. Other factors including the type of water and the presence of certain components in some fertilizers can influence phage viability and the trial's success [36]. Despite these challenges, many of the studies in Table 1 demonstrated a positive effect of phage treatment.

Table 1: Phage therapy trials of plant pathogens.

Pathogen	Host	Disease		References
Agrobacterium tumefaciens	Tomato	Crown gall	Bioassay with infected tissue—bacteriophage had no effect	[37]
Dickeya solani	Potato	Soft rot	Small effect seen when seed tubers were treated with phage prior to planting	[23]
Erwinia amylovora	Pome fruits	Fire blight	Phages isolated and characterised *in vitro*. Some testing *in planta*, promising results in combination with non-pathogenic carrier *Pantoea agglomerans*	[24, 38, 39]
Pectobacterium carotovorum subsp. *carotovorum*	Calla lily	Bacterial soft rot	Bacterial load reduction by phages but inhibition of killing by fertiliser solutions	[36]
Ralstonia solanacearum	Tobacco	Bacterial wilt	In a greenhouse trial pretreatment of plant roots with an avirulent strain and application of phage to the plants protected plants against bacterial wilt. No comparison was made to conventional chemical control methods	[40]
Ralstonia solanacearum	Tomato	Bacterial wilt	In a greenhouse trial pretreatment of tomato seedlings with φRSL1 prevented bacterial wilt in all plants, untreated plants all wilted. φRSL1 inhibited growth of bacteria but did not completely kill bacteria.	[41]
Streptomyces scabies	Potato	Potato scab	Phage treatment of seed tubers prior to planting reduced scab lesion coverage	[42]
Xanthomonas arboricola pv. *pruni*	Stonefruits	Bacterial spot	Application of phage to peach leaves prior to infection resulted in a 42% disease reduction compared to a nontreated control. Application of phage after infection had no effect	[43]
Xanthomonas axonopodis pv. *allii*	Onion	Xanthomonas leaf blight	Field and greenhouse trials of phage and plant activator provided equivalent protection to copper	[44]
Xanthomonas axonopodis pv. *vignaeradiatae*	Mungbean	Bacterial leaf spot	Synergistic effect of phage and streptomycin on mungbean seeds reduced seedling infection	[45]
Xanthomonas campestris pv. *juglandis*	Walnut	Walnut blight	Phage did not survive on walnut leaves in a greenhouse trial: pathogen was not included on leaves	[46]
Xanthomonas campestris pv. *pruni*	Peach	Leaf and fruit spot	A significant reduction in disease was seen in one out of three orchards with weekly application of a single phage	[47]
Xanthomonas campestris pv. *vesicatoria*	Tomato and Pepper	Bacterial spot	These studies have led to the successful development of a phage BCA (http://www.omnilytics.com/).	[48–51]
Xanthomonas citri subsp. *citri*	Citrus	Citrus canker	Mixed results in greenhouse and nursery trials when compared to copper bacteriocides	[25]
Xanthomonas fuscans subsp. *citrumelonis*	Citrus	Citrus bacterial spot	Mixed results in greenhouse and nursery trials when compared to copper bacteriocides	[25]

Each phage-phytopathogen-plant system has unique features and requires characterisation and optimisation of the phage biocontrol. External and environmental factors can play a role and cause variable results. For example, greenhouse-grown crops are in a more stable environment, whereas plants grown outside are exposed to more variable weather conditions and these factors vary between geographic locations. Phage survival and their persistence at the required site of action are affected by conditions such as pH, temperature, desiccation, rain, and UV. The most damaging factor appears to be UV irradiation in sunlight [52]. Interestingly, some plant extracts negatively affect the growth or viability of phage in vitro [53, 54] but it is unclear whether this is relevant during phage therapy on plants. Various approaches to address these challenges in phage biocontrol are available and will be discussed in the following sections.

Protective Formulations and Application Timing

In order to minimise UV damage, some researchers have investigated protective formulations. Few published studies have addressed stabilising agents, but one group found that certain combinations of sucrose, Casecrete NH-400, pregelatinised corn flour or skim milk, increased the persistence of phages active against Xanthomonas campestris pv. vesicatoria in greenhouse and field trials and improved treatment efficacy [51]. However, in a trial with phages against Xanthomonas axonopodis pv. citri and citrumelo, skim milk inhibited the action of the phages even though the phages persisted on the leaf surface for longer [25]. Balogh et al. also investigated the effect of the time of day of phage application and demonstrated that evening applications increased phage survival [51]. The lower UV intensity at dusk was attributed to this improved viral longevity. The frequency of phage application also appears to be specific to the particular phage-phytopathogen-plant system and the best results vary from daily to weekly application [25, 44, 48, 50, 51, 55]. Therefore, these conditions must be optimised for each newly developed phage BCA.

Coapplication with Other Control Strategies

Researchers have begun examining the effects of combining phage with existing or new control measures. Hypersensitive response and systemic acquired resistance plant activators have been tested against Xanthomonas spp. in two disease models in the presence or absence of phages. The combined approach provided disease control equal to, or greater than, either treatment alone [44, 48, 49] (Table 1) and comparable results were obtained with phage and copper hydroxide with mancozeb [48]. Recently, it was demonstrated that lipid-containing phages were the most susceptible to copper, whereas most dsDNA phages were unaffected [56]. This indicates that dsDNA phages are the best candidates for plant disease therapies in combination with copper. Phages can also be used in combination with streptomycin as they are not a direct target of antibiotics [45]. Cotherapy with copper or streptomycin could enable a "belt and braces" approach to minimise streptomycin/ copper or phage resistance. However, this approach is not possible in some parts of the world. For example, regulations that disallow the use of streptomycin for the control of plant diseases have been introduced in the EU [57]. Phage compatibility with other agrichemicals has not been thoroughly examined. Phages used would require testing for viability and persistence in the presence of any agrichemical treatment used simultaneously on the affected plants.

A range of bacterial and fungal BCAs that include pathogen antagonists have been developed and are commercially available. These include BlightBan A506 (Pseudomonas fluorescens A506; Nufarm Americas Inc), Blossom Bless (Pantoea agglomerans P10c; Gro-Chem NZ Ltd), and Superzyme (Bacillus subtilis, Trichoderma, and Pseudomonas putida; J H Biotech Inc). When used on their own, the protection provided by some of these products often varies and does not compare favourably with streptomycin, which is considered the benchmark [58]. These BCAs can complement conventional control strategies. For example, use of BCAs in the control of fire blight [59, 60] in apple trees reduced the number of streptomycin sprays required to provide the same amount of protection against E. amylovora infection in apple blossoms [58]. Growth of these BCAs is thought to be required for their action as competitors. Variable environmental conditions influence growth and the regulation of the production of secondary metabolites, such as antibiotics, which might account for some of the inconsistent responses to BCAs. One approach to improve plant protection is to incorporate phages into existing BCA products as part of the pest management plan.

Use of Carrier Bacteria

The abundance of host bacteria, in addition to the environmental factors mentioned above, causes fluctuations in phage numbers [61]. To improve phage persistence and efficacy the idea of coapplication of phages with nonpathogenic host bacteria to support phage replication has been raised [26, 62]. It is necessary to identify a suitable nonpathogenic carrier bacterium that does not affect the plant and to isolate phages that infect both the target phytopathogen and the carrier strain. Therefore, isolation of broad host range phages can also be beneficial for the purposes of biological control. Care is needed to use phages that do not infect plant beneficial bacteria in the phyllosphere or rhizosphere.

The use of competitive antagonists has been investigated in the control of fire blight for many years and can prevent or reduce infection [24, 26, 63]. An extensive panel of Erwinia amylovora phages have been isolated and characterised [24, 38, 39, 62, 64] and their use with carrier bacteria investigated [17, 26, 62]. In an early study a temperate phage that infected both a saprophyte and Erwinia amylovora provided increased protection in a pear slice bioassay [62]. However, current opinion is that temperate phages should be avoided for phage therapy. Recently, a Pantoea agglomerans carrier system was developed which amplified the phages and reduced disease in blossom bioassays [17, 26]. The application of lytic phages against Erwinia amylovora in conjunction with the Pantoea agglomerans carrier strain to potted apple trees reduced the incidence of fire blight to the same level as streptomycin [24]. Application of the carrier strain itself also helped prevent infection. In orchard trials, the coapplication of phages with carrier bacteria resulted in a significant reduction in disease incidence compared with the no phage control or the P. agglomerans carrier alone, and the protective effect was similar to that observed with streptomycin [17, 26].

Due to difficulties in isolating broad host range phages that infect suitable non-pathogenic strains, other strategies should be considered. For example, non-pathogenic mutant derivatives of the phytopathogen can be used as hosts for phage replication in the environment [40]. There are some risks if the pathogen is not fully attenuated or if reversion is possible. An alternative use for non-pathogenic or attenuated pathogenic hosts is for production of phages. Non-pathogenic hosts might enable safer and more economical phage amplification due to reduced purification requirements for removal of pathogenic bacteria.

Different Plant Disease Systems

The disease symptoms and location of infection within or on the plant can pose challenges for phage biocontrol. For example, Pseudomonas syringae pv. actinidiae and Erwinia amylovora spend some of the disease cycle inside the host plant following infection through plant openings or wounds [4, 65]. High numbers of bacteria can accumulate within cankers in the plant and are protected from any control agent applied to the outside of the plant that cannot penetrate to deeper tissues. One study has shown a preventative and beneficial effect on disease progression with phage treatment of citrus canker [25]. However, the ability of phage to act as a curative agent of cankers was not directly assessed. High-pressure injection of phage is a possible strategy that is under consideration. In cases where treatment is problematic, phage prophylaxis would be the preferred application at a time when the chance of infection is greatest [17].

Phages survive in soil for at least one month [66] but the type of soil, pH, moisture content, and nutrients all influence persistence and can affect phage bioavailability [22, 67]. There is a desire to use phages in soil against bacterial pathogens that infect plant tubers, such as Pectobacterium spp. and Dickeya spp. In a recent study, treatment of potato tubers with phages prevented soft rot disease caused by Dickeya solani in a controlled environment [23]. Only a small protective effect was seen when the tubers were planted and only when the tubers were allowed to dry completely after treatment. A mixture of phages against Dickeya solani and Pectobacterium atrosepticum would be useful as Adriaenssens et al. observed soft rot caused by both species during their trials [23].

Use of Filamentous Phages

The filamentous phages φRSS1 and φRSM3 have almost opposite effects on the virulence of Ralstonia solanacearum. Infection of Ralstonia solanacearum with φRSS1 increased twitching motility, EPS production, the expression of the virulence regulator phcA, and the rate of tomato plant wilting compared with a noninfected control [68]. In contrast, infection with φRSM3 decreased twitching, EPS and phcA expression, growth, and movement in tomato plant stems and caused less wilting [69]. This difference was due to a repressor in the genome of φRSM3 that is absent in φRSS1 [69]. Tomato plants exposed to φRSM3-infected bacterial cells increased the expression of genes to help the plant resist infection [28]. Indeed, pretreatment of tomato plants with φRSM3-infected bacteria prevented infection by noninfected cells and prevented bacterial wilt. The authors suggest that a mixture of φRSM3 and a lytic phage, such as φRSL1 [41], would be a good BCA [28].

APPLICATION

Safety, efficacy, intellectual property, and a market are important factors for the commercial success of phage control of phytopathogens. As phages are ubiquitous and are consumed with food and water without any negative effects, their safety as BCAs is not an issue. Phages for control of Listeria monocytogenes have FDA approval for use on food [70] and have been granted GRAS (generally recognised as safe) status. Interestingly, in the EU [57] and the USA LISTEX (Micreos Food Safety) [71] is considered organic, suggesting that organic growers could be a valuable market for phage biocontrol products. Furthermore, phages are used to treat bacterial infections in humans in Russia and Georgia [72, 73]. Clinical trials have tested the effectiveness of phages to treat ear infections [74] and have assessed the safety of phages for treating burn victims [75]. These strictly controlled medical trials demonstrate that phage therapy is safe and effective, indicating that phage control of phytopathogens will not cause adverse health problems. Phages are being developed as BCAs for control of human pathogens in animals and on food products, which has been reviewed elsewhere [76, 77] and in other articles in this special issue. Patent protection and intellectual property are important factors in commercialisation of phage BCAs. Despite the concept of phage therapy having existed for over 90 years, multiple companies have acquired patents and established commercial platforms. This has been thoroughly reviewed recently by Gill et al. [78]. In the agricultural sector Omnilytics has developed AgriPhage, a range of phage products for the control of Xanthomonas campestris pv. vesicatoria, for the treatment of bacterial spot of tomatoes and peppers, and Pseudomonas syringae pv. tomato, which is the causative agent of bacterial speck on tomatoes. We expect further growth in the area of phage control of plant pathogens, which is the least examined application of phage control and has the advantage of less regulatory and safety hurdles.

PHAGE RESISTANCE

Bacterial mechanisms of phage resistance are well understood and should be considered when designing a BCA to reduce resistance and/or to help develop alternative BCAs. Most stages during phage infection can be affected by resistance development (Figure 1) [19, 79]. Briefly, these mechanisms include prevention of phage adsorption, blocking

DNA entry, abortive infection, CRISPR/Cas systems, and restriction modification systems.

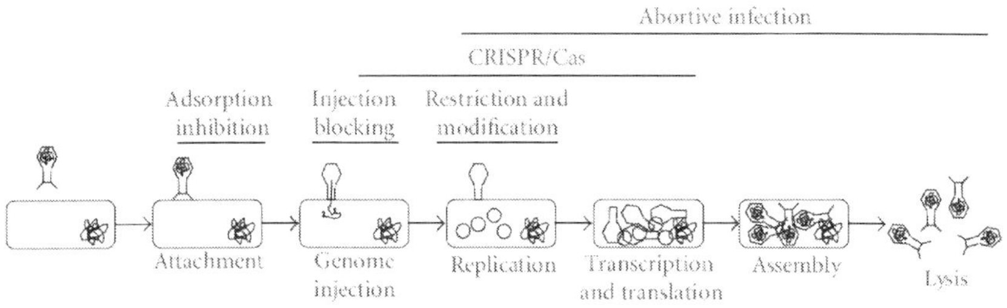

Figure 1: Bacteria can acquire phage resistance against most stages of the phage lifecycle. Fortunately, we can use our knowledge of these systems and the ability to evolve or isolate phage that infect resistant strains to minimise or avoid resistance (see text for details).

Receptors and Inhibition of Phage Adsorption

Cell surface receptors are essential for phage attachment. Receptor identification is important because receptor mutation is a common cause of phage resistance [80]. However, development of phage resistance can be beneficial. For example, mutants of Pectobacterium atrosepticum resistant to φS32 had mutations in LPS, which reduced their virulence in a potato tuber rot assay [81]. Likewise, Pectobacterium atrosepticum mutants resistant to φAT1 contained flagella mutations and were attenuated for motility and virulence [82]. The double-stranded RNA phages, φ6 and φ2954, that infect Pseudomonas syringae use Type IV pili for attachment [83–85]. Mutations in Type IV pili cause reduced survival in the phyllosphere and pathogenicity [86]. Phage φ6 also uses host cell phospholipids as secondary receptors during infection [87]. Because phages often recognize these important components (e.g., LPS and flagella), the resistant mutants are frequently less competitive or pathogenic. Therefore, careful choice of the phage cocktail can ensure that if resistant bacteria arise they will be attenuated. This approach has been utilised in phage therapy of E. coli infections in animal trials [88].

It is generally accepted that a phage mixture that uses different receptors is better for biocontrol because resistance to a carefully chosen range of phages cannot usually be acquired with a single point mutation [89, 90].

It is possible to use mutant bacterial hosts to enrich for the isolation of phages with alternative receptors. For example, a phage that targeted the flagellum was isolated using an LPS mutant host [82]. However, in most studies where multiple phages are used the exact receptors are unknown, limiting the potential benefit of such an approach. Cocktails are not always the most effective treatment. One study by Fujiwara et al. [41] characterised resistant strains generated by infection with three phages. Resistant strains were isolated for two phages, whereas no phage-resistant mutants were observed for a third phage. Use of this phage (φRSL1) was more effective in tomato plant assays and greenhouse experiments compared with either of the other phages used singularly or as a mixture. Discovering the receptors used by these phages might shed some light on these results. To overcome resistance, Omnilytics has developed a management plan, which involves monitoring the pathogen and updating the phage mixture if, or when, bacterial resistance emerges [91]. This involves the selection of host range (h-) mutants, which can be evolved to avoid resistance [55].

Intracellular Factors and Abortive Infection/ Toxin-Antitoxin Systems

Cytosolic factors are important for phage infection and their mutations can lead to resistance. Pseudomonas syringae phage φ2954 requires host glutaredoxin 3 (GrxC) for transcription of the third L segment of RNA and deletion of grxC led to resistance [92]. Phages with gene 1 mutations were easily selected that had overcome the loss of GrxC [92]. This indicates that phage can be isolated to overcome resistance caused by mutations of intracellular factors. Therefore, in theory, if φ2954 was to be used as a BCA, these escape mutants could be selected as part of the cocktail to avoid the impact of grxC mutants.

Other mechanisms interfere with phage reproduction such as the abortive infection (Abi) systems [19, 93]. Abi systems cause the "suicide" of infected cells and the inhibition of phage reproduction. Most Abi systems were identified in dairy bacteria but, recently, an Abi system, termed ToxIN, was isolated in Pectobacterium atrosepticum [94]. ToxIN inhibited infection by multiple phages and worked in different genera [94]. Theoretically, this broad efficacy and the presence of some of these systems on plasmids [94, 95] might pose a threat for phage as biocontrol agents. Fortunately, the only Abi system shown to function in a phytopathogen is ToxIN. ToxIN acts as a novel Type III protein-RNA toxin-antitoxin (TA) system [94, 96, 97]. TA systems consist of a toxic protein and an antitoxin and are found in most bacterial genomes [98]. Despite debated roles, TAs have been shown to provide resistance against phages [99]. Whether TA systems influence use of phage as biocontrol agents is at present unclear. Reassuringly, phage mutants can be isolated that avoid Abi/TA systems [97], demonstrating that resistance can be overcome if encountered.

CRISPR/Cas Resistance

Recently, roughly 40% of sequenced bacteria have been shown to possess an adaptable phage resistance system [100]. These systems contain a clustered regularly interspaced short palindromic repeat (CRISPR) array and CRISPR associated (Cas) proteins. CRISPR arrays acquire short stretches of nucleic acids (termed spacers) from invading phages. The arrays are transcribed and processed into small RNAs, which, with the assistance of Cas proteins, target and degrade spacer-complementary viral nucleic acids. In short, CRISPR/Cas provides a heritable memory of past invaders and elicits a sequence-specific immunity. The experimental analysis of CRISPR/Cas systems in phytopathogens is limited to Pectobacterium atrosepticum [101], Erwinia amylovora[102], and Xanthomonas oryzae [103]. Pectobacterium atrosepticum has three CRISPR arrays and the cas operon is expressed in planta and in vitro, indicating that phage resistance could be active during plant infection [101]. In Erwinia amylovora the CRISPR/Cas was used to study the evolutionary history of these strains and some spacers matched viral sequences but not to any sequenced Erwinia phages [102]. In Xanthomonas oryzae, sequence analyses suggested that the system had previously provided resistance against phage Xop411 but the phage had acquired a mutation to avoid the resistance system [103]. The role of CRISPR/Cas systems in plant pathogens is not well characterised but phage can be selected to avoid CRISPR resistance if, and when, it arises [104]. In summary, despite the presence of multiple resistance mechanisms, our understanding of these systems and the ability to easily select phage escape mutants and create intelligent cocktails can minimise any possible impact of resistance development in an effective therapy.

CONCLUSIONS AND FUTURE DIRECTIONS

As plant diseases continue to have a serious impact on food production worldwide, new approaches for control are sought. This has seen a resurgence of studies into the use of phage for prophylaxis and treatment of phytopathogens. As highlighted in this paper, multiple phage-phytopathogen-plant systems have been studied and promising results are beginning to emerge. However, although available, commercial application of phages to treat plant disease is still uncommon. Alternative strategies for phage-based control of plant pathogens are being developed. For example, one idea is to insert phage genes into plant genomes, especially for the control of systemic pathogens. To avoid the requirement of producing large amounts of phages or purified enzymes, transgenic tomato plants that express the CMP1 and CN77 endolysins in the xylem are being developed to kill invading C. michiganensis [33, 34]. Despite the possible advantages of this approach, the regulatory approval needed because of the use of transgenic tomato plants may present a challenge in certain countries and to the consumers. In conclusion, studies into phage BCAs will not only aid in tackling the problems of plant diseases but will also continue to shed light on the basic biology of phages and their pathogenic bacterial hosts.

ACKNOWLEDGMENTS

The authors thank members of the Fineran laboratory for useful discussions. Research in our laboratories on phage control of plant pathogens is supported by Zespri International Ltd and phage resistance in the Fineran laboratory is funded by the Marsden Fund and a Rutherford Discovery Fellowship (P. C. Fineran) from the Royal Society of New Zealand.

REFERENCES

1. P. Ronald, "Plant genetics, sustainable agriculture and global food security," Genetics, vol. 188, no. 1, pp. 11–20, 2011.

2. C. A. Gilligan, "Sustainable agriculture and plant diseases: an epidemiological perspective," Philosophical Transactions of the Royal Society B, vol. 363, no. 1492, pp. 741–759, 2008.
3. E. Stokstad, "Agriculture. Dread citrus disease turns up in California, Texas," Science, vol. 336, no. 6079, pp. 283–284, 2012.
4. M. Scortichini, S. Marcelletti, P. Ferrante Petriccione, and G. Firrao, "Pseudomonas syringae pv. actinidiae: a re-emerging, multi-faceted, pandemic pathogen," Molecular Plant Pathology, vol. 13, no. 7, pp. 631–640, 2012.
5. "Kiwifruit Vine Health," 2012, http://www.kvh.org.nz/.
6. D. A. Cooksey, "Genetics of bactericide resistance in plant pathogenic bacteria," Annual Review of Phytopathology, vol. 28, pp. 201–219, 1990.
7. P. S. McManus, V. O. Stockwell, G. W. Sundin, and A. L. Jones, "Antibiotic use in plant agriculture," Annual Review of Phytopathology, vol. 40, pp. 443–465, 2002.
8. D. W. Dye, "Control of Pseudomonas syringae with streptomycin," Nature, vol. 172, no. 4380, pp. 683–684, 1953.
9. F. Behlau, B. I. Canteros, G. V. Minsavage, J. B. Jones, and J. H. Graham, "Molecular characterization of copper resistance genes from Xanthomonas citri subsp. citri and Xanthomonas alfalfae subsp. citrumelonis," Applied and Environmental Microbiology, vol. 77, no. 12, pp. 4089–4096, 2011.
10. M. S. H. Hwang, R. L. Morgan, S. F. Sarkar, P. W. Wang, and D. S. Guttman, "Phylogenetic characterization of virulence and resistance phenotypes of Pseudomonas syringae," Applied and Environmental Microbiology, vol. 71, no. 9, pp. 5182–5191, 2005.
11. D. A. Cooksey, "Molecular mechanisms of copper resistance and accumulation in bacteria," FEMS Microbiology Reviews, vol. 14, no. 4, pp. 381–386, 1994.
12. J. M. Hirst, H. H. Riche, and C. L. Bascomb, "Copper accumulation in the soils of apple orchards near Wisbech," Plant Pathology, vol. 10, no. 3, pp. 105–108, 1961.
13. U. Pietrzak and D. C. McPhail, "Copper accumulation, distribution and fractionation in vineyard soils of Victoria, Australia," Geoderma,

vol. 122, no. 2–4, pp. 151–166, 2004.

14. N. Okabe and M. Goto, "Bacteriophages of Plant Pathogens," Annual Review of Phytopathology, vol. 1, pp. 397–418, 1963.
15. B. Balogh, J. B. Jones, F. B. Iriarte, and M. T. Momol, "Phage therapy for plant disease control," Current Pharmaceutical Biotechnology, vol. 11, no. 1, pp. 48–57, 2010. ·
16. J. B. Jones, L. E. Jackson, B. Balogh, A. Obradovic, F. B. Iriarte, and M. T. Momol, "Bacteriophages for plant disease control," Annual Review of Phytopathology, vol. 45, pp. 245–262, 2007.
17. M. Svircev, A.J. Castle, and S. M. Lehman, "Bacteriophages for control of phytopathogens in food production systems," in Bacteriophages in the Control of Food- and Waterborne Pathogens, P. M. Sabour and M.W. Griffiths, Eds., pp. 79–102, ASM Press, Washington, DC, USA, 2010.
18. Z. Klement, "Some new specific bacteriophages for plant pathogenic Xanthomonas spp," Nature, vol. 184, no. 4694, pp. 1248–1249, 1959.
19. N. K. Petty, T. J. Evans, P. C. Fineran, and G. P. C. Salmond, "Biotechnological exploitation of bacteriophage research," Trends in Biotechnology, vol. 25, no. 1, pp. 7–15, 2007. ·
20. D. A. Schofield, C. T. Bull, I. Rubio, W. P. Wechter, C. Westwater, and I. J. Molineux, "Development of an engineered "bioluminescent" reporter phage for the detection of bacterial blight of crucifers," Applied and Environmental Microbiology, vol. 78, no. 10, pp. 3592–3598, 2012.
21. M. R. J. Clokie and A. M. Kropinski, Eds., Bacteriophages Methods and Protocols, Volume 1: Isolation, Characterization, and Interactions, Humana Press, New York, NY, USA, 2009.
22. M. Svircev, S.M. Lehman, P. Sholberg, D. Roach, and A. J. Castle, "Phage biopesticides and soil bacteria: multilayered and complex interactions," in Biocommunication in Soil Microorganisms, G. Witzany, Ed., pp. 215–235, Springer, Berlin, Germany, 2011.
23. E. M. Adriaenssens, J. Van Vaerenbergh, D. Vandenheuvel et al., "T4-related bacteriophage LIMEstone isolates for the control of soft rot on potato caused by "Dickeya solani"," PLoS ONE, vol. 7, no. 3, Article ID e33227, 2012.

24. J. Boulé, P. L. Sholberg, S. M. Lehman, D. T. O'gorman, and A. M. Svircev, "Isolation and characterization of eight bacteriophages infecting Erwinia amylovora and their potential as biological control agents in British Columbia, Canada," Canadian Journal of Plant Pathology, vol. 33, no. 3, pp. 308–317, 2011.

25. B. Balogh, B. I. Canteros, R. E. Stall, and J. B. Jones, "Control of citrus canker and citrus bacterial spot with bacteriophages," Plant Disease, vol. 92, no. 7, pp. 1048–1052, 2008. ·

26. S. M. Lehman, Development of a bacteriophage-based biopesticide for fire blight [Ph.D. thesis], Brock University, St Catharines, Canada, 2007.

27. P. C. Fineran, N. K. Petty, and G. P. C. Salmond, "Transduction: host DNA transfer by bacteriophages," in The Encyclopedia of Microbiology, M. Schaechter, Ed., Elsevier, 2009.

28. H. S. Addy, A. Askora, T. Kawasaki, M. Fujie, and T. Yamada, "Utilization of filamentous phage □RSM3 to control bacterial wilt caused by Ralstonia solanacearum," Plant Disease, vol. 96, no. 8, pp. 1204–1208, 2012.

29. J. Borysowski, B. Weber-Dabrowska, and A. Górski, "Bacteriophage endolysins as a novel class of antibacterial agents," Experimental Biology and Medicine, vol. 231, no. 4, pp. 366–377, 2006.

30. M. J. Loessner, "Bacteriophage endolysins—current state of research and applications," Current Opinion in Microbiology, vol. 8, no. 4, pp. 480–487, 2005.

31. E. Echandi and M. Sun, "Isolation and characterization of a bacteriophage for the identification of Corynebacterium michiganense," Phytopathology, vol. 63, pp. 1398–1401, 1973.

32. F. D. Cook and H. Katznelson, "Isolation of bacteriophages for the detection of Corynebacterium insidiosum, agent of bacterial wilt of alfalfa," Canadian Journal of Microbiology, vol. 6, pp. 121–125, 1960.

33. J. Wittmann, R. Eichenlaub, and B. Dreiseikelmann, "The endolysins of bacteriophages CMP1 and CN77 are specific for the lysis of Clavibacter michiganensis strains," Microbiology, vol. 156, no. 8, pp. 2366–2373, 2010. ·

34. J. Wittmann, K.-H. Gartemann, R. Eichenlaub, and B. Dreiseikelmann,

"Genomic and molecular analysis of phage CMP1 from Clavibacter michiganensis subspecies michiganensis," Bacteriophage, vol. 1, no. 1, pp. 6–14, 2011.

35. H. W. Ackermann, D. Tremblay, and S. Moineau, "Long-term bacteriophage preservation," World Federation For Culture Collections Newsletter, vol. 38, pp. 35–40, 2004.
36. M. Ravensdale, T. J. Blom, J. A. Gracia-Garza, A. M. Svircev, and R. J. Smith, "Bacteriophages and the control of Erwinia carotovora subsp. carotovora," Canadian Journal of Plant Pathology, vol. 29, no. 2, pp. 121–130, 2007.
37. T. Stonier, J. McSharry, and T. Speitel, "Agrobacterium tumefaciens Conn. IV. Bacteriophage PB21 and its inhibitory effect on tumor induction," Journal of Virology, vol. 1, no. 2, pp. 268–273, 1967.
38. J. J. Gill, A. M. Svircev, R. Smith, and A. J. Castle, "Bacteriophages of Erwinia amylovora," Applied and Environmental Microbiology, vol. 69, no. 4, pp. 2133–2138, 2003.
39. E. L. Schnabel and A. L. Jones, "Isolation and characterization of five Erwinia amylovora bacteriophages and assessment of phage resistance in strains of Erwinia amylovora," Applied and Environmental Microbiology, vol. 67, no. 1, pp. 59–64, 2001.
40. H. Tanaka Negishi and H. Maeda, "Control of tobacco bacterial wilt by an avirulent strain of Pseudomonas solanacearum M4S and its bacteriophage," Annals of the Phytopathological Society of Japan, vol. 56, no. 2, pp. 243–246, 1990.
41. Fujiwara, M. Fujisawa, R. Hamasaki, T. Kawasaki, M. Fujie, and T. Yamada, "Biocontrol of Ralstonia solanacearum by treatment with lytic bacteriophages," Applied and Environmental Microbiology, vol. 77, no. 12, pp. 4155–4162, 2011.
42. F. McKenna, K. A. El-Tarabily, G. E. S. T. J. Hardy, and B. Dell, "Novel in vivo use of a polyvalent Streptomyces phage to disinfest Streptomyces scabies-infected seed potatoes," Plant Pathology, vol. 50, no. 6, pp. 666–675, 2001.
43. E. L. Civerolo and H. L. Kiel, "Inhibition of bacterial spot of peach foliage by Xanthomonas pruni bacteriophage," Phytopathology, vol. 59, pp. 1966–1967, 1969.
44. J. M. Lang, D. H. Gent, and H. F. Schwartz, "Management of

Xanthomonas leaf blight of onion with bacteriophages and a plant activator," Plant Disease, vol. 91, no. 7, pp. 871–878, 2007. ·

45. P. K. Borah, J. K. Jindal, and J. P. Verma, "Integrated management of bacterial leaf spot of mungbean with bacteriophages of Xav and chemicals," Journal of Mycology and Plant Pathology, vol. 30, no. 1, pp. 19–21, 2000.
46. D. L. McNeil, S. Romero, J. Kandula, C. Stark, A. Stewart, and S. Larsen, "Bacteriophages: a potential biocontrol agent against walnut blight (Xanthomonas campestris pv. juglandis)," New Zealand Plant Protection, vol. 54, pp. 220–224, 2001.
47. Saccardi, E. Gambin, M. Zaccardelli, G. Barone, and U. Mazzucchi, "Xanthomonas campestris pv. pruni control trials with phage treatments on peaches in the orchard," Phytopathologia Mediterranea, vol. 32, pp. 206–210, 1993.
48. Obradovic, J. B. Jones, M. T. Momol, B. Balogh, and S. M. Olson, "Management of tomato bacterial spot in the field by foliar applications of bacteriophages and SAR inducers," Plant Disease, vol. 88, no. 7, pp. 736–740, 2004.
49. Obradovic, J. B. Jones, M. T. Momol et al., "Integration of biological control agents and systemic acquired resistance inducers against bacterial spot on tomato," Plant Disease, vol. 89, no. 7, pp. 712–716, 2005. ·
50. J. E. Flaherty, J. B. Jones, B. K. Harbaugh, G. C. Somodi, and L. E. Jackson, "Control of bacterial spot on tomato in the greenhouse and field with H-mutant bacteriophages," HortScience, vol. 35, no. 5, pp. 882–884, 2000.
51. Balogh, J. B. Jones, M. T. Momol et al., "Improved efficacy of newly formulated bacteriophages for management of bacterial spot on tomato," Plant Disease, vol. 87, no. 8, pp. 949–954, 2003.
52. F. B. Iriarte, B. Balogh, M. T. Momol, L. M. Smith, M. Wilson, and J. B. Jones, "Factors affecting survival of bacteriophage on tomato leaf surfaces," Applied and Environmental Microbiology, vol. 73, no. 6, pp. 1704–1711, 2007. ·
53. H. Chantrill, C. E. Coulthard, L. Dickinson, G. W. Inkley, W. Morris, and A. H. Pyle, "The action of plant extracts on a bacteriophage of Pseudomonas pyocyanea and on influenza A virus," Journal of

General Microbiology, vol. 6, no. 1-2, pp. 74–84, 1952.

54. Delitheos, E. Tiligada, A. Yannitsaros, and I. Bazos, "Antiphage activity in extracts of plants growing in Greece," Phytomedicine, vol. 4, no. 2, pp. 117–124, 1997.
55. J. E. Flaherty, B. K. Harbaugh, J. B. Jones, G. C. Somodi, and L. E. Jackson, "H-mutant bacteriophages as a potential biocontrol of bacterial blight of geranium," HortScience, vol. 36, no. 1, pp. 98–100, 2001.
56. J. Li Dennehy, "Differential bacteriophage mortality on exposure to copper," Applied and Environmental Microbiology, vol. 77, no. 19, pp. 6878–6883, 2011.
57. "Commission decision of 30 January 2004 concerning the non-inclusion of certain active substances in Annex I to Council Directive 91/414/EEC and the withdrawl of authorisations for plant protection products containing these substances," Official Journal of the European Union, 2004, http://eur-lex.europa.eu/LexUriServ/LexUriServ.do?uri=OJ:L:2004:037:0027:0031:EN:PDF.
58. G. W. Sundin, N. A. Werner, K. S. Yoder, and H. S. Aldwinckle, "Field evaluation of biological control of fire blight in the Eastern United States," Plant Disease, vol. 93, no. 4, pp. 386–394, 2009. ·
59. Billing, "Fire blight. Why do views on host invasion by Erwinia amylovora differ?" Plant Pathology, vol. 60, no. 2, pp. 178–189, 2011.
60. Palacio-Bielsa, M. Roselló, P. Llop, and M. M. López, "Erwinia spp. from pome fruit trees: similarities and differences among pathogenic and non-pathogenic species," Trees, vol. 26, no. 1, pp. 13–29, 2012.
61. P. Gómez and A. Buckling, "Bacteria-phage antagonistic coevolution in soil," Science, vol. 332, no. 6025, pp. 106–109, 2011. ·
62. J. M. Erskine, "Characteristics of Erwinia amylovora bacteriophage and its possible role in the epidemiology of fire blight," Canadian Journal of Microbiology, vol. 19, no. 7, pp. 837–845, 1973.
63. M. J. Hattingh, S.V. Beer, and E. W. Lawson, "Scanning electron microscopy of apple blossoms colonized by Erwinia amylovora and E. herbicola," Phytopathology, vol. 76, pp. 900–904, 1986.

64. S. M. Lehman, A. M. Kropinski, A. J. Castle, and A. M. Svircev, "Complete genome of the broad-host-range Erwinia amylovora phage φEa21-4 and its relationship to Salmonella phage f elix O1," Applied and Environmental Microbiology, vol. 75, no. 7, pp. 2139–2147, 2009. ·

65. J. L. Norelli, A. L. Jones, and H. S. Aldwinckle, "Fire blight management in the twenty-first century: using new technologies that enhance host resistance in apple," Plant Disease, vol. 87, no. 7, pp. 756–765, 2003.

66. N. W. Assadian, G. D. Di Giovanni, J. Enciso, J. Iglesias, and W. Lindemann, "The transport of waterborne solutes and bacteriophage in soil subirrigated with a wastewater blend," Agriculture, Ecosystems and Environment, vol. 111, no. 1–4, pp. 279–291, 2005. ·

67. M. Kimura, Z. J. Jia, N. Nakayama, and S. Asakawa, "Ecology of viruses in soils: past, present and future perspectives," Soil Science and Plant Nutrition, vol. 54, no. 1, pp. 1–32, 2008. ·

68. H. S. Addy, A. Askora, T. Kawasaki, M. Fujie, and T. Yamada, "The filamentous phage □RSS1 enhances virulence of phytopathogenic Ralstonia solanacearum on tomato," Phytopathology, vol. 102, no. 3, pp. 244–251, 2012.

69. H. S. Addy, A. Askora, T. Kawasaki, M. Fujie, and T. Yamada, "Loss of virulence of the phytopathogen Ralstonia solanacearum through infection by phiRSM filamentous phages," Phytopathology, vol. 102, no. 5, pp. 469–477, 2012. L. Bren, "Bacteria-eating virus approved as food additive," FDA Consumer, vol. 41, no. 1, 2007.

70. "Regulatory position ListexTM," 2012, http://www.micreosfoodsafety.com/en/listex-regulatory.aspx.

71. M. Kutateladze and R. Adamia, "Bacteriophages as potential new therapeutics to replace or supplement antibiotics," Trends in Biotechnology, vol. 28, no. 12, pp. 591–595, 2010. ·

72. Loc-Carrillo and S. T. Abedon, "Pros and cons of phage therapy," Bacteriophage, vol. 1, no. 2, pp. 111–114, 2011. A. Wright, C. H. Hawkins, E. E. Änggård, and D. R. Harper, "A controlled clinical trial of a therapeutic bacteriophage preparation in chronic otitis due to antibiotic-resistant Pseudomonas aeruginosa; A preliminary

report of efficacy," Clinical Otolaryngology, vol. 34, no. 4, pp. 349–357, 2009. ·

73. M. Merabishvili, J. P. Pirnay, G. Verbeken et al., "Quality-controlled small-scale production of a well-defined bacteriophage cocktail for use in human clinical trials," PLoS ONE, vol. 4, no. 3, Article ID e4944, 2009.

74. R. J. Atterbury, "Bacteriophage biocontrol in animals and meat products," Microbial Biotechnology, vol. 2, no. 6, pp. 601–612, 2009.

75. J. Mahony, O. McAuliffe, R. P. Ross, and D. van Sinderen, "Bacteriophages as biocontrol agents of food pathogens," Current Opinion in Biotechnology, vol. 22, no. 2, pp. 157–163, 2011. ·

76. J. J. Gill, T. Hollyer, and P. M. Sabour, "Bacteriophages and phage-derived products as antibacterial therapeutics," Expert Opinion on Therapeutic Patents, vol. 17, no. 11, pp. 1341–1350, 2007. ·

77. S. J. Labrie, J. E. Samson, and S. Moineau, "Bacteriophage resistance mechanisms," Nature Reviews Microbiology, vol. 8, no. 5, pp. 317–327, 2010. ·

78. T. R. Blower, T.J. Evans, P.C. Fineran, I.K. Toth, I.J. Foulds, and G. P. C. Salmond, "Phage-receptor interactions and phage abortive infection: potential biocontrol factors in a bacterial plant pathogen," in Biology of Molecular Plant-Microbe Interactions, H. Antoun et al., Ed., pp. 1–7, International Society for Molecular Plant-Microbe Interactions, St Paul, Minn, USA, 2010.

79. T. J. Evans, A. Ind, E. Komitopoulou, and G. P. C. Salmond, "Phage-selected lipopolysaccharide mutants of Pectobacterium atrosepticum exhibit different impacts on virulence," Journal of Applied Microbiology, vol. 109, no. 2, pp. 505–514, 2010. ·

80. T. J. Evans, A. Trauner, E. Komitopoulou, and G. P. C. Salmond, "Exploitation of a new flagellatropic phage of Erwinia for positive selection of bacterial mutants attenuated in plant virulence: towards phage therapy," Journal of Applied Microbiology, vol. 108, no. 2, pp. 676–685, 2010.

81. K. Vidaver, R. K. Koski, and J. L. Van Etten, "Bacteriophage phi6: a lipid containing virus of Pseudomonas phaseolicola," Journal of Virology, vol. 11, no. 5, pp. 799–805, 1973.

82. J. S. Semancik, A. K. Vidaver, and J. L. Van Etten, "Characterization of a segmented double helical RNA from bacteriophage □6," Journal of Molecular Biology, vol. 78, no. 4, pp. 617–625, 1973.

83. X. Qiao, Y. Sun, J. Qiao, F. Di Sanzo, and L. Mindich, "Characterization of Φ2954, a newly isolated bacteriophage containing three dsRNA genomic segments," BMC Microbiology, vol. 10, article 55, 2010. ·

84. S. S. Hirano and C. D. Upper, "Bacteria in the leaf ecosystem with emphasis on Pseudomonas syringae—a pathogen, ice nucleus, and epiphyte," Microbiology and Molecular Biology Reviews, vol. 64, no. 3, pp. 624–653, 2000.

85. V. Cvirkaite-Krupovič, M. M. Poranen, and D. H. Bamford, "Phospholipids act as secondary receptor during the entry of the enveloped, double-stranded RNA bacteriophage φ6," Journal of General Virology, vol. 91, no. 8, pp. 2116–2120, 2010. ·

86. H. W. Smith and M. B. Huggins, "Successful treatment of experimental Escherichia coli infections in mice using phage: its general superiority over antibiotics," Journal of General Microbiology, vol. 128, no. 2, pp. 307–318, 1982.

87. Y. Tanji, T. Shimada, M. Yoichi, K. Miyanaga, K. Hori, and H. Unno, "Toward rational control of Escherichia coli O157:H7 by a phage cocktail," Applied Microbiology and Biotechnology, vol. 64, no. 2, pp. 270–274, 2004. ·

88. Y. Tanji, T. Shimada, H. Fukudomi, K. Miyanaga, Y. Nakai, and H. Unno, "Therapeutic use of phage cocktail for controlling Escherichia coli O157:H7 in gastrointestinal tract of mice," Journal of Bioscience and Bioengineering, vol. 100, no. 3, pp. 280–287, 2005. ·

89. Agriphage, 2004, http://www.omnilytics.com/products/agriphage/agriphage_info/agriphage_overview.html.

90. J. Qiao, X. Qiao, Y. Sun, and L. Mindich, "Role of host protein glutaredoxin 3 in the control of transcription during bacteriophage ϕ2954 infection," Proceedings of the National Academy of Sciences of the United States of America, vol. 107, no. 13, pp. 6000–6004, 2010.·

91. M. C. Chopin, A. Chopin, and E. Bidnenko, "Phage abortive infection in lactococci: variations on a theme," Current Opinion in

Microbiology, vol. 8, no. 4, pp. 473–479, 2005. ·

92. P. C. Fineran, T. R. Blower, I. J. Foulds, D. P. Humphreys, K. S. Lilley, and G. P. C. Salmond, "The phage abortive infection system, ToxIN, functions as a protein-RNA toxin-antitoxin pair," Proceedings of the National Academy of Sciences of the United States of America, vol. 106, no. 3, pp. 894–899, 2009. ·
93. T. R. Blower, F. L. Short, F. Rao et al., "Identification and classification of bacterial Type III toxin-antitoxin systems encoded in chromosomal and plasmid genomes," Nucleic Acids Research, vol. 40, no. 13, pp. 6158–6173, 2012.
94. T. R. Blower, X. Y. Pei, F. L. Short et al., "A processed noncoding RNA regulates an altruistic bacterial antiviral system," Nature Structural and Molecular Biology, vol. 18, no. 2, pp. 185–190, 2011.
95. T. R. Blower, P. C. Fineran, M. J. Johnson, I. K. Toth, D. P. Humphreys, and G. P. C. Salmond, "Mutagenesis and functional characterization of the RNA and protein components of the toxIN abortive infection and toxin-antitoxin locus of Erwinia," Journal of Bacteriology, vol. 191, no. 19, pp. 6029–6039, 2009. ·
96. Y. Yamaguchi, J. H. Park, and M. Inouye, "Toxin-antitoxin systems in bacteria and archaea," Annual Review of Genetics, vol. 45, pp. 61–79, 2011.
97. L. Short, T. R. Blower, and G. P. Salmond, "A promiscuous antitoxin of bacteriophage T4 ensures successful viral replication," Molecular Microbiology, vol. 83, no. 4, pp. 665–668, 2012.
98. S. Al-Attar, E. R. Westra, J. Van Der Oost, and S. J. J. Brouns, "Clustered regularly interspaced short palindromic repeats (CRISPRs): the hallmark of an ingenious antiviral defense mechanism in prokaryotes," Biological Chemistry, vol. 392, no. 4, pp. 277–289, 2011. ·
99. R. Przybilski, C. Richter, T. Gristwood, J. S. Clulow, R. B. Vercoe, and P. C. Fineran, "Csy4 is responsible for CRISPR RNA processing in Pectobacterium atrosepticum," RNA Biology, vol. 8, no. 3, pp. 517–528, 2011. ·
100. Rezzonico, T. H. M. Smits, and B. Duffy, "Diversity, evolution, and functionality of clustered regularly interspaced short palindromic repeat (CRISPR) regions in the fire blight pathogen Erwinia

amylovora," Applied and Environmental Microbiology, vol. 77, no. 11, pp. 3819–3829, 2011. ·

101. E. Semenova, M. Nagornykh, M. Pyatnitskiy, I. I. Artamonova, and K. Severinov, "Analysis of CRISPR system function in plant pathogen Xanthomonas oryzae," FEMS Microbiology Letters, vol. 296, no. 1, pp. 110–116, 2009. H. Deveau, R. Barrangou, J. E. Garneau et al., "Phage response to CRISPR-encoded resistance in Streptococcus thermophilus," Journal of Bacteriology, vol. 190, no. 4, pp. 1390–1400, 2008.

CHAPTER

10

CHARACTERIZATION OF NOVEL BACTERIOPHAGES FOR BIOCONTROL OF BACTERIAL BLIGHT IN LEEK CAUSED BY PSEUDOMONAS SYRINGAE PV. PORRI

Sofie Rombouts[1,2], Anneleen Volckaert[3], Sofie Venneman[4], Bart Declercq[5], Dieter Vandenheuvel[6], Camille N. Allonsius[6], Cinzia Van Malderghem[2], Ho B. Jang[1], Yves Briers[1,7], Jean P. Noben8, Jochen Klumpp[9], Johan Van Vaerenbergh[2], Martine Maes[2,10] and Rob Lavigne[1]*

[1]Laboratory of Gene Technology, Department of Biosystems, KU Leuven, Leuven, Belgium

[2]Unit Plant— Crop Protection, Institute for Agricultural and Fisheries Research, Merelbeke, Belgium

[3]Vegetable Research Centre (PCG), Kruishoutem, Belgium

[4]Research Station for Vegetable Production, Sint-Katelijne-Waver, Belgium

[5]Inagro, Rumbeke, Belgium

[6]Laboratory of Applied Microbiology and Biotechnology, Department of Bioscience Engineering, University of Antwerp, Antwerpen, Belgium

[7]Laboratory of Applied Biotechnology, Department of Applied Biosciences, Ghent University, Ghent, Belgium

[8]School of Life Sciences, Biomedical Research Institute and Transnational University Limburg, Hasselt University, Diepenbeek, Belgium

[9]Institute of Food, Nutrition and Health, ETH Zurich, Zurich, Switzerland

[10]Lab. of Microbiology, Department of Biochemistry and Microbiology, Ghent University, Gent, Belgium

Keywords: Pseudomonas syringae pv. porri, leek bacterial blight, phage therapy, KIL-like viruses, phylogenomics

Pseudomonas syringae pv. *porri*, the causative agent of bacterial blight in leek (*Allium porrum*), is increasingly frequent causing problems in leek cultivation. Because of the current lack of control measures, novel bacteriophages were isolated to control this pathogen using phage therapy. Five novel phages were isolated from infected fields in Flanders (vB_PsyM_KIL1, vB_PsyM_KIL2, vB_PsyM_KIL3, vB_PsyM_KIL4, and vB_PsyM_KIL5), and were complemented with one selected host range mutant phage (vB_PsyM_KIL3b). Genome analysis of the phages revealed genome sizes between 90 and 94 kb and an average GC-content of 44.8%. Phylogenomic networking classified them into a novel clade, named the "KIL-like viruses," related to the *Felixounalikevirus* genus, together with phage phiPsa374 from *P. syringae* pv. *actinidiae*. *In vitro* characterization demonstrated the stability and lytic potential of these phages. Host range analysis confirmed heterogeneity within *P. syringae* pv. *porri*, leading to the development of a phage cocktail with a range that covers the entire set of 41 strains tested. Specific bioassays demonstrated the *in planta* efficacy of phages vB_PsyM_KIL1, vB_PsyM_KIL2, vB_PsyM_KIL3, and vB_PsyM_KIL3b. In addition, two parallel field trial experiments on three locations using a phage cocktail of the six phages showed variable results. In one trial, symptom development was attenuated. These data suggest some potential for phage therapy in controlling bacterial blight of leek, pending optimization of formulation and application methods.

INTRODUCTION

In recent years, an increase in the prevalence of leek bacterial blight was noted in Flanders, Belgium. The disease is caused by the bacterial pathogen *Pseudomonas syringae* pv. *porri*, which was first reported in the United Kingdom by Lelliott (1952). Two decades later, the species was isolated in New Zealand by Hale (1975) who further described the pathogen. Later, the bacteria was classified as a new pathovar, *P. syringae* pv. *porri*, based on extensive research by Samson et al. in France in 1998. Currently, the disease has spread globally and is reported in The Netherlands (Janse, 1982; van Overbeek et al., 2010), Italy (Varvaro, 1983), the United States (Koike et al., 1999), Australia (Noble et al., 2006), Greece (Glynos and Alivizatos, 2006), Japan (Goto, 1972), and Korea (Myung et al., 2011, 2012). Typical symptoms include leaf curling and yellowing of the middle vein in young plants and water soaked spots on older leaves and flowering stems (Samson et al., 1998; Noble et al., 2006). Leek (*Allium porrum*) is the major host, but the pathogen has also been diagnosed on onion (*Allium cepa*) and shallot (*Allium cepa* var. *aggregatum*; Noble et al., 2006). The disease is known to be transmitted by seed (Koike et al., 1999; Ikene et al., 2003), but crop waste also plays a role in contaminating new leek plants in the field (van Overbeek et al., 2010).

The recent rise in leek bacterial blight is consistent with a growing number of Flemish farmers obtaining leek transplants from plant nurseries. High plant densities in those nurseries, combined with plant manipulations such as irrigation and mowing, promote dissemination of the pathogen among the transplants (Koike et al., 1999). In a previous study, the causative agent of recent leek blight epidemics in Flanders was examined, leading to the proposed subdivision of the strains into two groups, with small genomic differences (Rombouts et al., 2015). Knowledge on this intra-pathovar diversity is important to determine the specificity of phages isolated and characterized within this study.

To date, disease management in leek production mainly consists of prevention. Clean seeds, a strict hygiene in plant nurseries and the use of more tolerant varieties are important control measures. However, once the first symptoms of bacterial blight appear, no solution is available because of the ban on streptomycin derivatives to control bacterial plant diseases in the European Union. The use of copper-based agrichemicals is also undesirable in view of ecotoxicity and bacterial resistance development (Cooksey, 1994; Pietrzak and McPhail, 2004; Hwang et al., 2005). Therefore, the use of phage therapy constitutes an attractive alternative. Bacteriophages are viruses that specifically infect bacteria, their replication resulting in the lysis of their bacterial host and the release of newly formed viral particles. Several reviews have previously been published, highlighting the possibilities and limitations of phage therapy in plant disease control (Gill and Abedon, 2003; Jones et al.,

2007; Balogh et al., 2010). In agriculture, phage therapy research has been conducted for pathogens in other crops including "*Dickeya solani*" (Adriaenssens et al., 2012), *Erwinia amylovora* (Gill et al., 2003; Boulé et al., 2011), *Pectobacterium carotovorum* (Ravensdale et al., 2007), *Ralstonia solanacearum* (Fujiwara et al., 2011), *Xanthomonas axonopodis* pvs. *citri* and *citrumelo* (Balogh et al., 2008), and *P. syringae* pv. *actinidiae*(Frampton et al., 2014). To date, phage research resulted in a limited number of commercial phage-based products for agricultural use. These products are for control of two tomato pathogens, *Xanthomonas campestris* pv. *vesicatoria* and *P. syringae* pv. *tomato* (AgriPhage; Flaherty et al., 2000; Balogh et al., 2003) and against potato rot (Biolyse, APS Biocontrol Ltd.).

Phages infecting *P. syringae* pv. *porri* have not yet been described. In this manuscript, we focus on the isolation of novel phages which cover the bacterial diversity present in Flemish leek production. Extensive *in vitro* characterization of the phages was followed by bio-assays and field trials to evaluate the potential of these phages for the biocontrol of bacterial blight in leek.

MATERIALS AND METHODS

Bacterial Strains and Growth Conditions

The bacterial strains used in this study are listed in Table 1. *P. syringae* pv. *porri* strains were isolated from fields in Flanders and covered the existing diversity. Their identification and characterization was previously published (Rombouts et al., 2015). These isolates were supplemented with P55, an isolate from soil from the study of van Overbeek et al. (2010), and reference strains of *P. syringae* pv. *porri* from the Collection Française de Bactéries Phytopathogènes (CFBP; Beaucouzé, France), including the type strain and a set of phylogenetic related strains. Bacteria were grown in lysogeny broth with reduced salt concentration (LB_{rs}) (0.5 g/l NaCl) at 26°C, or on LB_{rs} plates supplemented with 1.5% agar. LB_{rs} with 0.7% agar (soft agar) was used for the plate overlays.

Table 1: *P. syringae* pv. *porri* **strains and related** *P. syringae* pathovars with their phage sensitivity and BOX-PCR profiles.

Bacterial strain[a]	Plant origin	Origin[b]	Year[c]	Identification	KIL 1	KIL 2	KIL 3	KIL 4	KIL 5	KIL 3b
•GBBC 1427	*Allium porrum*	BE	2012	*P.s.* pv. *porri*	–	–	–	+	+	–
•GBBC 1428	*Allium porrum*	BE	2012	*P.s.* pv. *porri*	–	–	–	+	+	–
•GBBC 1438	*Allium porrum*	BE	2012	*P.s.* pv. *porri*	–	–	–	+	+	–
•GBBC 1444	*Allium porrum*	BE	2012	*P.s.* pv. *porri*	–	–	–	+	+	–
•LMG 28496	*Allium porrum*	BE	2012	*P.s.* pv. *porri*	–	–	–	+	+	–
×GBBC 715	*Allium porrum*	BE	2001	*P.s.* pv. *porri*	+	+	+	+	+	+
×GBBC 722	*Allium porrum*	BE	2001	*P.s.* pv. *porri*	+	+	+	+	+	+
×GBBC 728	*Allium porrum*	BE	2002	*P.s.* pv. *porri*	+	+	+	+	+	+
×GBBC 747	*Allium porrum*	BE	2002	*P.s.* pv. *porri*	+	+	+	+	+	+
×GBBC 1088	*Allium porrum*	MA	2011	*P.s.* pv. *porri*	+	+	+	+	+	+
×GBBC 1089	*Allium porrum*	MA	2011	*P.s.* pv. *porri*	+	+	+	+	+	+
×GBBC 1090	*Allium porrum*	MA	2011	*P.s.* pv. *porri*	–	–	–	–	–	+
×GBBC 1113	*Allium porrum*	BE	2003	*P.s.* pv. *porri*	+	+	+	+	+	+
×GBBC 1165	*Allium porrum*	BE	2004	*P.s.* pv. *porri*	+	+	+	+	+	+
×GBBC 1166	*Allium porrum*	BE	2004	*P.s.* pv. *porri*	+	+	+	+	+	+
×GBBC 1170	*Allium porrum*	BE	2004	*P.s.* pv. *porri*	+	+	+	+	+	+
×GBBC 1184	*Allium porrum*	BE	2004	*P.s.* pv. *porri*	+	+	+	+	+	+
×GBBC 1255	*Allium porrum*	BE	2005	*P.s.* pv. *porri*	+	+	+	+	+	+
×GBBC 1256	*Allium porrum*	BE	2005	*P.s.* pv. *porri*	–	–	–	–	–	+
×GBBC 1267	*Allium porrum*	BE	2005	*P.s.* pv. *porri*	+	+	+	+	+	+
×GBBC 1269	*Allium porrum*	BE	2005	*P.s.* pv. *porri*	–	–	–	–	+	+
×GBBC 1272	*Allium porrum*	BE	2005	*P.s.* pv. *porri*	+	+	+	+	+	+
×GBBC 1273	*Allium porrum*	BE	2005	*P.s.* pv. *porri*	+	+	+	+	+	+
×GBBC 1277	*Allium porrum*	BE	2006	*P.s.* pv. *porri*	+	–	–	–	–	+
×GBBC 1286	*Allium porrum*	BE	2006	*P.s.* pv. *porri*	+	+	+	+	+	+
×GBBC 1311	*Allium porrum*	BE	2007	*P.s.* pv. *porri*	+	+	+	+	+	+
×GBBC 1424	*Allium porrum*	BE	2012	*P.s.* pv. *porri*	+	+	+	+	+	+
×GBBC 1426	*Allium porrum*	BE	2012	*P.s.* pv. *porri*	+	+	+	+	+	+
×GBBC 1433	*Allium porrum*	BE	2012	*P.s.* pv. *porri*	+	+	+	+	+	+
×GBBC 1434	*Allium porrum*	BE	2012	*P.s.* pv. *porri*	+	+	+	+	+	+
×GBBC 1435	*Allium porrum*	BE	2012	*P.s.* pv. *porri*	+	+	+	+	+	+
×GBBC 1452	*Allium porrum*	BE	2012	*P.s.* pv. *porri*	+	+	+	+	+	+
×GBBC 1459	*Allium porrum*	BE	2013	*P.s.* pv. *porri*	+	+	+	+	+	+
×GBBC 1462	*Allium porrum*	BE	2013	*P.s.* pv. *porri*	+	+	+	+	+	+
×GBBC 1893	*Allium porrum*	NL	2013	*P.s.* pv. *porri*	+	+	+	+	+	+
×GBBC 1894	*Allium porrum*	NL	2013	*P.s.* pv. *porri*	+	+	+	+	+	+
×LMG 28495	*Allium porrum*	BE	2011	*P.s.* pv. *porri*	+	+	+	+	+	+
×P55[d]	soil of leek field	NL	2010	*P.s.* pv. *porri*	+	–	–	–	–	+
×CFBP 1908[PT]	*Allium porrum*	FR	1978	*P.s.* pv. *porri*	+	+	+	+	+	+
×CFBP 1687	*Allium porrum*	GB	1949	*P.s.* pv. *porri*	+	+	+	+	+	+
×CFBP 1770	*Allium porrum*	NZ	1973	*P.s.* pv. *porri*	+	+	+	+	+	+
CFBP 3228[PT]	*Oryza sativa*	JP	1983	*P.s.* pv. *oryzae*	–	–	–	–	–	–
CFBP 1634[PT]	*Coffea arabica*	BR	1958	*P.s.* pv. *garcae*	–	–	+	–	–	+
CFBP 1674[PT]	*Avena sativa*	–	1958	*P.s.* pv. *striafaciens*	–	–	–	–	–	–
CFBP 2216[T]	*Avena sativa*	GB	1958	*P.s.* pv. *coronafaciens*	–	–	–	–	–	–
CFBP 4117[PT]	*Zizania aquatica*	US	1983	*P.s.* pv. *zizaniae*	–	–	–	–	–	–

The two symbols (• and ×) indicate two BOX-PCR pattern groups. The arrow gives the position of the discriminative band. Light and dark green: lysis after 24 and 48 h, respectively and yellow: lysis from without. [a]GBBC: culture collection of plant pathogenic bacteria at ILVO; CFBP, Collection Française de Bactéries Phytopathogènes: INRA Angers; LMG, Belgian Coordinated Collections of Microorganisms at the Laboratory of Microbiology of Ghent University with [T] as type strains and [PT] as pathovar reference strains. [b]Geographical origin. [c]Year of isolation. [d]received from van Overbeek et al. (2010).

Bacteriophage Isolation, Amplification, and Purification

Phages were isolated from soil samples taken in 2011 and 2012, from the same fields from which the *P. syringae* pv. *porri* strains were isolated. To enrich for phages, a bacterial host (LMG 28495 or LMG 28596) was grown in 25 ml LB_{rs} at 26°C up to exponential growth phase and 5 g of soil sample was added. After overnight incubation, 250 µl chloroform was added and incubation was continued for 1 h to lyse all bacterial cells. Next, the mixture was centrifuged (30 min, 4500 rpm) in a Sorvall Legend RT+ centrifuge with swing-out 4-place rotor, type 75006445 (Thermo Scientific, Waltham, MA, USA). The supernatant was filtered through a Millex® syringe filter with a 0.45 µm pore size (Merck Millipore Ltd.) and 10 µl was spotted on a soft agar layer that contained the bacterial host. Lysis zones were picked up with sterile toothpicks and suspended in phage buffer (10 mM Tris-HCl; pH 7.5; 10 mM $MgSO_4$; 150 mM NaCl). These suspensions were plated by pooling 250 µl overnight bacterial host culture, 100 µl phage suspension and 4 ml LB_{rs} overlay agar, according to the overlay agar technique (Adams, 1959). After overnight incubation at 26°C, single plaques were picked up again. Three successive single plaque isolations were performed to achieve pure phage isolates. Based on their DNA restriction pattern (data not shown) different phages were selected for further analysis.

One bacteriophage (KIL3b) was obtained after a co-cultivation experiment. This experiment consisted of the infection of an exponentially growing liquid culture of *P. syringae* pv. *porri* strain CFBP 1687 with phage KIL3 at a multiplicity of infection (MOI) of 0.01. For 4 weeks, 40 µl of the liquid culture was transferred to a tube with 4 ml fresh LB_{rs} medium every 2 to 3 days and incubated at room temperature. Supernatant of every two-day-old liquid culture was spot tested against the bacterial strain P55 to select for a broadened lytic activity as none of the other phages were able to lyse this strain.

Phages were amplified by plating 10^5 plaque forming units (pfu) per plate together with their bacterial host on a soft agar overlay plate. After overnight incubation at 26°C, the soft agar was scraped off and suspended in phage buffer. Soft agar and cell debris were subsequently removed by centrifugation (20 min, 4000 rpm) in a Sorvall Legend RT+ centrifuge. The supernatant was filtered over a 0.22 µm membrane Millex® syringe filter (Merck Millipore Ltd.), after which polyethylene glycol (PEG) 8000 (Sigma-Aldrich) containing 1 M NaCl was added to the suspension to a final concentration of $8\%_{w/v}$. After minimum 3 h incubation at 4°C, phages were precipitated by centrifugation (30 min, 4600 rpm) and the pellet was dissolved in 1 ml phage buffer.

Phage purification was performed using CsCl-gradient ultracentrifugation. A phage suspension (15 ml) with a minimum concentration of 10^{10} pfu/ml was layered on a CsCl step gradient (1.33, 1.45, 1.50, and 1.70 CsCl g/cm^3) and centrifuged (3 h, 140.000 × g) with a Beckman Coulter Optima L-90K ultracentrifuge (rotor type SW28) (Brea, CA, USA). All phages were collected at the interface between the 1.45 and 1.50 g/cm^3 density layer and dialyzed three times for 30 min against 300 volumes of phage buffer.

Electron Microscopy

To obtain transmission electron microscopy (TEM) images of the bacteriophages, dilutions of the samples were spotted on carbon coated grids (Quantifoil, Großlöbichau, Germany) after glow-discharge, and negatively stained with 2% uranyl acetate. A Philips CM12 microscope was used at 120 kV acceleration voltage. Images were produced using a Gatan Orius 1 k camera.

Host Range Analysis and General Characterization

Phage host range was tested by spotting a phage suspension on plates with a soft agar layer supplied with a specific bacterial isolate. For all the phages, three different concentrations, ranging from 10^2 to 10^6 pfu/ml, were tested to determine infectivity. The bacterial strains used in the host range assay are listed in Table 1.

Adsorption experiments were performed according to Adriaenssens et al. (2012). Briefly, bacteria and phage were mixed at an MOI of 0.0001. Next, samples were taken every minute and bacteria were immediately lysed by adding chloroform. The upper phase was titred to determine the number of non-adsorbed phages.

Killing curves at different MOIs were generated by infecting a bacterial culture at an optical density at 600 nm (OD_{600}) of 0.3. The OD_{600} of infected and uninfected cultures was monitored every 20 min during 6 h. OD_{600} results were the average of three independent biological repeats.

Phage stability was tested by incubating a phage suspension of 10^6 pfu/ml in phage buffer at different temperatures (−20, 4, 16, 30, 37, and 50°C) and at different pH ranging from 1 to 13 (universal buffer consisting of 150 mM KCl, Janssen Chimica, Geel, Belgium; 10 mM KH_2PO_4, VWR International, Leuven, Belgium; 10 mM sodium citrate, Acros Organics; 10 mM H_3BO_3, Acros Organics; adjusted to pH 1–13 with NaOH, or HCl). Infectivity of the phages at different temperatures was tested by incubating soft agar plates with spots of different phage concentrations at room temperature (±21°C), 26 and 30°C and counting the infective phage titers.

The frequency of bacterial resistance was determined using an adaptation of a previously developed method (Beale, 1948). Using the agar overlay method, phage and bacteria were plated at MOI 1 and MOI 10, in order to lyse all bacteria on the plate. After 72 h of incubation, emerging colonies were counted and about 20 colonies were cultured and re-tested for resistance against the phages in Table 2. Phage-resistant isolates were plated on Pseudomonas agar F (PAF, Becton Dickinson) supplemented with glycerol to check for fluorescence under UV-light and analyzed with a multiplex PCR containing specific primers (Ineke Wijkamp, Rijkzwaan, personal communication) to confirm their *P. syringae* pv. *porri* identity.

Table 2: Bacterial host of the new bacteriophages and phage genome characteristics determined by bioinformatic analysis.

Phage name	GenBank accession number	Bacterial host	Genome length (bp)	GC content (%)	# ORFs	# tRNAs	Terminators	Bacterial promoters
KIL1	KU130126	CFBP 1687	90695	44.86	159	5	18	57
KIL2	KU130127	CFBP 1687	92466	44.79	163	9	17	59
KIL3	KU130128	CFBP 1687	92089	44.74	161	5	17	61
KIL4	KU130129	LMG 28496	92825	44.89	167	9	18	60
KIL5	KU130130	LMG 28496	93384	44.97	169	9	18	59
KIL3b	KU130131	CFBP 1687	92099	44.72	161	5	17	61

Lysogeny in these resistant bacteria was tested as previously described (Petty et al., 2006). Supernatant from overnight cultures of resistant bacteria were collected and tested for the spontaneous release of phage particles by spotting 5 μl drops onto soft agar containing the different bacterial hosts. The supernatants from CFBP 1687 and LMG 28496 were used as negative controls, and pure phage solutions (10^6 pfu/ml) were used as positive controls for plaque formation.

Genome and Proteome Analysis

DNA Isolation and Sequencing

Contaminating bacterial DNA was removed from phage lysates by DNase treatment. Upon inactivation of the DNase by the addition of EDTA, phage particles were disrupted by adding SDS and proteinase K. The released phage DNA was purified by standard phenol-chloroform extraction and subsequent ethanol precipitation according to Sambrook and Russell (2001). DNA concentration and purity was verified with the NanoDrop ND-1000 spectrophotometer (Thermo Fisher Scientific Inc.). DNA integrity and genome size was evaluated on an 0.8% agarose gel. High throughput sequencing was done using the Illumina MiSeq platform. A 2*250 bp paired-end library was prepared for each sample, tagged with a unique adapter sequence. The quality of each library preparation was controlled using the Agilent Bioanalyzer. All library preps were equally pooled and sequenced. After processing, the reads of every library prep were assembled in a single contig with a general coverage over 1000x using the CLC Bio Genomics Workbench de novo assembly algorithm (version 7.5.1) (Aarhus, Denmark).

"In silico" Analysis

The genomes were scanned for open reading frames (ORFs) with GeneMark.hmm and GeneMarkS software (Lukashin and Borodovsky, 1998; Besemer et al., 2001). Shine-Dalgarno sequences were verified manually upstream of each annotated ORF. Functional annotation was carried out by comparing translated ORFs in a Blastp analysis (Altschul et al., 2005) against the nonredundant GenBank protein database and with the Protein Homology/analogY Recognition Engine v2.0 (PHYRE2) (Kelley and Sternberg, 2009). tRNAs were detected with the programs ARAGORN and tRNAscanSE (Lowe and Eddy, 1997; Laslett and Canback, 2004). Rho-independent terminators were predicted with ARNold, a search program that combines two complementary programs: RNAMotif and Erpin (Gautheret and Lambert, 2001; Macke et al., 2001). Probable promoter sequences were identified by looking for conserved intergenic motifs in the 100 bp upstream of every ORF with MEME and PHIRE (Lavigne et al., 2004; Bailey et al., 2009).

Protein Clustering and Phylogenetic Analysis

Homologous proteins, deduced from their DNA sequences, were examined in the phages KIL1, KIL2, KIL3, KIL4, KIL5, KIL3b, and a number of fully sequenced bacteriophages using HMMER (version3.0; http://hmmer.janelia.org), together with the ACLAME search (version0.4; http://aclame.ulb.ac.be; as of June 2015). In this manner, we retrieved a total of 4632 predicted protein sequences represented in the genomes of the 26 relevant phages. We performed pairwise similarity comparisons for each predicted protein by using the ACLAME database with the database of "viruses" and an E-value < 0.001 (Lima-Mendez et al., 2008). The remaining proteins which could not be assigned into any ACLAME protein family were defined as the unclassified protein families (UPFs) as previously described (Jang et al., 2013). The protein families that were common to all 32 representative phage genomes were identified and their phylogenetic relationships were inferred. After multiple sequence alignment of the conserved proteins using CLUSTALW2, non-informative positions were excluded with the BMGE software package (Criscuolo and Gribaldo, 2010). The two alignments were then concatenated into a FASTA file and a phylogenetic tree was built with MEGA5 (version 5.2.1; http://www.megasoftware.net/; Tamura et al., 2011) using a Jones–Taylor–Thornton (JTT) model. To estimate the robustness of the trees, we used the maximum-likelihood algorithm provided with bootstrap support (n = 1000 replicates).

Proteome Analysis

Structural proteins of phages KIL3 and KIL5 were identified by SDS-PAGE gel electrophoresis, isolation of gel bands, and subsequent trypsinization and ESI-MS/MS as described previously (Lavigne et al., 2009a). The resulting data were analyzed using Mascot (version 2.3.01) and Sequest (version 1.2.0.208), against a database containing all possible ORFs based on the DNA sequence of the phages.

Phage Therapy Experiments

Bio-Assay

In planta activity of the phages was tested by injecting phage and bacterial suspensions into leek leaves. Leek plants of the disease susceptible cultivar Krypton (Nunhems) were grown in separate pots until full development of three leaves. Next, 0.1 ml of bacterial suspension was injected with a syringe into the middle of the leaf. Subsequently, 3 cm above the bacterial injection place 0.1 ml of phage suspension (10^9pfu/ml) was injected (Supplementary Figure 1). In the first bio-assay all 41 *P. syringae* pv. *porri* strains (10^7 CFU/ml) were inoculated in leek leaves (three inoculated leaves/strain) and all phage types (10^9pfu/ml) were co-inoculated with a *P. syringae* pv. *porri* representative (eight inoculations/phage-bacteria combination). This experiment was repeated using the disease susceptible cultivar Striker (Bejo) and a lower bacterial concentration (10^6 CFU/ml). Phages KIL1, KIL2, KIL3, and KIL3b (10^9 pfu/ml) were co-inoculated with bacterial strain CFBP 1687 and phages KIL4 and KIL5 with LMG 28496 (10 inoculations/phage-bacteria combination). Plants were covered for 48 h with a plastic bag to maintain humid conditions and kept at a temperature of 25°C. Lesion lengths were measured after 10 days. Injections with only bacteria, phage or buffer served as controls.

Field Trials

In a first trial, leek transplants were treated with phages before they were planted in an infested field. First, the field was infested by spraying a solution of CFBP 1687 with a concentration of 10^6 CFU/ml on the soil at a rate of 1000 l/ha, at 1.5 bar and with 0.01% Tween® 20 as surfactant. The next day, before planting, leek plants were submerged in a solution containing a mixture of the six different phages, each at a concentration of 10^7 pfu/ml. Control plants were planted without treatment. Four blocks of the phage-treated and non-treated plants were randomly distributed over the field, with each block containing 300 plants. At multiple time points after planting, disease incidence (number of damaged plants) as well as disease severity (% of leaf surface affected) was measured for 20–50 plants per block. This trial was performed simultaneously on three different locations in Flanders: Kruishoutem, PCG (N 50.94337°, E 3.52710°), Sint-Katelijne Waver, PSKW (N 51.078120°, E 4.528180°), and Beitem, Inagro (N 50.901508°, E 3.124464°).

In a second trial, leek transplants were planted in a non-infested field. Two months after planting, bacterial infection was performed with strain

CFBP 1687. A suspension with a concentration of 10^6CFU/ml was spread over the plants at a rate of 1000 l/ha, at 1.5 bar and with 0.01% Tween® 20 as surfactant. The next day, a mix of the six different phages, each at a concentration of 10^9 pfu/ml and supplemented with 0.01% Tween® 20, was sprayed over the plants, in a way that each plant was covered with on average 10^8 pfu of each phage after spraying. Plants were covered with plastic for 48 h and assessed as described. This trial was also performed in parallel on the three different locations in Flanders mentioned above.

Statistical Analysis of Data

Statistical analyses were performed with IBM SPSS Statistics 20. For the bio-assays, normality of the data was assessed with the Kolmogorov-Smirnov and Shapiro-Wilk test at a significance level of 0.05. Homogeneity of variances was tested using the Levene's test. When the data were distributed normally and homoscedasticity was proven, an one-way ANOVA test was performed to check for differences. If the tests showed significant differences, the data were further analyzed with a *post-hoc* Tukey HSD test. To test for differences between groups of not normally-distributed data, a Kruskal-Wallis non-parametric test was used. When differences were present, Mann-Whitney U non-parametric test was used to compare the groups, paired two by two. For the field trials, the statistical program ARM (version 2015.3) was used. This program, which is designed specifically for field trials, uses the Bartlett›s test to test the homogeneity of variances and the Kurtosis and Skewness tests to check normality. One-way ANOVA with the *post-hoc*Duncan test was used for normal distributed and homoscedastic data. As a non-parametric test, the Friedman test was applied to check for differences between several variables.

RESULTS

Phage Isolation and TEM

Phages were isolated from soil samples taken in 2011 and 2012 from the same fields from which *P. syringae* pv. *porri* isolates were obtained (Table 2). After enrichment of the phages present in the soil, their capacity to lyse *P. syringae* pv. *porri* strain CFBP 1687 was tested. Based on their differential DNA restriction pattern (data not shown) phages were selected for further analysis. By this method, five novel phages were discovered for *P. syringae* pv. *porri* and the phages were named KIL1, KIL2, KIL3, KIL4, and KIL5 (referring to the involved research institutes KU Leuven and ILVO) with scientific names vB_PsyM_KIL1 etc. as proposed by Kropinski et al. (2009). The host strains used for amplification and characterization of each of these phages are indicated in Table 2.

TEM images of a phage representative revealed that they belong to the *Myoviridae* family with morphological similarity to the Pseudomonas phage *PB1* (genus *Pbunalikevirus*), as shown for phage KIL1 in Figure 1. The other analyzed phages have an identical morphology.

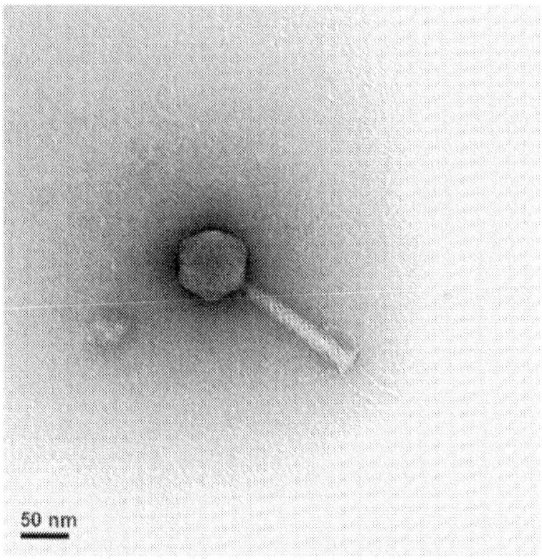

Figure 1: EM picture of phage KIL1.

Host Range Analysis

To investigate the specificity of the phages, a host range analysis was performed (Table 1), by using standard serial dilution spotting assays to distinguish infection from lysis from without phenomena, as described previously (Adriaenssens et al., 2012). Some of the bacterial isolates could not be infected by our five phage isolates from soil, therefore the host-range mutant phage KIL3b was developed. The results indicate that the bacterial strains can be divided in three groups. One group consists of five field isolates from the year 2012 and was only infected by phages KIL4 and KIL5. A second group of strains was infected by all the investigated phages and a third group of strains only showed lysis after 48 h and by a limited number of phages. None of the pathovars related to *P. syringae* pv. *porri* was infected by the phages demonstrating their specificity for pathovar *porri*. However, *P. syringae* pv. *garcae* (CFBP 1634) showed "lysis from without" at phage concentrations of 10^6 pfu/ml. In addition, none of the phages infected all *P. syringae* pv. *porri* strains tested but the combined host range of the phages covers all 41 isolates tested. A previous BOX-PCR analysis on DNA of all the bacterial strains used in the host range analysis provided information about their genomic diversity and identified two groups within *P. syringae* pv. *porri*(Rombouts et al., 2015). These two groups were confirmed by phage profiling, suggesting that differences in the bacterial genome are responsible for the observed host range. Two phages, KIL3 and KIL5 were selected for further characterization.

General Characteristics

To assess the infection parameters, adsorption experiments were performed for phage KIL3. On average, 84% of phages were irreversibly adsorbed to the host cell after 1 min and 99% after 6 min with an adsorption constant k [$k = (2.3/(B*t))*\log(P_0/P)$, with B the bacterial titer at time zero, P the phage titer and t the time] at 1 min of 7.52×10^{-9} ml/min. Compared to other *Myoviridae* phages, this is slower than reported for phage T4 (2.4×10^{-9} ml/min) but comparable to the adsorption constants of phages LIMEstone1 (9.53×10^{-9} ml/min) and LIMEstone2 (2.05×10^{-8} ml/min) infecting *Dickeya* (Kasman et al., 2002; Adriaenssens et al., 2012).

Killing curves generated by infecting an exponentially growing cultures of *P. syringae* pv. *porri* CFBP 1687 and LMG 28496 with phages KIL3 and KIL5, respectively, at different MOI demonstrated their virulence (Supplementary Figure 2). Optical density of the bacterial cultures at 600 nm decreased after 80 and 100 min for phage KIL3 and KIL5, respectively. KIL5 showed a later but steeper decline reaching $OD_{600} < 0.1$ after 180 min instead of 220 min for KIL3. No rise in OD_{600} appeared within the 5 h of monitoring indicating that resistance did not develop at that point.

The influence of temperature and pH on the viability was tested. As shown for KIL3 and KIL5 (representatives of the two phage clades in the cocktail), the phages were stable between 4 and 37°C in phage buffer for 24 h, but after 24 h of incubation at 50°C, a two and one $\log_{10}$ unit decrease was noted for phage KIL3 and KIL5, respectively. After freezing, all viable KIL3 phages were lost, while the titer of KIL5 decreased by three log10 units. Phages were stable from pH 4 to 12 for 24 h.

An analysis of the optimal infection temperature revealed that none of the phages are able to infect their host when grown at temperatures of 30°C. Phages KIL1, KIL2, KIL3, KIL4, and KIL5 were able to infect at temperatures of 26°C and room temperature (±21°C), and KIL3b only infected at room temperature.

For two phage-host interactions, frequency of bacterial resistance development was determined. An exponentially growing culture of strain CFBP 1687 was infected with KIL3 at MOI 1 and after 72 h incubation, an average of 110 phage-resistant bacteria appeared on plate. To exclude the possibility of contamination, the identity of 12 colonies was verified with PCR-analysis using specific primers, confirming their identity as *P. syringae* pv. *porri*. This resulted in a resistance frequency of 1.83×10^{-6}. A host screen with those resistant bacterial derivatives revealed that five of them were resistant to the six phages, a phenomenon called cross-resistance. The other isolates could still be infected by the host-range mutant phage KIL3b. The same experiment was performed with phage KIL4. After 72 h incubation on average 200 resistant colonies appeared and PCR-analysis was used to confirm the identity of 12 of them, resulting in a calculated resistance frequency of 3.33×10^{-6}. Again, cross-resistance was observed for part of the strains (four out of 12), the other resistant strains were still infected by KIL3b.

A spot test with supernatant from an overnight culture of those resistant isolates on strains CFBP 1687 and LMG 28496 suggested there was no induction of lysogens under the conditions tested. In general, as all phages produced clear plaques there was no indication for lysogeny.

Genome and Proteome Analysis

Genome Analysis

The genomes of all six phages were sequenced using the Illumina MiSeq platform and single contigs were obtained. Information about the genome characteristics of the phages is summarized in Table 2.

ORFs were identified and similarity at the protein level was verified by Blastp analysis. In addition, a putative function could be assigned to some ORFs without homologs using PHYRE2. All tRNAs were found in the same region spanning 2.7 kb in front of the large terminase subunit, a part of the packaging machinery (Figure 2). This is consistent with the genome organization of the related PAK_P1-like phages (Debarbieux et al., 2010; Henry et al., 2015). When comparing the tRNAs present in the six genomes, phages KIL1, KIL3, and KIL3b encode the same five tRNAs. KIL2, KIL4, and KIL5 also have these five tRNAs in their genomes, supplemented with four additional tRNAs. Depending on the phage, 17 or 18 rho-independent terminators were retained after manual verification and were located across the entire genome. Probable promoter sequences were identified by looking for conserved motifs in the 100 bp upstream region of every ORF. Using MEME, a motif resembling a typical *E. coli* bacterial promoter sequence (TTGACA-N_{17-18}-TATAAT) was found in all phages. As in the most closely related *P. syringae*phage PhiPsa374, the 41 bp conserved motif surrounding the bacterial promotor sequences in PAK_P1-like viruses was not encountered thereby confirming its specificity for *P. aeruginosa* phages (Henry et al., 2015). No toxin genes, virulence genes or genes related to lysogeny were discovered in the phage genomes, indicating their suitability for phage therapy.

A remarkable feature of all KIL-like phages is the presence of a homopolymeric G-stretch of variable length at one or two locations in the genome. Illumina reads at those positions contain 6–13 Gs. Both stretches are found in intergenic regions, yet not clearly associated to known promoter elements. A similar feature was reported previously in PB1-like phages, in a sequence coding for a baseplate protein and in the ORF coding for a DNA adenine methylase in Bordetella phages BPP-1, BIP-1, and BMP-1 (Liu et al., 2004; Ceyssens et al., 2009).

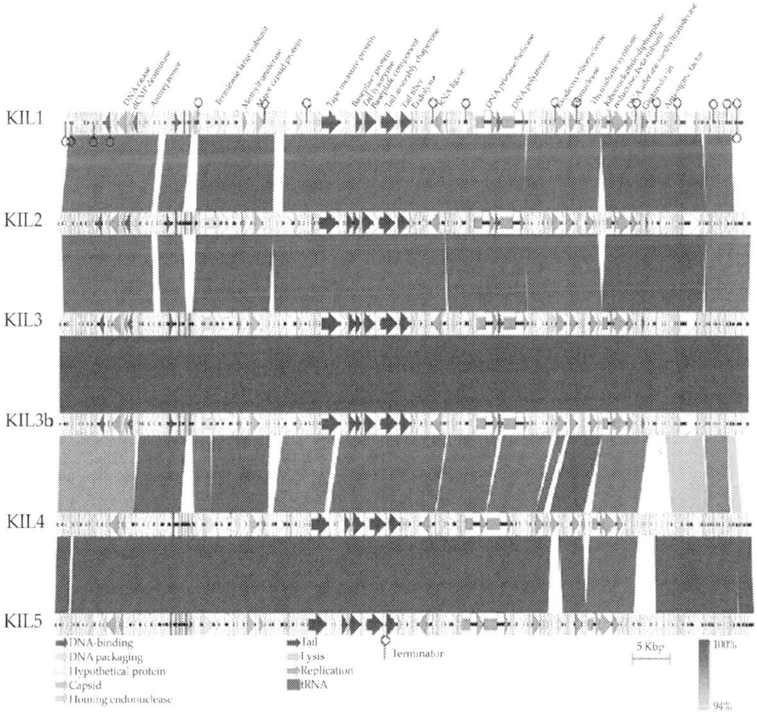

Figure 2: Representation of the genome organization of the six KIL-like phages and homology (blastn) between them.

It was proposed that the resulting frameshift serves as a control point for expression levels of the protein. Another remarkable region in the KIL-like phage genomes is a long A-stretch varying from 14 to 33 nucleic acids in length. In the most related phage phiPsa374 those homopolymeric stretches are not present. A phylogenetic analysis of the six phages based on their genome sequences confirms the subdivision of the phages into two groups, correlated to their host range. The first group constitutes of phage KIL1, KIL2, KIL3, and KIL3b, the second group contains phages KIL4 and KIL5. Between the groups, some differences in the ORFs are predicted, mostly among unknown genes. The difference between the genomes of phage KIL3 and KIL3b, which is a host range mutant of the former, is limited to the number of repeats located in the homopolymeric A- and G-stretches. No clear biological function could be attributed to this phenomenon.

The Phylogenetic and Proteome-Based Relationships

To further frame these phages in their comparative genomic context, phylogenetic trees were constructed based on the concatenated datasets of two structural genes encoding the major capsid protein and the baseplate protein. These two proteins are shared among the genomes of four clades: the FelixO1-like viruses, PAK_P1-like viruses, and KPP10-like viruses within the *Felixounalikevirus* genus and the rV5-like viruses, as well as phages CR3, ICP1, and our six phage isolates. Each protein was assigned to "Family:virus:314" and "Family:virus:3622" by ACLAME-based clustering. Phylogenetic analysis revealed the four clades comprising the KPP10-, PAK_P1-, FelixO1, and rV5-like phages, respectively (Left, Figure 3). Most clades branch deeply and are well supported with bootstrap support values of more than 90%. Within the *Felixounalikevirus* genus, the *Pseudomonas* phages KIL1, KIL2, KIL3, KIL4, KIL5, and KIL3b are grouped into a distinct clade, together with a *Pseudomonas* phage phiPsa374.

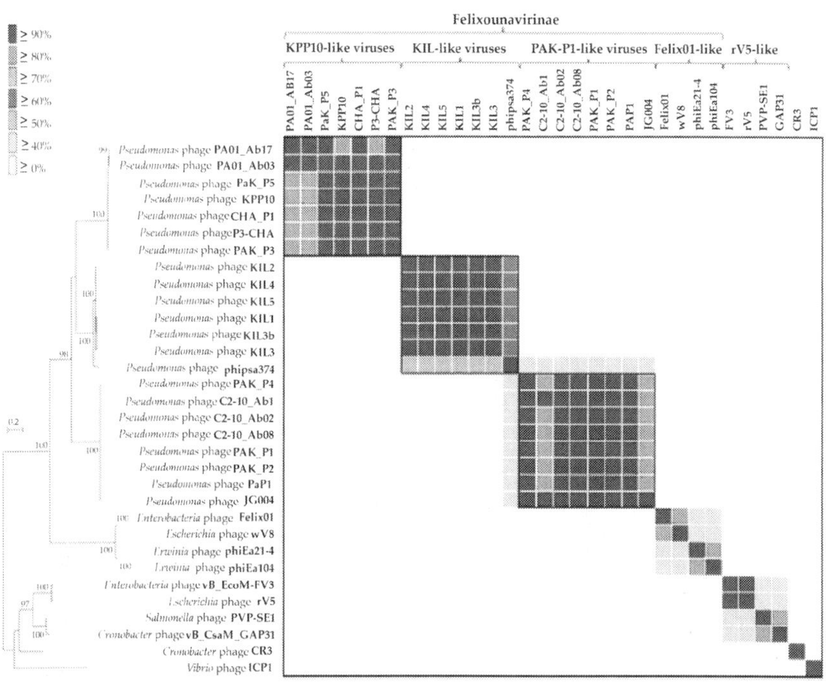

Figure 3: Phylogenetic and proteome-based relationships of phages KIL1, KIL2, KIL3, KIL4, KIL5, and KIL3b. Left, maximum-likelihood tree was con-

structed upon a concatenation of two structural proteins (major capsid protein and baseplate protein) that are common to the 32 phages. The numbers at the branch represent the bootstrap values. Edges with bootstrap values above 75% are represented. The scale bar indicates the number of substitution per site. Right, in the matrix view, the columns and rows correspond to the phages. Each cell indicates the percentage of shared protein families between two phages. Color tags are used to visualize the genomic relatedness based on the proportion of shared protein families.

To investigate the genome relatedness based on the proportion of conserved homologous proteins between two phages, we performed pairwise comparisons of the 32 phage genomes, which allowed us to assign 5622 protein sequences to 1322 homologous protein families. A cut-off value of 40% of homologous proteins was used to assign phage genomes to a putative genus (Lavigne et al., 2008, 2009b). The pattern of proteome-based grouping was highly similar to our phylogenetic results, in which KIL1, KIL2, KIL3, KIL4, KILP5, KIL3b, and phiPsa374 form a conserved group with ≥50% homologous proteins in common to each other's genome (Right, Figure 3). These seven phages are also related to three groups of the *Felixounalikevirus* phages, sharing more than 20% of their homologous proteins, with the PAK_P1-like phages being the closest relatives (average 34% shared proteins). Notably, of these, phiPsa374 shared the highest overlap percentages of the proteomic content (≥40%) with the PAK_P1-like phages.

Proteome Analysis

Analysis of the proteome of phages KIL3 and KIL5 (representatives of the two clades) by ESI-MS/MS led to the identification of 51 and 67 proteins, respectively. For KIL3 and KIL5, respectively 27 proteins and 26 had a predicted function based on homology to other phage proteins. In both phages, most of the detected proteins are early phage proteins or structural phage proteins. The use of non CsCl-purified phages might explain the presence of the non-structural phage proteins since they can remain in the phage lysate after bacterial lysis. Alternately, these proteins may be considered as candidate proteins which may be co-injected during infection. The most abundant proteins based on the number of unique peptides recovered are the tape measure protein, ORF 46 (no predicted function) and the major capsid protein in phage KIL3. Similarly, the most abundant proteins of phage KIL5 are the capsid protein, ORF 46 (no predicted function) and the tail fiber protein (ORF 76). As was seen by Henry et al. (2015) for the related phage PAK_P3, the ORF upstream of the capsid protein had the second highest relative abundance (total number of spectra / Molecular Weight) next to the capsid protein itself, and this in both phages. Although the function of this protein could not be predicted, an association with the capsid protein is suspected. The differences in the number of proteins identified for the two phages could be explained by a difference in protein concentration as the number of peptides recovered for proteins of KIL5 were in general higher than those for KIL3. The identification of these peptides by mass spectrometry confirms the *in silico* ORF predictions. One structural protein, a tail lysozyme (ORF 70), could not be detected in KIL3; however, the corresponding protein in KIL5 (ORF 71) was identified. All other predicted structural proteins could be experimentally confirmed by mass spectrometry.

Phage Therapy

Bio-Assay on Leek Leaves

To investigate whether the phages were also capable of lysing their bacterial host in the plant environment, bio-assays were performed on leek leaves.

In a first bio-assay, the bacterial concentration was chosen to be 10^7 CFU/ml, the amount of bacteria necessary to produce clear symptoms. For the phages, a relatively high concentration of 10^9 pfu/ml was used to prove efficacy. All 41 *P. syringae* pv. *porri* strains were inoculated and showed comparable virulence, presented on the left side of the boxplot (Figure 4). Strains CFBP 1687 and CFBP 1770 were chosen as representatives of *P. syringae* pv. *porri* and were used as bacterial host in combination with the different phages. Lesion lengths on the phage treated leaves were compared to the lesion lengths induced by all *P. syringae* pv. *porri* strains used in this study. In general, each phage reduced the mean lesion length compared to the non-phage treated leaves. The reduction in lesion length was significant for phages KIL1, KIL3, and KIL3b with p-values of 0.003, 0.005, and 0.004, respectively. Phages KIL2, KIL4, and KIL5 reduced the lesion length, but were not able to prevent bacterial infection completely.

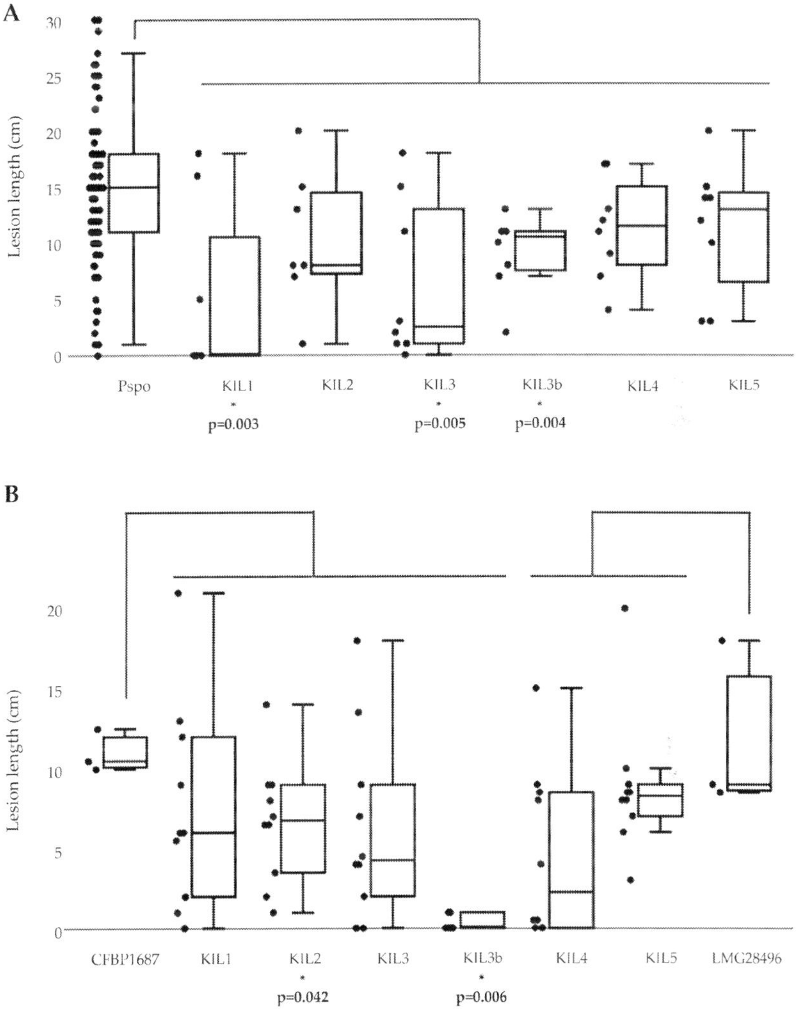

Figure 4: (A) Lesion lengths in the first bio-assay in which all 41 *P. syringae* pv. *porri* strains (10^7CFU/ml) were inoculated in leek leaves (three inoculated leaves/strain) and all phage types (10^9pfu/ml) were co-inoculated with a *P. syringae* pv. *porri* representative (eight inoculations/phage-bacteria combination). (B) Lesion lengths in the second bio-assay in which KIL1, KIL2, KIL3, and KIL3b (10^9 pfu/ml) were co-inoculated with bacterial strain CFBP 1687 (10^6 CFU/ml) and phages KIL4 and KIL5 with LMG 28496 (10 inoculations/phage-bacteria combination).

In a second bio-assay, the effect of the phages on a lower bacterial concentration of 10^6 CFU/ml was tested using ten leaves per phage. CFBP 1687 was used in combination with phages KIL1, KIL2, KIL3, and KIL3b, phages KIL4 and KIL5 were co-inoculated with bacterial strain LMG 28496. Again, lesion lengths of the phage-treated leaves were compared to the lengths produced by their respective host without phage. As expected, lesion lengths were smaller in all treatments compared to the previous assay with higher bacterial concentrations. Only phages KIL2 and KIL3b decreased lesion length significantly compared to their bacterial control, with p-values of respectively 0.042 and 0.006. Leaves treated with phage KIL3b showed no symptoms for the most part. The results of this second bio-assay confirm that phages KIL4 and KIL5 are less suitable to reduce lesion length significantly. Only phage KIL3b showed significant reduction in both assays, demonstrating its *in planta* antibacterial effect. The effect of phages KIL1, KIL2, and KIL3 varied between the assays. Nevertheless, it was considered prudent to maintain all phages in a cocktail for subsequent testing in field trials.

Field Trials

In a first field trial set-up, the potential of phages to protect leek transplants against bacterial infection at planting was tested. Transplants were submerged in a phage solution containing the six phages before planting them in an infested field. This experiment was performed on three different locations in Flanders in the year 2014 (Table 3).

Table 3: Percentage symptomatic plants after 3 months for the field trial at three different locations where 1200 transplants were submerged in phage solution (10^7pfu/ml) before planting in a field infested with CFBP 1687.

Phage treatment	PCG	PSKW	Inagro
No	92.5	15.0	23.5
Yes	82.0	18.0	18.0

In the trial at PCG, leek plants of cultivar Kenton (Nunhems) were used. The first symptoms developed 25 days after planting in phage-treated and non-treated plants. Two months later, percentages of diseased plants increased to 82% for the phage-treated plants and 92.5% for the control plants. Later on, symptoms disappeared again in both groups and percentages of diseased plants were again lower 4 months after planting. The same trend was apparent when scoring the percentage of leaf surface affected.

The same experiment was performed in PSKW with cultivar Harston (Nunhems). Two and three months after planting, the disease incidence was measured. Percentages of symptomatic plants varied from 29% for the phage-treated plants and 19% for the control plants after 2 months to 18 and 15% after 3 months, respectively. No significant differences were observed.

In the experiment at Inagro, leek plants of cultivar Harston (Nunhems) were used. No disease incidence was recorded 2 months after planting. After 3 months, measurements reported an infection rate of 23.5% of control plants and 18% for the phage treated plants.

In all three experiments, no significant differences in disease occurrence between treated and untreated plants could be noticed. In the trials performed at PSKW and Inagro, disease incidence remained low indicating that infection was not successful. Conclusions on the efficacy of phage treatment could therefore not be made. From the trial results of PCG we can conclude that phage treatment was not successful in protecting the leek plants from infection when planted in an infested field. However, monitoring at an earlier time-point could be better to detect an effect of the phages. It is also possible that more frequent phage applications are necessary to protect the plants against infection.

In a second field trial, phages were sprayed on artificially infected plants, to test for their potential as a crop protection agent in a later growth stage. First, bacteria were sprayed onto the plants 3 months after planting. The day after, a phage cocktail with the six phages present in a concentration of 10^9 pfu/ml each was sprayed onto the plants. Results are shown in Table 4.

Table 4: Percentage symptomatic plants after 1 (PCG and PSKW) and 2 months (Inagro) for the field trial with *P. syringae* pv. *porri* **CFBP 1687 (10^6 CFU/ml) and the phage cocktail (10^9** pfu/ml) being sprayed over 1000 leek plants per treatment.

Treatment	PCG	PSKW	Inagro
Untreated	6.0	32.0	19.5
Phage	11.0	38.0	24.0
Bacteria	63.0	42.0	21.5
Bacteria + phage	38.5	30.0	19.0

In the trial at PCG, again cultivar Kenton (Nunhems) was used and phage treatment was repeated four times every 2 weeks. The first disease assessment was performed 19 days after infection and demonstrated a higher infection rate at the bacteria treated plant in relation to the other plants, meaning that infection was successful. One month after infection, infection rate in the bacteria-treated plants increased and a difference trend could be measured for the phage-treated object in comparison with the non-treated infected object with, respectively, 38.5 and 63% symptomatic plants. In subsequent measurements this trend continued although the differences became less apparent. The same trend was demonstrated by the data displaying the affected leaf surface. At all time-points, infection was slightly higher in the phage-treated control compared to the untreated plants. The trial demonstrated that phages could decrease symptom development but were not able to completely stop infection. A higher phage concentration and more frequent applications can possibly lead to a better disease control.

The same trial was performed at PSKW with plants of the cultivar Harston (Nunhems). Phage treatments were applied every week for 8 weeks with half the dose of phages compared to the trial of Kruishoutem, meaning that a concentration of 0.5×10^9 pfu/ml of each phage was used. Disease incidence was measured one, 2 and 3 months after infection but no differences between the plants could be measured. Even the untreated and infected plants showed the same level of symptoms, indicating the possibility that symptoms were masked by natural infection with another pathogen.

The trial performed at Inagro used plants of the cultivar Harston (Nunhems). Five phage applications were carried out: one a week before infection, one the evening of the day during which bacterial infection was performed, and three more weekly applications after infection. Two months after infection disease incidence was measured showing no differences between the four treatments. Even the uninfected and untreated control and the uninfected phage-treated control displayed an infection of respectively 19.5 and 24% indicating the presence of a natural infection that could not be affected by our phages.

Results of the three trials demonstrate that only at PCG conducive conditions for infection with *P. syringae* pv. *porri* were present. From that trial we can conclude that phage application can lead to a reduction in bacterial disease incidence but cannot completely prevent it. Other factors such as natural infections with other pathogens make it difficult to predict the effect and reliability of phage therapy in field trials.

DISCUSSION

The increase of bacterial blight of leek and the related economical losses attracted our interest in the problem a few years ago leading to an extensive investigation of the causative agent, *P. syringae* pv. *porri*. Previous research divided the Flemish isolates into two groups based on their BOX-PCR fingerprints (Rombouts et al., 2015). Genomic differences among *P. syringae* pv. *porri* strains were also described by Noble et al. (2006), who showed minor differences based on profiles generated with IS50-primer amplification and RFLP of a 16S rDNA fragment. In contrast, these authors as well as Koike et al. (1999) and van Overbeek et al. (2010) did not discriminate different genotypes among their isolates from California, Australia and the Netherland when applying different repetitive-element-PCRs (BOX, REP, and ERIC). This overall bacterial diversity was kept in mind when looking for novel phages capable of lysing *P. syringae* pv. *porri*. Host range analysis of the five phages isolated from soil and the H-mutant phage confirmed the BOX-PCR grouping of bacterial strains, suggesting that differences at the bacterial genome are responsible for the observed host range. Previous genome analysis of a bacterial strain of each group demonstrated that variation was mostly situated in prophage content and mobile genetic elements (Rombouts et al., 2015). No specific element could be found that linked bacterial diversity with the phage host range. In general, genomic differences can affect phage infection at different stages. If those genome differences result in an altered phage receptor, different receptor binding proteins (RBP) capable of recognizing the altered receptor of the five bacterial strains in group one could be present among the phages KIL4 and KIL5. Changes in host range due to a spontaneous mutation in a tail fiber protein (replacement of a positively charged lysine by an uncharged asparagine) were previously reported for two related phages, *Pseudomonas aeruginosa* phages PaP1 and JG004 (Le et al., 2013). However, the exact role of the variation and mutation observed here does not lead to a straightforward hypothesis and requires further investigation. In spite of the overall bacteriophage diversity, all phages we found against *P. syringae* pv. *porri* belong to the *Myoviridae* family. Genome sequence analysis further classified them into a new clade within the *Felixounavirinae* genus.

The phage characteristics analyzed in this study show that all six phages are suitable for use in phage therapy. First, they are strictly virulent, meaning that infection with the phages results in bacterial lysis minimizing the risk of horizontal transfer of pathogenicity genes as is the risk with lysogenic phages (Penadés et al., 2015). Furthermore, by demonstrating a temperature stability from 4 to 37°C and pH stability from pH 4 to 12, they are capable of surviving in the plant environment when applied. However, a disadvantage is their limitation to infect only bacteria grown at temperatures of 28°C or lower. This means that in summer, when temperatures occasionally rise above 28°C bacteria may not be killed by the phages. Probably, the bacterial receptor necessary for phage infection is not expressed at higher temperatures. Temperature dependent infection was previously reported for phage φS1 which is not able to infect *Pseudomonas fluorescens* cells grown at 37°C (Sillankorva et al., 2004). It has been shown that expression of bacterial genes can be temperature dependent. For example, the expression of flagellar components and the surfactant syringafactin is reduced in *P. syringae* strain B728a at a temperature of 30°C, resulting in reduced swimming and swarming motility (Hockett et al., 2013). Host range analysis demonstrates that none of the phages is capable of lysing all tested *P. syringae* pv. *porri* strains, but a combination of different phages in a phage cocktail can cover all strains used. The results indicate that there is some resistance development among *P. syringae* pv. *porri* strain CFBP 1687 when challenged with the bacteriophages KIL3 or KIL4. Although the kill curves indicate that in the first 5 h there is no resistance development, after 72 h an average resistance frequency of around 2×10^{-6} was noted. When phages are applied in a phage cocktail, this resistance development is less likely to occur (Gill and Hyman, 2010; Barbosa et al., 2013). The possibility of resistance through lysogenic conversion was excluded, thereby confirming their suitability for phage therapy for this parameter. In addition, development of the host-range mutant phage KIL3b demonstrates that host-range expanding adaptions are possible to counter bacterial resistance, especially because this phage was still able to infect more than half of the resistant bacteria.

To investigate evolutionary relationships of KIL1, KIL2, KIL3, KIL4, and KIL5 with their close relatives, phylogenetic analysis was combined with a whole genome proteomic approach (Lavigne et al., 2008, 2009b). The overall phylogenetic relationships between the concatenated sequences of the major capsid protein and the baseplate proteins of the *Felixounalikevirus* phages are consistent with previous taxonomic description (Henry et al., 2015). The phages KIL1, KIL2, KIL3, KIL4, and KIL5 as well as phiPsa374 can clearly be grouped into a new phylogenetic clade. Their location within the *Felixounalikevirus* genus suggests that the sequences of two structural proteins are highly conserved within their own group and appear to be most closely related to those of the "PAK_P1-like viruses" and "KPP1-like viruses." In addition, our proteome-based analysis, which is more sensitive and accurate for taxonomic classification for phage genomes (Lavigne et al., 2008), supports the phylogenetic relationships of the *Felixounalikevirus* phages. Given that shared homologs between the phage genomes can be used as an indication of phage members in the same genus (Lavigne et al., 2008, 2009b), we proposed a possible division of the *Felixounalikevirus* genus into the four subclades, which include the "KPP10-like viruses" (phages PAO1 Ab17, PAO1_Ab03, PAK_P5, KPP10, CHA_P1, P3-CHA, and PAK_P3), "PAK_P1-like viruses" (PAK_P4, C2-10_Ab1, C2-10_Ab02, C2-10_Ab-8, PAK_P1, PaP1, and JG004), "FelixO1-like viruses" (FelixO1, wV8, phiEa21-4, and phiEa104), and "KIL-like viruses" (KIL1, KIL2, KIL3, KIL4, KILP5, KIL3b, and phiPsa374). Interestingly, although phage phiPsa374 has been suggested as a member of the PAK_P1-like clade (Frampton et al., 2014; Henry et al., 2015), our results indicate the genome of phiPsa374 aligns with more shared protein families to the KIL-like phages.

No toxin genes, virulence genes or genes related to lysogeny could be discovered in the phage genomes at this time, indicating their suitability for phage therapy. Efficacy of the phages in the plant environment was analyzed with bio-assays in leek leaves. Significant results were only obtained for phages KIL1, KIL2, KIL3, and KIL3b but the broader host range of phages KIL4 and KIL5 led to their incorporation in the phage cocktail for the field trials. When treating transplants with a phage cocktail before planting in an infested field, no significant differences were observed. Spraying of phages on artificially infected plants led to a slightly reduced symptom development in one of the three trials, yet again without statistical significance. Since two trials suffered from natural infections, their impact has most likely masked the effect of the treatments. Although isolation and characterization of phages is reported for several plant pathogens, results of other field trials are scarce. A recent overview has recently been published by Czajkowski (Czajkowski, 2016). A field trial with phages against *Dickeya solani* on potato also reports of phage therapy resulting in minor differences in disease severity (Adriaenssens et al., 2012). In experiments with phages against *X. axonopodis* pvs *citri* and *citrumelo*, significant control of bacterial spot was reported, yet treatment with copper-mancozeb was more effective (Balogh et al., 2008). As other researchers mentioned, open field (and greenhouse) applications with phages encounter many challenges such as UV-light, desiccation, application method, and the need of a large quantities of phages. Especially the phyllosphere is a harsh environment for phages, making leaf diseases difficult to treat (Gill and Abedon, 2003; Goodridge, 2004; Iriarte et al., 2012). This study indicates that phage therapy is able to reduce bacterial blight symptoms in leek, but to improve the efficacy of the phages in the field, phage persistence in the plant phyllosphere should be improved by the use of protective formulations, addition of non-pathogenic phage-propagating bacterial strains or adapting timing and frequency of application as suggested previously (Jones et al., 2007; Balogh et al., 2010).

ACKNOWLEDGMENTS

The authors would like to thank Dr. LS van Overbeek of Wageningen UR (The Netherlands) for providing strain P55. The first author would like to thank Mr. Thijs van den Brande for his help with some of the figures and Mr. Johan Dick and Mrs. Cindy Elisabeth for taking care of the plants. The research was funded by the Agency for Innovation by Science and Technology (IWT, project nr.100881) with co-funding of Boerenbond, INGRO cvba, LAVA cvba, and the seed companies Bejo, Enza, Nunhems, Hazera, Rijk Zwaan, and Syngenta.

REFERENCES

1. Adams, M. H. (1959). *Bacteriophages.* New York, NY: Interscience Publishers Inc.
2. Adriaenssens, E. M., Van Vaerenbergh, J., Vandenheuvel, D., Dunon, V., Ceyssens, P. J., De Proft, M., et al. (2012). T4-Related Bacteriophage LIMEstone Isolates for the Control of Soft Rot on Potato Caused by '*Dickeya Solani.*' *PLoS ONE* 7:e33227. doi: 10.1371/journal.pone.0033227
3. Altschul, S. F., Wootton, J. C., Gertz, E. M., Agarwala, R., Morgulis, A., Schäffer, A. A., et al. (2005). Protein database searches using compositionally adjusted substitution matrices. *FEBS J.* 272, 5101–5109. doi: 10.1111/j.1742-4658.2005.04945.x
4. Bailey, T. L., Boden, M., Buske, F., Frith, M., Grant, C. E., Clementi, L., et al. (2009). MEME suite: tools for motif discovery and searching. *Nucleic Acids Res.* 37, 202–208. doi: 10.1093/nar/gkp335
5. Balogh, B., Canteros, B. I., Stall, R. E., and Jones, J. B. (2008). Control of citrus canker and citrus bacterial spot with bacteriophages. *Plant Dis.* 92, 1048–1052. doi: 10.1094/PDIS-92-7-1048
6. Balogh, B., Jones, J. B., Iriarte, F. B., and Momol, M. T. (2010). Phage therapy for plant disease control. *Curr. Pharm. Biotechnol.* 11, 48–57. doi: 10.2174/138920110790725302
7. Balogh, B., Jones, J. B., Momol, M. T., Olson, S. M., Obradovic, A., King, P., et al. (2003). Improved efficacy of newly formulated bacteriophages for management of bacterial spot on tomato. *Plant Dis.* 87, 949–954. doi: 10.1094/PDIS.2003.87.8.949

8. Barbosa, C., Venail, P., Holguin, A. V., and Vives, M. J. (2013). Co-evolutionary dynamics of the Bacteria Vibrio Sp. CV1 and phages V1G, V1P1, and V1P2: implications for phage therapy. *Microb. Ecol.* 66, 897–905. doi: 10.1007/s00248-013-0284-2
9. Beale, G. H. (1948). A method for the measurement of mutation rate from phage sensitivity to phage resistance in *Escherichia coli. J. Gen. Microbiol.* 2, 131–142. doi: 10.1099/00221287-2-2-131
10. Besemer, J., Lomsadze, A., and Borodovsky, M. (2001). GeneMarkS: a self-training method for prediction of gene starts in microbial genomes. implications for finding sequence motifs in regulatory regions. *Nucleic Acids Res.* 29, 2607–2618. doi: 10.1093/nar/29.12.2607
11. Boulé, J., Sholberg, P. L., Lehman, S. M., O›gorman, D. T., and Svircev, A. M. (2011). Isolation and characterization of eight bacteriophages infecting *Erwinia amylovora* and their potential as biological control agents in British Columbia, Canada. *Can. J. Plant Pathol.* 33, 308–317. doi: 10.1080/07060661.2011.588250
12. Ceyssens, P. J., Miroshnikov, K., Mattheus, W., Krylov, V., Robben, J., Noben, J. P., et al. (2009). Comparative analysis of the widespread and conserved PB1-like viruses infecting *Pseudomonas aeruginosa. Environ. Microbiol.* 11, 2874–2883. doi: 10.1111/j.1462-2920.2009.02030.x
13. Cooksey, D. A. (1994). Molecular mechanisms of copper resistance and accumulation in bacteria. *FEMS Microbiol. Rev.* 14, 381–386. doi: 10.1111/j.1574-6976.1994.tb00112.x
14. Criscuolo, A., and Gribaldo, S. (2010). BMGE (Block Mapping and Gathering with Entropy): a new software for selection of phylogenetic informative regions from multiple sequence alignments. *BMC Evol. Biol.* 10:210. doi: 10.1186/1471-2148-10-210
15. Czajkowski, R. (2016). Bacteriophages of soft rot enterobacteriaceae - a minireview. *FEMS Microbiol. Lett.* 363:fnv230. doi: 10.1093/femsle/fnv230
16. Debarbieux, L., Leduc, D., Maura, D., Morello, E., Criscuolo, A., Grossi, O. L., et al. (2010). Bacteriophages can treat and prevent *Pseudomonas aeruginosa* lung infections. *J. Infect. Dis.* 201, 1096–1104. doi: 10.1086/651135

17. Flaherty, J. E., Jones, J. B., Harbaugh, B. K., Somodi, G. C., and Jackson, L. E. (2000). Control of bacterial spot on tomato in the greenhouse and field with H-mutant bacteriophages. *HortScience* 35, 882–884.

18. Frampton, R. A., Taylor, C., Holguín Moreno, A. V., Visnovsky, S. B., Petty, N. K., Pitman, A. R., et al. (2014). Identification of bacteriophages for biocontrol of the kiwifruit canker phytopathogen *Pseudomonas syringae* pv. actinidiae. *Appl. Environ. Microbiol.* 80, 2216–2228. doi: 10.1128/AEM.00062-14

19. Fujiwara, A., Fujisawa, M., Hamasaki, R., Kawasaki, T., Fujie, M., and Yamada, T. (2011). Biocontrol of *Ralstonia solanacearum* by treatment with lytic bacteriophages. *Appl. Environ. Microbiol.* 77, 4155–4162. doi: 10.1128/AEM.02847-10

20. Gautheret, D., and Lambert, A. (2001). Direct RNA motif definition and identification from multiple sequence alignments using secondary structure profiles. *J. Mol. Biol.* 313, 1003–1011. doi: 10.1006/jmbi.2001.5102

21. Gill, J., and Abedon, S. T. (2003). Bacteriophage ecology and plants. *APSnet*. doi: 10.1094/APSnetFeature-2003-1103. [Epub ahead of print].

22. Gill, J. J., and Hyman, P. (2010). Phage choice, isolation, and preparation for phage therapy. *Curr. Pharm. Biotechnol.* 11, 2–14. doi: 10.2174/138920110790725311

23. Gill, J. J., Svircev, A. M., Smith, R., and Castle, A. J. (2003). Bacteriophages of *Erwinia amylovora*. *Appl. Environ. Microbiol.* 69, 2133–2138. doi: 10.1128/AEM.69.4.2133-2138.2003

24. Glynos, P. E., and Alivizatos, A. S. (2006). Leek rot caused by *Pseudomonas syringae* pv. *porri*. *Ann. Benaki Phytopathol. Inst.* 20, 82–87.

25. Goodridge, L. D. (2004). Bacteriophage biocontrol of plant pathogens: fact or fiction? *Trends Biotechnol.* 22, 384–385. doi: 10.1016/j.tibtech.2004.05.007

26. Goto, M. (1972). Bacterial leaf blight of onion in Japan. *Plant Dis. Rep.* 56, 490–493.

27. Hale, C. N. (1975). Bacteriosis of leek in New Zealand. *J. Agric. Res.* 18, 251–254. doi: 10.1080/00288233.1975.10423640

28. Henry, M., Bobay, L. M., Chevallereau, A., Saussereau, E., Ceyssens, P. J., and Debarbieux, L. (2015). The search for therapeutic bacteriophages uncovers one new subfamily and two new genera of *Pseudomonas*-infecting *Myoviridae*. *PLoS ONE* 10:e0117163. doi: 10.1371/journal.pone.0117163

29. Hockett, K. L., Burch, A. Y., and Lindow, S. E. (2013). Thermo-regulation of genes mediating motility and plant interactions in *Pseudomonas syringae*. *PLoS ONE* 8:e59850. doi: 10.1371/journal.pone.0059850

30. Hwang, M. S. H., Morgan, R. L., Sarkar, S. F., Wang, P. W., and Guttman, D. S. (2005). Phylogenetic characterization of virulence and resistance phenotypes of *Pseudomonas syringae*. *Appl. Environ. Microbiol.* 71, 5182–5191. doi: 10.1128/AEM.71.9.5182-5191.2005

31. Ikene, I., Grondeau, C., Bureau, C., and Samson, R. (2003). "*Pseudomonas* leek blight: study of seed transmission," in *Pseudomonas Syringae and Related Pathogens*, eds N. S. lacobellis, A. Collmer, S. W. Hutcheson, J. W. Mansfield, C. E. Morris, J. Murillo, N. W. Schaad, D. E. Stead, G. Surico, and M. S. Ullrich (Dordrecht: Springer), 93–95. doi: 10.1007/978-94-017-0133-4_10

32. Iriarte, F. B., Obradovic, A., Wernsing, M. H., Jackson, L. E., Balogh, B., Hong, J. A., et al. (2012). Soil-based systemic delivery and phyllosphere *in vivo* propagation of bacteriophages: two possible strategies for improving bacteriophage persistence for plant disease control. *Bacteriophage* 2, 215–224. doi: 10.4161/bact.23530

33. Jang, H. B., Fagutao, F. F., Nho, S. W., Park, S. B., Cha, I. S., Yu, J. E., et al. (2013). Phylogenomic network and comparative genomics reveal a diverged member of the KZ-related group, marine *Vibrio* phage JM-2012. *J. Virol.* 87, 12866–12878. doi: 10.1128/JVI.02656-13

34. Janse, J. D. (1982). Enkele nieuw waargenomen aantastingen van *Pseudomonas syringae*. *Verslagen En Mededelingen Plantenziektenkundige Dienst.* 158, 34–35.

35. Jones, J. B., Jackson, L. E., Balogh, B., Obradovic, A., Iriarte, F. B., and Momol, M. T. (2007). Bacteriophages for plant disease control. *Annu. Rev. Phytopathol.* 45, 245–262. doi: 10.1146/annurev.

phyto.45.062806.094411

36. Kasman, L. M., Kasman, A., Westwater, C., Dolan, J., Schmidt, M. G., and Norris, J. S. (2002). Overcoming the phage replication threshold: a mathematical model with implications for phage therapy. *J. Virol.* 76, 5557–5564. doi: 10.1128/JVI.76.11.5557-5564.2002
37. Kelley, L. A., and Sternberg, M. J. E. (2009). Protein structure prediction on the web: a case study using the phyre server. *Nat. Protoc.* 4, 363–371. doi: 10.1038/nprot.2009.2
38. Koike, S. T., Barak, J. D., Hendersen, D. M., and Gilbertson, R. L. (1999). Bacterial blight of leek: a new disease in California caused by *Pseudomonas syringae*. *Plant Dis.* 83, 165–170. doi: 10.1094/PDIS.1999.83.2.165
39. Kropinski, A. M., Prangishvili, D., and Lavigne, R. (2009). Position paper: the creation of a rational scheme for the nomenclature of viruses of bacteria and archaea. *Environ. Microbiol.* 11, 2775–2777. doi: 10.1111/j.1462-2920.2009.01970.x
40. Laslett, D., and Canback, B. (2004). ARAGORN, a program to detect tRNA genes and tmRNA genes in nucleotide sequences. *Nucleic Acids Res.* 32, 11–16. doi: 10.1093/nar/gkh152
41. Lavigne, R., Ceyssens, P. J., and Robben, J. (2009a). "Chapter 14 phage proteomics: applications of mass spectrometry," in *Bacteriophages Methods and Protocols*, Vol. 2, eds M. R. J. Clokie and A. Kropinski (Dordrecht: Humana Press), 239–251.
42. Lavigne, R., Darius, P., Summer, E. J., Seto, D., Mahadevan, P., Nilsson, A. S., et al. (2009b). Classification of *Myoviridae* bacteriophages using protein sequence similarity. *BMC Microbiol.* 9:224. doi: 10.1186/1471-2180-9-224
43. Lavigne, R., Seto, D., Mahadevan, P., Ackermann, H. W., and Kropinski, A. M. (2008). Unifying classical and molecular taxonomic classification: analysis of the podoviridae using BLASTP-based tools. *Res. Microbiol.* 159, 406–414. doi: 10.1016/j.resmic.2008.03.005
44. Lavigne, R., Sun, W. D., and Volckaert, G. (2004). PHIRE, a deterministic approach to reveal regulatory elements in bacteriophage genomes. *Bioinformatics* 20, 629–635. doi: 10.1093/

bioinformatics/btg456

45. Le, S., He, X., Tan, Y., Huang, G., Zhang, L., Lux, R., et al. (2013). Mapping the tail fiber as the receptor binding protein responsible for differential host specificity of *Pseudomonas aeruginosa* bacteriophages PaP1 and JG004. *PLoS ONE* 8:e68562. doi: 10.1371/journal.pone.0068562

46. Lelliott, R. A. (1952). A new bacterial disease of leeks. *Plant Pathol.* 1, 84–85. doi: 10.1111/j.1365-3059.1952.tb00035.x

47. Lima-Mendez, G., Van Helden, J., Toussaint, A., and Leplae, R. (2008). Reticulate representation of evolutionary and functional relationships between phage genomes. *Mol. Biol. Evol.* 25, 762–777. doi: 10.1093/molbev/msn023

48. Liu, M., Gingery, M., Doulatov, S. R., Liu, Y., Hodes, A., Baker, S., et al. (2004). Genomic and genetic analysis of *Bordetella* bacteriophages encoding reverse transcriptase-mediated tropism-switching cassettes. *Society* 186, 1503–1517. doi: 10.1128/jb.186.5.1503-1517.2004

49. Lowe, T. M., and Eddy, S. R. (1997). tRNAscan-SE: a program for improved detection of transfer RNA genes in genomic sequence. *Nucleic Acids Res.* 25, 955–964. doi: 10.1093/nar/25.5.0955

50. Lukashin, A. V., and Borodovsky, M. (1998). GeneMark.hmm: new solutions for gene finding. *Nucleic Acids Res.* 26, 1107–1115. doi: 10.1093/nar/26.4.1107

51. Macke, T. J., Ecker, D. J., Gutell, R. R., Gautheret, D., Case, D. A., and Sampath, R. (2001). RNAMotif, an RNA secondary structure definition and search algorithm. *Nucleic Acids Res.* 29, 4724–4735. doi: 10.1093/nar/29.22.4724

52. Myung, I. S., Joa, J. H., and Shim, H. S. (2011). Bacterial leaf spot of onion caused by *Pseudomonas syringae* pv. *porri*, a new disease in Korea. *Plant Dis.* 95, 1311. doi: 10.1094/PDIS-03-11-0250

53. Myung, I. S., Lee, Y. K., and Shim, H. (2012). Bacterial blight of shallot, caused by *Pseudomonas syringae* pv. *porri*, a new disease in Korea. *Plant Pathol. J.* 28, 454. doi: 10.5423/PPJ.DR.08.2011.0149

54. Noble, D. H., Cother, E. J., Hailstones, D. L., Flack, M., Oxspring, L., and Hall, B. (2006). Characterisation of *Pseudomonas*

*syringae*strains associated with a leaf disease of leek in Australia. *Eur. J. Plant Pathol.* 115, 419–430. doi: 10.1007/s10658-006-9033-4

55. Penadés, J. R., Chen, J., Quiles-Puchalt, N., Carpena, N., and Novick, R. P. (2015). Bacteriophage-mediated spread of bacterial virulence genes. *Curr. Opin. Microbiol.* 23, 171–178. doi: 10.1016/j.mib.2014.11.019

56. Petty, N. K., Foulds, I. J., Pradel, E., Ewbank, J. J., and Salmond, G. P. C. (2006). A generalized transducing phage (phiIF3) for the genomically sequenced *Serratia marcescens* strain Db11: a tool for functional genomics of an opportunistic human pathogen. *Microbiology* 152, 1701–1708. doi: 10.1099/mic.0.28712-0

57. Pietrzak, U., and McPhail, D. C. (2004). Copper accumulation, distribution and fractionation in vineyard soils of Victoria, Australia. *Geoderma* 122, 151–166. doi: 10.1016/j.geoderma.2004.01.005

58. Ravensdale, M., Blom, T. J., Gracia-Garza, J. A., Svircev, A. M., and Smith, R. J. (2007). Bacteriophages and the control of *Erwinia carotovora* subsp. carotovora. *Can. J. Plant Pathol.* 29, 121–130. doi: 10.1080/07060660709507448

59. Rombouts, S., Van Vaerenbergh, J., Volckaert, A., Baeyen, S., De Langhe, T., Declercq, B., et al. (2015). Isolation and characterization of *Pseudomonas syringae pv.* porri from Leek in Flanders. *Eur. J. Plant Pathol.* 144, 185–198. doi: 10.1007/s10658-015-0761-1

60. Sambrook, J., and Russell, D. (2001). *Molecular Cloning: A Laboratory Manual.* Cold Spring Harbor, NY: Laboratory Press.

61. Samson, R., Shafik, H., Benjama, A., and Gardan, L. (1998). Description of the bacterium causing blight of leek as *Pseudomonas syringae* pv. *porri* (pv. Nov.). *Phytopathology* 88, 844–850. doi: 10.1094/PHYTO.1998.88.8.844

62. Sillankorva, S., Oliveira, R., Vieira, M. J., Sutherland, I., and Azeredo, J. (2004). *Pseudomonas fluorescens* infection by bacteriophage Φs1: the influence of temperature, host growth phase and media. *FEMS Microbiol. Lett.* 241, 13–20. doi: 10.1016/j.femsle.2004.06.058

63. Tamura, K., Peterson, D., Peterson, N., Stecher, G., Nei, M., and Kumar, S. (2011). MEGA5: molecular evolutionary genetics

analysis using maximum likelihood, evolutionary distance, and maximum parsimony methods. *Mol. Biol. Evol.* 28, 2731–2739. doi: 10.1093/molbev/msr121

64. van Overbeek, L. S., Nijhuis, E. H. M., Koenraadt, H., Visser, J., and van Kruistum, G. (2010). The role of crop waste and soil in *Pseudomonas syringae* pv. *porri* infection of leek (*Allium Porrum*). *Appl. Soil Ecol.* 46, 457–463. doi: 10.1016/j.apsoil.2010.09.004
65. Varvaro, L. (1983). Una batteriosi del porro *Allium porrum* in Puglia. *Phytophatol. Mediterr.* 22, 86–88.

CHAPTER

11

CONTROL OF PIERCE'S DISEASE BY PHAGE

Mayukh Das[1,2], Tushar Suvra Bhowmick[1,2], Stephen J. Ahern[1,2], Ry Young[2,3], Carlos F. Gonzalez[1,2]

[1]Department of Plant Pathology and Microbiology, Texas A&M University, College Station, Texas, United States of America,

[2]Center for Phage Technology, Texas A&M University, College Station, Texas, United States of America,

[3]Department of Biochemistry and Biophysics, Texas A&M University, College Station, Texas, United States of America

ABSTRACT

Pierce's Disease (PD) of grapevines, caused by *Xylella fastidiosa* subsp. *fastidiosa* (*Xf*), is a limiting factor in the cultivation of grapevines in the US. There are presently no effective control methods to prevent or treat PD. The therapeutic and prophylactic efficacy of a phage cocktail composed of four virulent (lytic) phages was evaluated for control of PD. *Xf* levels in grapevines were significantly reduced in therapeutically or prophylactically treated grapevines. PD symptoms ceased to progress one week post-therapeutic treatment and symptoms were not observed in prophylactically treated grapevines. Cocktail phage levels increased in grapevines in the presence of the host. No *in planta* phage-resistant *Xf* isolates were obtained. Moreover, *Xf* mutants selected for phage resistance *in vitro* did not cause PD symptoms. Our results indicate that phages have great potential for biocontrol of PD and other economically important diseases caused by *Xylella*.

INTRODUCTION

Pierce's Disease (PD) caused by the xylem-limited, gamma-proteobacterium *Xylella fastidiosa*subsp. *fastidiosa* (*Xf*) [1], is a severe threat to the wine industry in the United States. *Xf* is transmitted by xylem sap-feeding insect vectors and colonizes the xylem vessels of host plants by movement through bordered pits to neighboring vessels. The formation of biofilms within the vessels results in plugging and disease [2]. Existing disease control methods are often only partially successful. Other than rouging of infected grapevines, the most widely used current practice relies on controlling the vectors through widespread application of systemic insecticides such as neonicotinoids that are absorbed by plants and transferred throughout the vascular system [3,4]. Neonicotinoids are neuro-active insecticides that derive their toxicity by acting mainly agonistically on nicotinic acetylcholine receptors on the post-synaptic membrane [5,6,7]. The neonicotinoid pesticide, imidacloprid, is one of the most widely used insecticides in the world and is recommended for the control of the glassy winged sharpshooter (*Homalodisca coagulate*), a major vector of *Xf* [3,5]. However, recent concerns of the role of neonicotinoids in honey bee colony collapse disorder [8] has resulted in a two year ban of three neonicotinoid insecticides (clothianidin, imidacloprid and thiamethoxam) by the European Commission because of the acute and chronic effects on bee colony survival and development [9]. In 2013, the US Environmental Protection Agency issued a notification to registrants on registered pesticides containing imidacloprid, dinotefuran, clothianidin or thiamethoxam advising of strengthen pollinator protective labeling on neonicotinoid products [10].

Clearly there is a need to develop alternative methods for control PD. Current approaches include the development of transgenic grapevines and rootstocks resistant to PD [11,12,13,14] and the use of avirulent *Xf* strains to protect susceptible grapevines [15]. We submit that the used of bacteriophages (phages) offers an environmentally-friendly, effective, and sustainable biocontrol method for PD.

Phages are the most abundant and ubiquitous genetic entity on earth [16,17]. Besides ubiquity, two major properties of phages, specificity and exponential propagation, make them attractive as antibacterial agents. Specifically, a cocktail of phages exhibiting broad host range activity that reflects a diversity of receptors would maximize the potency of the treatment and minimize the possibility for development of resistance. Well-developed phage cocktails can act as a natural biocontrol for bacterial infections, targeting a pathogenic bacterium without affecting the animal or plant host or their commensal microflora. Guidelines to the development of phage-based biocontrol strategies are yet to be formalized but it is already generally accepted that temperate phages should not be candidates for therapeutic use because of their potential capacity for specialized or generalized transduction of pathogenesis determinants [18,19].

In this study, a cocktail of phages previously shown to be incapable of lysogeny [20] was evaluated for therapeutic and prophylactic treatment of PD in *Vitis vinifera* in greenhouse studies. Considering the practical application of the phage biocontrol system, we also addressed the potential for the development of phage-resistant mutants. This is the first report for the use of phages to treat or prevent PD of grapevines.

MATERIALS AND METHODS

Bacterial Strains, Phages and Inoculum Preparation

All bacterial strains used in this study are listed. *Xf* strain Temecula 1 (*Xf*-T1) was used in greenhouse experiments. 50 *Xf* isolates obtained from grapevine (*V. vinifera*) samples expressing PD symptoms from vineyards in Santa Clara, Napa, Sonoma counties, CA and Uvalde County, TX were used in molecular typing and phage host range testing. Cultures were maintained on PW-M agar (PW-MA) medium as described by Ahern *et al.* (2014) [20,21]. For grapevine inoculations, five-day-old cultures of the *Xf* isolates were used to make bacterial suspensions in P-buffer [15]. Phages Sano, Salvo, Prado and Paz were used in this study [20]. High-titer phage lysates (10^{10} PFU/ml) were prepared and titered as described by Ahern *et al.* (2014) [20]. Lysates were stored at 4°C.

Ethics Statement

A United States Department of Agriculture-Animal and Plant Health Inspection Service (USDA-APHIS) Permit was issued for interstate transfer of isolates and/or plant material collected in locations. Transfer of bacterial isolates and plant material was authorized by USDA-APHIS permit P526P-11-02320. California isolates were collected on land owned or leased by Ridge Vineyards in Santa Clara, Napa and Sonoma counties, CA. Texas isolates were collected on land in Uvalde County, TX owned by the Texas A&M AgriLife Research and Extension Service. No endangered or protected species were sampled. No animals were involved in this study.

Phage Host Range And Host Simple Sequence Repeat (SSR) Analysis

SSR analysis was performed to identify the genetic diversity within the same 50 *Xf* isolates using 11 of the SSR loci described by Lin *et al.* (2005) [22]. *Xf*-T1, *X. fastidiosa* subsp. *sandyi* (strain Ann-1) and *X. fastidiosa* subsp. *multiplex* (strain Dixon) were also used for SSR analysis. The host range of cocktail phages was determined by the serial dilution spot assay on overlays as described by Ahern *et al.* [20] using the panel of 50 *Xf* isolates as hosts.

Grapevine Growth Conditions

Dormant *V. vinifera* cv. Cabernet Sauvignon clone 08 on 1103P rootstock were purchased from Vintage Nurseries (Wasco, California, USA). Grapevines were planted in 7-gallon pots using 101 Sunshine Mix 1 (Sun Gro Horticulture, Vancouver, British Columbia, Canada). Grapevines were grown in a greenhouse on a 16 h light (26°C, 300–400 $\mu E\ m^{-2}\ s^{-1}$)/8 h dark (18°C) cycle supplemented with illumination from sodium vapor lamps. Grapevines were watered every other day with tap water and fertilized with Peter's General Purpose 20-20-20 fertilizer and micronutrients every 15 d. Grapevines were progressively pruned to provide uniform plants with two unbranched solitary shoots. Lateral shoots and buds were removed. For each grapevine, two cordons were staked and allowed to grow until each cordon was ~80 cm in length before grapevines were used for experiments (Fig 1A).

Figure 1: Grapevines, points of inoculation and harvest method.

(A) Trained grapevine exhibiting two cordons. POI indicated with red arrows. (B) Grapevine root, trunk and cordon segments were numbered from POI (0), below (-) or above (+) in ~13 cm segments. Root portion was divided into three segments and numbered as R0, R-1, or R-2.

Grapevine Inoculation with Bacteria and Phage

Three independent experiments in triplicate were conducted to determine phage efficacy using identical protocols. One-way analyses of variance (ANOVAs) were performed as described below. No significant difference ($P > 0.05$) was found among the three experiments using the above analysis, we therefore report the results from a single representative trial.

The trial was conducted using a phage cocktail (Sano, Salvo, Prado and Paz) to evaluate phage therapeutic and prophylactic efficacy.

For disease control and to evaluate pathogen movement, 9 grapevines inoculated with *Xf*-T1 were assayed in triplicate immediately after inoculation (0 min) and at weeks 8, and 12 post-pathogen inoculation. Individual cordons were inoculated between the second and the third node on opposite sites (two points/cordon) with 40 μl of bacterial suspension (1×10^9 CFU/ml) using the needle inoculation technique as described by Hopkins (2005) [15] (Fig 1A). Control grapevines were mock inoculated with P-buffer using the same protocol. Grapevines were not watered 4 d prior inoculation to insure uptake of the inoculation.

To determine phage distribution and the effect of the phage cocktail on grapevines, 21 grapevines were treated with cocktail (1×10^{10} PFU/ml) as described above. Grapevines treated only with phage were assayed in triplicate immediately after treatment and at every two weeks for 12 weeks.

In the therapeutic phage treatment, 15 grapevines were inoculated with *Xf*-T1 as above. At week 3 post-pathogen inoculation, *Xf*-T1 inoculated 15 grapevines were treated with 40 μl of a phage cocktail suspension (1×10^{10} PFU/ml) using the same inoculation protocol and technique as above. *Xf* and phage levels were evaluated (in triplicate grapevines) at the time of inoculation (0 min), 6, 8, 10 and 12 weeks, as described below.

In the prophylactic phage treatment, 15 grapevines were treated with phage cocktail using the same inoculation protocol as above, and at week 3 post-phage treatment grapevines were inoculated with *Xf*-T1 as above. Grapevines were assayed in triplicate as described below immediately after treatment (0 min) and at weeks 6, 8, 10, and 12 post treatment. All grapevines were evaluated for disease symptom development twice weekly for 12 weeks. Grapevines were scored as positive for PD symptoms when a single leaf exhibited symptoms. Grapevines inoculated with P-buffer were assayed to confirm the absence of *Xf*and/or phage by quantitative real time PCR (qRT-PCR), as described below.

Sample Collection and Processing

Grapevine samples were designated, collected and processed as shown in Fig 1B. Briefly, duplicate cordons and root system from each grapevine were excised from the trunk and divided into 5–6 segments (~13 cm each). Each segment was homogenized using a 20 × 115 mm generator (PRO Scientific, CT, USA) in 15 ml of P-buffer, filtered through sterile cheesecloth to remove plant debris and centrifuged (10,000 × *g* at 4°C for 15 min). Pellets were resuspended into 1 ml of sterile double distilled water (ddH$_2$O) for bacterial DNA extraction for qRT-PCR or plating to selective media. Supernatants were used for phage DNA extraction for qRT-PCR or direct plating to quantify PFU (see below).

qRT-PCR

Grapevine tissue extracts were assayed for phage and *Xf* by qRT-PCR. The propidium monoazide (PMA) protocol as described by Nocker *et al.* (2006) was used to detect viable cells of *Xf* [23].

Phage-Resistant *Xf* Mutants

To screen for the occurrence of phage-resistant *Xf* mutants *in planta*, isolates were obtained from both for therapeutically and prophylactically treated grapevines at 12 or 9 weeks post-*Xf*-T1 inoculation, respectively. All isolates were confirmed at the species and subspecies level using PCR analysis as described by Hernandez-Martinez *et al.* (2006) [24] and were screened for phage sensitivity by the serial dilution spot assay on overlays and soft agar overlay method as described by Ahern *et al.* (2014) [20]. Additionally, *in vitro* phage-resistant mutants of *Xf*-T1 were selected for using soft agar overlay method as described by Ahern *et al.* (2014) [20] with modifications. Briefly, 100μl of *Xf*-T1 (1 × 10^8 CFU/ml) and 100μl (1 × 10^8 PFU/ml) of each cocktail phage were individually mixed in 7 ml of PW-MA soft agar and poured on PW-MA and incubated at 28°C. Plates were evaluated for growth of single colonies for up to 10 d. Colonies were streak purified three times and confirmed at the species and subspecies level using PCR analysis as described above. Phage sensitivity was determined by the serial dilution spot assay on overlays and soft agar overlay method as described by Ahern *et al.* (2014) [20].

Screening of Twitching Motility

Twitching motility of *Xf*-T1 and derivatives was examined as previously described by Ahern *et al.* (2014) [20].

Movement and Pathogenicity of Phage-Resistant *Xf* Mutants

Phage-resistant mutants were evaluated for ability to move through grapevines and cause PD symptoms. 15 grapevines each were inoculated with *Xf*-T1 or phage-resistant *Xf*-T1 mutants as described above. Segments from triplicate grapevines were assayed at week 4, 6, 8, and 10 post-inoculation and assayed by qRT-PCR as described above. Grapevines were evaluated for disease symptoms weekly for 12 weeks.

Statistical Analysis

The mean and standard deviation (SD) were calculated from the levels of phages Sano, Salvo Prado, and Paz, and/or *Xf*-T1 obtained from triplicate grapevine samples. ANOVAs were performed using split-plot design to test for significant variation between vine segments across treatments for three independent experiments. Tukey's Honestly Significant Difference (HSD) test and Student's *t* test were used for pairwise comparisons in order to test for significant differences between PFUs of grapevines treated only with cocktail verses therapeutically treated grapevines with cocktail or PFUs of grapevines treated only with cocktail verses grapevines prophylactically treated with cocktail. The same tests were used to assess the significant difference between CFUs of grapevines inoculated only with *Xf*-T1verses grapevines treated therapeutically with cocktail or CFUs of grapevines inoculated only with *Xf*-T1 verses grapevines prophylactically treated with cocktail. Significant differences are noted by asterisks (*) in figures. ANOVAs and HSD were performed using Statistix 9.0 (http://www.statistix.com/). Student's *t* test was performed using R-software version 2.15.3 (http://www.r-project.org).

RESULTS

Phage Host Range and SSR Analysis

The host range of *Xf* phages Sano, Salvo, Prado and Paz [20] was evaluated using a collection of 10 Texas and 40 California *Xf* isolates representative of 20 SSR typing groups. There was no observed correlation between phage sensitivity and SSR groups. However, phages Sano, Salvo, Prado and Paz formed plaques on 25, 46, 48 and 31 of the 50 *Xf* isolates and on 11, 20, 20 and 16 SSR groups, respectively. The observed overlap indicated that the phages comprising the cocktail had broad host range activity individually and collectively. The four cocktail phages were also able to form plaques on three *X. fastidiosa*isolates from coffee plants (*Coffee arabica*) exhibiting coffee leaf scorch symptoms and on *X. fastidiosa* strains Ann-1 and Dixon causing oleander and almond leaf scorch, respectively [20].

Disease development and movement of *Xf* in grapevines

Characteristic leaf scorching symptoms were first observed on individual leaves of *Xf*-T1 inoculated, non-cocktail treated, grapevines at week 3–4. The leaf symptoms further developed into typical banding pattern of dark red to orange (Fig 2A–2E). Grapevines inoculated with only *Xf*-T1 exhibited full grapevine PD symptoms by week 8 and 12 post-inoculation (Fig 2A and 2B). Control grapevines inoculated with P-buffer [20] did not exhibit PD symptoms (Fig 2F). qRT-PCR of extracts from duplicate cordons representing triplicate samples of the grapevines inoculated with *Xf*-T1 showed movement of the pathogen throughout the grapevine segments assayed (Fig 3A and 3B, White bars). *Xf*-T1 was detected at the points of inoculation (POI) in cordons (~10^3 CFU/ gm of tissue (gt)), the lower segments (including root systems) at low concentrations (~10^2–10^3 CFU/gt) and at higher levels (maxima ~10^5 CFU/gt) in the upper segments of the trunks at week 8 post-inoculation (Fig 3A, White bars). At week 12 post-inoculation, *Xf*-T1 was detected in the root systems at low concentrations (~10^2 CFU/gt) and in upper segments of both cordons at higher concentrations (~10^3–10^6 CFU/gt; Fig 3B, White bars). Control grapevines (P-buffer inoculated) were confirmed as *Xf* free by qRT-PCR.

Figure 2: Symptoms of *Xf* **in phage non-treated and phage-treated grapevines.** (A,B) *Xf* disease control grapevines inoculated with *Xf*-T1. Symptoms at week 8 (A) and 12 (B) post-pathogen inoculation. (C–E) Grapevine leaves exhibiting PD symptoms. (F) Control grapevine inoculated with P-buffer at week 8 (no symptoms). (G) Control grapevine treated only with cocktail phage at week 12 (no symptoms). (H,I) *Xf*-T1 inoculated and phage cocktail treated grapevines. Symptoms at week 8 (H) and week 12 (I) post-pathogen inoculation.

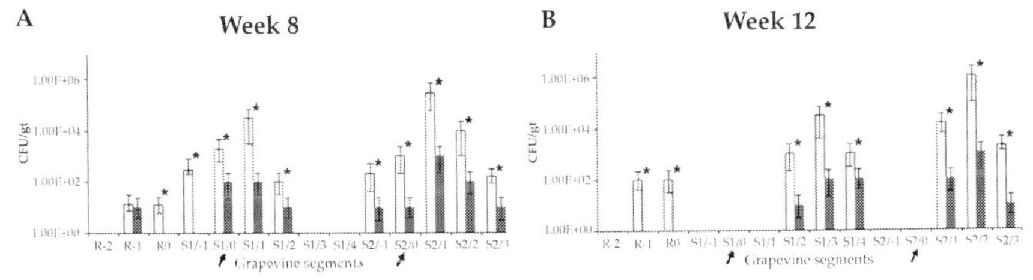

Figure 3: Disease development, movement of *Xf* **and therapeutic efficacy of phage cocktail.** (A,B) Quantitative levels of *Xf*-T1 at week 8 (A) and 12 (B) post-inoculation of grapevines with only *Xf*-T1 (White bars) (symptoms Fig 2A and 2B, respectively) or *Xf*-T1 and treated with phage cocktail at week 3 post-*Xf*-T1 inoculation (Grey bars) (symptoms Fig 2H and 2I, respectively). POI (indicated with arrows) and grapevine segments numbered as POI (0), below (-) or above (+) in ~13 cm segments. Root divided into three segments and numbered as R0, R-1, or R-2. Segments of similar proximal distance from the graft point from triplicate grapevines were assayed using qRT-PCR to determine the mean CFU/gt ± s.d. Each bar represents s.d. Significant differences are noted by asterisks (*). qRT-PCR line plots obtained for *Xf*-T1 had R^2values of greater than 0.9 and efficiencies of 157%.

Phage Distribution and Persistence in Grapevines

It was of interest to determine the distribution of cocktail phages in the absence of a host in grapevines, since no prior reports existed. Phage levels in grapevines treated only with cocktail (three grapevines with two cordons each) were quantitatively assessed to determine phage distribution over a 12 week period (Fig 4). The highest concentration of cocktail phages was detected in segments at the POI of cordons (~10^5–10^6 PFU/gt) at week 2 post-treatment, with lower levels detected (~10^2–10^3 PFU/gt) in the roots and trunk segments ~12 cm above and below the POI (Fig 4A). During week 4, further distribution of phages was detected at ~25 cm above and below the POI, with detection (maxima ~10^2 PFU/gt) only at ~38 cm above the POI by week 6 (Fig 4B and 4C). Phage levels reached maximum distribution at week 8 (~50 cm above and below the POI) and then declined during weeks 10–12 in the absence of a host to an average of 10^1–10^2 PFU/gt in each cordon (Fig 4D–4F). However, the cocktail phages (~4×10^1 PFU/gt) were still detected in petioles collected at ~25–38 cm above the POI at week 12. Grapevines treated only with cocktail did not exhibit disease symptoms at any time during the test period (Fig 2G), nor were cocktail phages detected by qRT-PCR in P-buffer-inoculated grapevines.

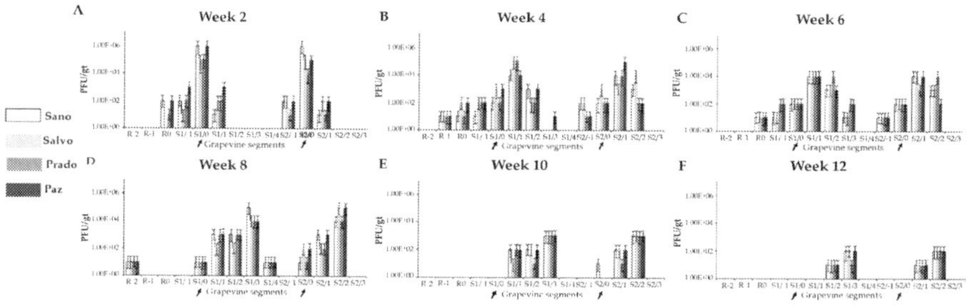

Figure 4: Distribution and persistence of cocktail phages in grapevines. (A–F) Distribution of individual cocktail phage (color coded to left) in grapevines was determined at week 2 (A), 4 (B), 6 (C), 8 (D), 10 (E) and 12 (F) post-inoculation. POI (indicated with arrows) and grapevine segments numbered as POI (0), below (-) or above (+) in ~13 cm segments. Root divided into three segments and numbered as R0, R-1, or R-2. Segments of similar proximal distance from the graft point from triplicate grapevines were assayed using qRT-PCR to determine the mean PFU/gt ± s.d. Each bar represents s.d. Standard qRT-PCR line plots obtained for cocktail phage Sano, Salvo, Prado and Paz had R^2 values of greater than 0.9 and efficiencies of 129%, 120%, 127% and 123%, respectively.

Therapeutic efficacy

To assess potential therapeutic efficacy of the phage cocktail, *Xf*-T1 infected grapevines were treated with phage cocktail at 3 weeks post-pathogen inoculation. At week 8 and 12 (week 5 and 9 post-cocktail treatment, respectively) the level of *Xf*-T1 was significantly lower (10–1000 fold, $P < 0.05$) in treated grapevines (Fig 3A and 3B, Grey bars) as compared to non-treated grapevines (Fig 3A and 3B, White bars). Non-therapeutically treated grapevines showed leaf scorching symptoms as early as week 4, with all grapevines exhibiting typical PD symptoms by week 7–8. In contrast, the cocktail treated grapevines did not further develop symptoms after week 4 (1 week post-cocktail treatment) and were asymptomatic for the duration of the experiment (Fig 2H and 2I). The therapeutic effect was correlated with a 10–200 fold ($P < 0.05$) increase in the level of the cocktail phages during week 6 through 12 (Fig 5A–5D,

Crosshatched bars) as compared to controls that were treated only with phage cocktail (Fig 5A–5D, Grey bars).

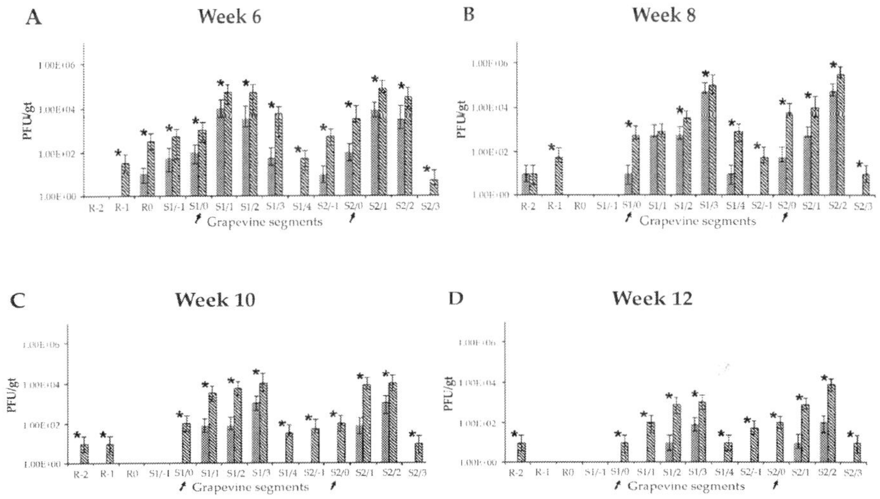

Figure 5: Persistence and replication of cocktail phages in therapeutic study. (A–D) Average quantitative levels of cocktail phages in grapevines treated only with cocktail (Grey bars) and in *Xf*-T1 inoculated and cocktail treated (3 week post-*Xf*-T1 inoculation) grapevines (Crosshatched bars) at week 6 (A), 8 (B), 10 (C) and 12 (D) post-*Xf*-T1 inoculation. POI (indicated with arrows) and grapevine segments numbered as POI (0), below (-) or above (+) in ~13 cm segments. Root divided into three segments and numbered as R0, R-1, or R-2. Segments of similar proximal distance from the graft point from triplicate grapevines were assayed to determine the mean PFU/gt ± s.d. of the four phages in cocktail. Each bar represents s.d. Significant differences are noted by asterisks (*).

Prophylactic Efficacy

Additional studies were conducted with grapevines treated with the phage cocktail and then inoculated with *Xf*-T1 at week 3 post-treatment to determine the prophylactic efficacy of the phage cocktail. In these grapevines, pathogen levels reached a maxima of ~10^3 CFU/gt (Fig 6A and 6B, Grey bars) in the segments of the grapevines assayed at week 8 and 12 post-phage cocktail treatment, compared to a maxima of ~10^6 CFU/gt (Fig 6A and 6B, White bars) in non-prophylactically treated grapevines. Moreover, prophylactically treated grapevines showed no

PD symptoms at week 8 or 12 (data not shown). As above, the levels of cocktail phages were 10–200 fold higher ($P < 0.05$) in prophylactically treated grapevines compared to cocktail only treated grapevines, and correlated with the prophylactic efficacy.

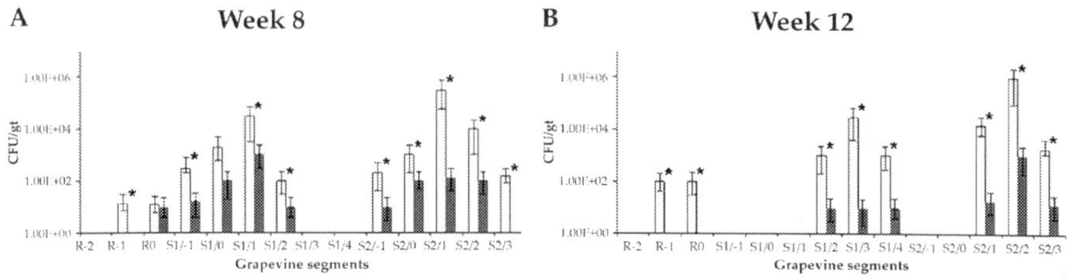

Figure 6: ***Xf*** **levels in non-treated and prophylactically treated grapevines.** (A,B) Quantitative levels of *Xf* at week 8 (A) and 12 (B) post-inoculation in grapevines inoculated only with *Xf*-T1 (White bars) or phage cocktail treated and inoculated with *Xf*-T1 at week 3 post-cocktail treatment (Grey bars). POI (indicated with arrows) and grapevine segments numbered as POI (0), below (-) or above (+) in ~13 cm segments. Root divided into three segments and numbered as R0, R-1, or R-2. Segments of similar proximal distance from the graft point from triplicate grapevines were assayed using qRT-PCR to determine the mean CFU/gt ± s.d. Each bar represents s.d. Significant differences are noted by asterisks (*).

Phage-Resistant *Xf* Mutants and Twitching Motility

At the termination of both therapeutic and prophylactic greenhouse studies at week 12, extracts from grapevine segments from each treatment were plated to determine if surviving *Xf*-T1 isolates were resistant to cocktail phages. Plating of plant extracts from therapeutically or prophylactically treated grapevines yielded an average of 4×10^2 CFU/gt and 3×10^2 CFU/gt, respectively (Figs 3B and 6B, respectively). Representative isolates (20 ea.) from each cordon of each of three grapevines from both therapeutic and prophylactic treatments were confirmed as *Xf* and retained full sensitivity to the cocktail phages (data not shown).

Since phage-resistant *Xf* mutants were not found *in planta*, spontaneous *Xf*-T1 mutants resistant to each of the four cocktail phages were selected *in vitro*. All mutants (43 isolates) were confirmed as *Xf* to subspecies level by PCR and were resistant to all four cocktail phages (data not shown). Of eight representative phage-resistant *Xf*-T1 derivatives tested, only one XF15.51 had lost twitching motility. Analysis of the type IV pilus gene clusters of XF15.51 revealed an IS1327 (PD1333; NC_004556.1) insertion in codon 24 of *pilQ*(PD1691; AAO29529.1), which encodes a protein in the outer membrane complex required for the extrusion and retraction of type IV pili [25]. XF15.51 and two of the phage-resistant mutants that retained twitching motility (XF15.7 and XF15.11) exhibited an absolute adsorption defect for all four cocktail phages, whereas the adsorption rate constants for the four phages to the parental *Xf*-T1 were ~4×10^{-12} ml cell^{-1} min^{-1}, as previously observed (data not shown) [20]. The simplest interpretation of these data is that the four cocktail phages share a requirement for both the type IV pilus, lacking in XF15.1 and an unidentified secondary receptor, defective in the other seven mutants. There was no detectable difference in the growth rate of the phage-resistant mutants as compared to the parental *Xf*-T1 strain in PW broth or solid medium (data not shown).

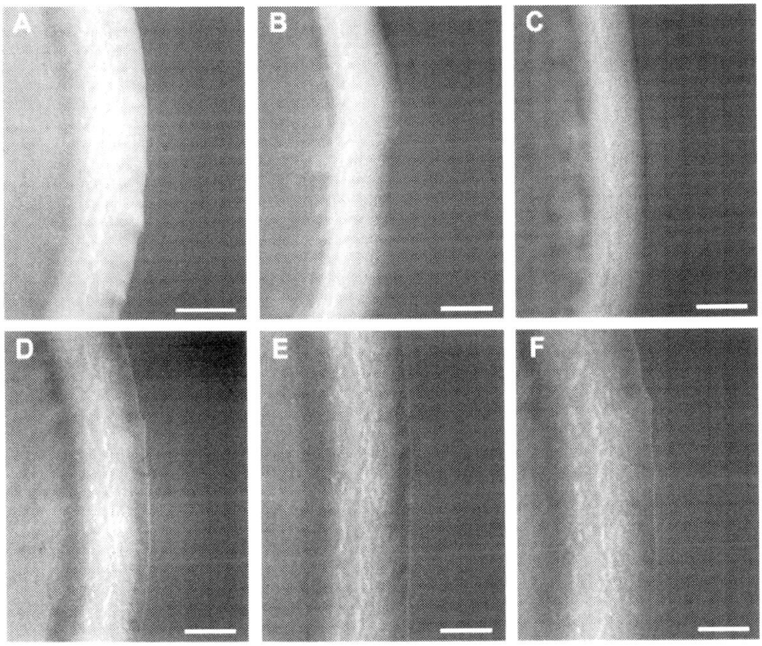

Figure 7: Colony morphology as indicator of twitching motility of wild-type *Xf*-T1 and phage-resistant derivatives. Colonies of the indicated isolates

were grown for 5 d (28°C) on the appropriate solid medium modified with 1.2% agar. Colonies with peripheral fringe are twitch positive. (A) wild-type *Xf*-T1 (B) *Xf* Temecula 1Δ*pilA* (C) XF15.51 (D) XF15.7 (E) XF15.11 (F) XF15.37. Bars, 0.1 mm.

Pathogenicity and Movement of *Xf* Phage-Resistant Mutants in Grapevines

The phage-resistant *Xf* mutants were tested in grapevines to determine pathogenicity and movement. Grapevines inoculated these derivatives did not develop PD symptoms over a 12 week post-inoculation period, whereas grapevines inoculated with the parental strain (*Xf*-T1) exhibited typical PD symptoms by week 7–8 (data not shown). Triplicate grapevines inoculated with each mutant were segmented and each segment was assayed by qRT-PCR at week 4, 6, 8, and 10 post-inoculation. The non-motile XF15.51 did not exhibit movement in grapevines beyond the POI, since functional type IV pili are required for movement and virulence (data not shown) [26]. In contrast, the other phage-resistant *Xf* that retained twitching motility colonized and showed movement throughout the grapevines (~38 cm above and below the POI) similar to wild type (*Xf*-T1). This suggests the secondary receptor or receptors used by these phages are essential for virulence.

DISCUSSION

Presently, there are no effective control strategies for PD. The most widely used measure for control of PD is the application of neonicotinoid pesticides that have been reported to adversely affect the winterization of honey bee colonies and contributing to colony collapse disorder [8]. In June 2013, a mass die-off of native pollinators attributed to applications of neonicotinoid insecticides led Oregon state authorities to adopt permanent restrictions on dinotefuran and imidacloprid [27]. Not only do neonicotinoids pose a threat to pollinators, but also they have been found widely in fruits and vegetables and have shown excitatory effects on cultured cerebellar neurons from neonatal rats, suggesting possible neurotoxicity in developing mammalian brains [28,29].

In this study, we demonstrated the therapeutic and prophylactic efficacy for the application of a cocktail composed of four virulent (lytic) *Xf* phages. The phage cocktail reduced the pathogen levels in grapevines and prevented PD symptom development. This is the first report of the application of a phage-based biocontrol strategy for the control of PD and may offer an alternative to the use of broad-acting insecticides for the control of vector-transmitted plant diseases caused by *Xylella* species.

The results show that the phages were able to be distributed in the grapevines after initial introduction, since they were detected in petioles 50 cm above the POI and in root systems 50 cm below the POI in grapevines treated only with phage. In the absence of the host, phage levels declined over a 12 week period (Fig 4). In contrast, in therapeutically and prophylactically treated grapevines phage levels increased 10–200 fold ($P < 0.05$) when a permissive host was present (Fig 5). To our knowledge, this is the first report of distribution, persistence and multiplication of phages in a plant vascular system.

A major concern regarding the use of phages as therapeutic agents is the development of phage resistance [18,30,31]. To address this concern, a therapeutic cocktail consisting of four lytic phages exhibiting differential host ranges was used in this study. In multiple *in planta*experiments, we have not detected phage-resistant mutants (data not shown). In fact, we have isolated phage from petioles of grapevines exhibiting PD symptoms [20]. *Xf* is known to persist in grapevines at low levels without causing disease [32], and thus the presence of *Xf* at low levels in the presence of low levels of phage would not be unexpected, since phage-host interactions depend on the concentrations of both viruses and cells [33]. However, phage-resistant derivatives obtained *in vitro* and tested in our system, were all avirulent, irrespective of retention of twitching motility. We have previously reported that the four cocktail phages are lytic and required the presence of type IV pili for adsorption [20]. It was therefore not unexpected to obtain phage-resistant mutants such as XF15.51, with an IS insertion in *pilQ*, and that these mutants would be avirulent *in planta*. However, all other phage-resistant mutants tested were also avirulent, despite the presence of functional type IV pili. This suggests that a secondary receptor is required for phage infection and that this secondary receptor is essential for virulence. Analysis of these mutants to identify the genes responsible is underway.

The implementation of a phage-based biocontrol system composed of a cocktail of lytic phages is a strategy for the treatment or prevention of PD. Furthermore, the target specific nature of the cocktail phages for *Xf* is not harmful to humans, animals, plants or associated beneficial microflora. *In vitro* experiments confirmed that *Xf* subspecies causing almond [20], oleander [20] or coffee leaf scorch were sensitive to the four cocktail phages. The developed phages may offer a treatment for the olive-quick decline syndrome recently reported in Italy that is attributed to *Xylella* [34]. The application of phages as biocontrol agents not only offers the wine industry an effective and environmentally safe method to control PD in grapevines, but also has the potential to control other diseases caused by *Xylella*.

ACKNOWLEDGMENTS

We thank Tsutomu Nishida for his support throughout this project. We thank Mr. Paul Draper and the staff at Ridge Vineyards for their involvement and support of the project. We thank John S. Hartung, USDA-ARS, for providing the *Xf* coffee isolates and Jason J. Gill for his critical reading of the manuscript.

REFERENCES

1. Janse JD, Obradovic A. Xylella fastidiosa: Its biology, diagnosis, control and risks. J Plant Pathol. 2010;92: S1.35–S1.48.
2. Newman KL, Almeida RP, Purcell AH, Lindow SE. Use of green fluorescent strain for analysis of Xylella fastidiosa colonization of Vitis vinifera. Appl Environ Microbiol. 2003;69: 7319–27. pmid:14660381
3. Byrne FJ, Toscano NC. Understanding the dynamics of neonicotinoid insecticidal activity against the glassy winged sharpshooter: development of target thresholds in grapevines. In Proceedings, 2010 Pierce's Disease Research Symposium, pp. 33–35. California Department of Food and Agriculture, San Diego, CA.
4. Starner K, Goh KS. Detections of Imidacloprid in surface waters of three agricultural regions of California, USA, 2010–2011. Bull Environ Cont Toxicol. 2012;88: 316–321. pmid:22228315
5. Van Dijk TC, Van Staalduinen MA, Van der Sluijs JP. Macro-

Invertebrate Decline in Surface Water Polluted with Imidacloprid. Plos One. 2013;8: e62374. pmid:23650513

6. Buckingham SD, Lapied B, Le Corronc H, Grolleau F, Sattelle DB. Imidacloprid actions on insect neuronal acetylcholine receptors. J Exp Biol. 1997;200: 2685–2692. pmid:9326496
7. Matsuda K, Buckingham SD, Kleier D, Rauh JJ, Sattelle DB. Neonicotinoids: insecticides acting on insect nicotinic acetylcholine receptors. Trends Pharmacol Sci. 2001;22: 573–580. pmid:11698101
8. Lu C, Warchol KM, Callahan RA. Sub-lethal exposure to neonicotinoids impaired honey bees winterization before proceeding to colony collapse disorder. Bull Insectol. 2014;67: 125–130.
9. Commission Implementing Regulation (EU) No. 485/2013 of 24 May 2013 amending Implementing Regulation (EU) No. 540/2011, as regards the conditions of approval of the active substances clothianidin, thiamethoxam and imidacloprid, and prohibiting the use and sale of seeds treated with plant protection products containing those active substances. Official Journal of the European Union L 139, 25 May 2013, pp. 12–26.
10. U.S. Environmental Protection Agency. Registered products containing imidacloprid, dinotefuran, clothianidin or thiamethoxam. 22 July 2013. Available: http://www2.epa.gov/sites/production/files/2013-11/documents/bee-july2013-letter.pdf.
11. Gilchrist D, Lincoln J. Pierce's disease control and bacterial population dynamics in wine grape varieties grafted on rootstocks expressing anti-apoptotic sequences. In Proceedings, 2010 Pierce's Disease Research Symposium, pp. 180–186. California Department of Food and Agriculture, San Diego, CA.
12. Kirkpatrick B. Identification and utilization of cold temperature induced grapevine metabolites to manage Pierce's disease. In Proceedings, 2010 Pierce's Disease Research Symposium, pp. 191–195. California Department of Food and Agriculture, San Diego, CA.
13. Lindow SE. Control of Pierce's disease using pathogen signal molecules. In Proceedings, 2010 Pierce's Disease Research Symposium, pp. 118–133. California Department of Food and Agriculture, San Diego, CA.

14. Aguero CB, Uratsu SL, Greve C, Powell AL, Labavitch JM, Meredith CP, et al. Evaluation of tolerance to PD and Botrytis in transgenic plants of Vitis vinifera L. expressing the pear PGIP gene. Mol Plant Path. 2005;6: 43–51. pmid:20565637
15. Hopkins DL. Biological control of Pierce's disease in the vineyard with strains of Xylella fastidiosa benign to grapevine. Plant Dis. 2005;89: 1348–1352.
16. Hendrix RW. Bacteriophages: Evolution of the Majority. Theor Popul Biol. 2002;61: 471–480. pmid:12167366
17. Suttle CA. Viruses in the sea. Nature. 2005;437: 356–361. pmid:16163346
18. Gill JJ, Hyman P. Phage choice, isolation, and preparation for phage therapy. Curr Pharm Biotechnol. 2010;11: 2–14. pmid:20214604
19. Kropinski AM. Phage therapy—Everything old is new again. Can J Infect Dis Med Microbiol. 2006;17: 297–306. pmid:18382643
20. Ahern SJ, Das M, Bhowmick TS, Young R, Gonzalez CF. Characterization of novel virulent broad-host-range phages of Xylella fastidiosa and Xanthomonas. J Bacteriol. 2014;196: 459–71. pmid:24214944
21. Summer EJ, Enderle CJ, Ahern SJ, Gill JJ, Torres CP, Appel DN, et al. Genomic and Biological Analysis of Phage Xfas53 and Related Prophages of Xylella fastidiosa. J Bacteriol. 2010;192: 179–190. pmid:19897657
22. Lin H, Civerolo EL, Hu R, Barros S, Francis M, Walker MA. Multilocus Simple Sequence Repeat Markers for Differentiating Strains and Evaluating Genetic Diversity of Xylella fastidiosa. Appl Environ Microbiol. 2005;71: 4888–4892. pmid:16085890
23. Nocker A, Cheung C, Camper AK. Comparison of propidium monoazide with ethidium monoazide for differentiation of live vs. dead bacteria by selective removal of DNA from dead cells. J Microbiol Meth. 2006;67: 310–20.
24. Hernandez-Martinez R, Costa HS, Dumenyo CK, Cooksey DA. Differentiation of strains of Xylella fastidiosa grape, almonds, and oleander using a multiprimer PCR assay. Plant Dis. 2006;90: 1382–1388.

25. Meng Y, Li Y, Galvani CD, Hao G, Turner JN, Burr TJ, et al. Upstream migration of Xylclla fastidiosa via pilus-driven twitching motility. J Bacteriol. 2005;187: 5560–5567. pmid:16077100
26. Li Y, Hao G, Galvani CD, Meng Y, De La Fuente L, Hoch HC, ct al. Type I and type IV pili of Xylella fastidiosa affect twitching motility, biofilm formation and cell-cell aggregation. Microbiology. 2007;153: 719–726. pmid:17322192
27. What's the problem with neonicotinoids? Retail nursery and garden center IPM news. March 2014. Available: http://uccemg.com/files/185197.pdf.
28. Chen M, Tao L, McLean J, Lu C. Quantitative Analysis of Neonicotinoid Insecticide Residues in Foods: Implication for Dietary Exposures. J Agric Food Chem. 2014;62: 6082–6090. pmid:24933495
29. Kimura-Kuroda J, Komuta Y, Kuroda Y, Hayashi M, Kawano H. Nicotine-like effects of the neonicotinoid insecticides acetamiprid and imidacloprid on cerebellar neurons from neonatal rats. Plos One. 2012;7: e32432. pmid:22393406
30. Smith HW, Huggins MB. Effectiveness of phages in treating experimental Escherichia coli diarrhoea in calves, piglets and lambs. J Gen Microbiol. 1983;129: 2659–2675. pmid:6355391
31. Levin BR, Bull JJ. Population and evolutionary dynamics of phage therapy. Nat Rev Microbiol. 2004,2: 166–173. pmid:15040264
32. Almeida RPP, Purcell AH. Biological traits of Xylella fastidiosa strains from grapes and almonds. Appl Environ Microbiol. 2003;69: 7447–7452. pmid:14660397
33. Loc-Carrillo C, Abedon ST. Pros and cons of phage therapy. Bacteriophage. 2011;1: 111–114. pmid:22334867
34. European Food Safety Authority. EFSA issues urgent advice on plant bacteria Xylella fastidiosa. 26 November 2013.

CHAPTER

12

GENOMIC, PROTEOMIC AND MORPHOLOGICAL CHARACTERIZATION OF TWO NOVEL BROAD HOST LYTIC BACTERIOPHAGES ΦPD10.3 AND ΦPD23.1 INFECTING PECTINOLYTIC PECTOBACTERIUM SPP. AND DICKEYA SPP.

Robert Czajkowski[1], Zofia Ozymko[1], Victor de Jager[2], Joanna Siwinska[1], Anna Smolarska[1], Adam Ossowicki[3], Magdalena Narajczyk[4], Ewa Lojkowska[1]

[1]Laboratory of Plant Protection and Biotechnology, Department of Biotechnology, Intercollegiate Faculty of Biotechnology, University of Gdansk and Medical University of Gdansk, Gdansk, Poland,

[2]Netherlands Institute of Ecology (NIOO-KNAW), Wageningen, the Netherlands,

[3]Laboratory of Biological Plant Protection, Department of Biotechnology, Intercollegiate Faculty of Biotechnology, University of Gdansk and Medical University of Gdansk, Gdansk, Poland,

[4]Laboratory of Electron Microscopy, Faculty of Biology, University of Gdansk, Gdansk, Poland

ABSTRACT

Pectinolytic *Pectobacterium* spp. and *Dickeya* spp. are necrotrophic bacterial pathogens of many important crops, including potato, worldwide. This study reports on the isolation and characterization of broad host lytic bacteriophages able to infect the dominant *Pectobacterium*spp. and *Dickeya* spp. affecting potato in Europe *viz. Pectobacterium carotovorum* subsp. *carotovorum* (Pcc), *P. wasabiae* (Pwa) and *Dickeya solani* (Dso) with the objective to assess their potential as biological disease control agents. Two lytic bacteriophages infecting stains of Pcc, Pwa and Dso were isolated from potato samples collected from two potato fields in central Poland. The ΦPD10.3 and ΦPD23.1 phages have morphology similar to other members of the *Myoviridae* family and the *Caudovirales* order, with a head diameter of 85 and 86 nm and length of tails of 117 and 121 nm, respectively. They were characterized for optimal multiplicity of infection, the rate of adsorption to the Pcc, Pwa and Dso cells, the latent period and the burst size. The phages were genotypically characterized with RAPD-PCR and RFLP techniques. The structural proteomes of both phages were obtained by fractionation of phage proteins by SDS-PAGE. Phage protein identification was performed by liquid chromatography-mass spectrometry (LC-MS) analysis. Pulsed-field gel electrophoresis (PFGE), genome sequencing and comparative genome analysis were used to gain knowledge of the length, organization and function of the ΦPD10.3 and ΦPD23.1 genomes. The potential use of ΦPD10.3 and ΦPD23.1 phages for the biocontrol of *Pectobacterium* spp. and *Dickeya* spp. infections in potato is discussed.

INTRODUCTION

Soft rot *Enterobacteriaceae* (SRE): *Pectobacterium* spp. and *Dickeya* spp. are ubiquitous necrotrophic bacterial pathogens of agriculturally important crops worldwide; they can be isolated from plants, soil, and water as well as from the surface of and within insects [1]. Due to the fact that SRE can seriously affect plant health and crop yield and because they are widely spread in the environment, these bacteria are ranked among the top ten most significant bacterial pathogens in agriculture [2].

In potato, the main soft rot *Enterobacteriaceae* causing blackleg, which affects the growing plant, and tuber soft rot of potato in storage and transit are *Pectobacterium carotovorum* subsp. *carotovorum* (Pcc), *Pectobacterium atrosepticum* (Pba) [3], *P. carotovorum* subsp. *brasiliense* (Pcb) [4], *P. wasabiae* (Pwa) [5] and several *Dickeya* spp. among which *D. solani* (Dso) has been recognized recently as the most important [5–7]. In Europe, blackleg and soft rot disease may result in relatively high losses in potato crops not only in the field and in storage but also indirectly in seed potato production due to declassification and rejection of seed lots [8].

In temperate climatic zones, the symptoms caused by the different SRE in potato are often difficult to discriminate [6, 9]. SRE bacteria are able not only to establish a systematic infection but also ultimately to kill the host plant [7, 9]. They characteristically produce large quantities of different extracellular plant cell wall degrading enzymes (e. g. cellulases, pectinases, proteases, pectate lyases, polygalacturonases, pectin methylesterases phospholipases) that allow the bacteria to infiltrate vascular compartments and result in a maceration (rotting) of plant host tissues [2, 6, 9].

Of all *Pectobacterium* spp. infecting potato, Pcc has the widest host range globally, whereas Pba is associated mainly with potato grown under temperate climate [10]. Recently, a highly aggressive bacterium Pcb was found to cause symptoms in potato in tropical and temperate regions [4, 11] and Pwa found for the first time in horse radish in Japan [5, 12] and was later on potato in New Zeeland, South Africa, Canada and several European countries [13, 14]. In contrast, *Dickeya* spp. can affect a number of plant species in different temperature zones [6], however, until recently it has been associated with blackleg and tuber soft rot only in tropical and subtropical regions. Until 2005, in temperate climates *Dickeya* spp. were considered of less importance and only weakly virulent strains of *D. dianthicola* were isolated from blackleg diseased plants in Western and Northern Europe [15]. However, due to the recent introduction of a new virulent *Dickeya* species—Dso—to the potato ecosystem [16], the situation has changed and now infection of potato by Dso is dominant in Europe, leading to even higher economic losses than in the past [17, 18].

Since the prohibition of mercury-based compounds as a pesticide and the general disapproval in Europe of antibiotics utilization in agriculture, attempts to control SRE in potato have been ineffective due to the lack of suitable tools and strategies [8]. Currently, the management of the SRE in potato is therefore based mainly on hygienic practices during plant cultivation and tuber storage as well as on seed certification based on the use of healthy (axenic) initial propagative material in seed production, limited number of generations and rejection of highly infected crops during multiplication [8]. So far, these measures have been partially successful but have not led to the total control of soft rot and blackleg pathogens [6, 19].

Bacteriophages (phages) have been proposed as potential biological control agents against plant pathogenic bacteria. They have been evaluated for different pathogens e. g. *Erwinia amylovora*, *Xanthomonas pruni*, *Pseudomonas tolaasii*, *Streptomyces scabies* and *Ralstonia solanacearum* (for review see [20]). Phages were also experimentally tested to control *Pectobacterium* spp. [21] and *Dickeya* spp. [22, 23] in potato and other crops with relative success but only limited attempts have been made to characterize these lytic bacteriophages [22–24].

Despite numerous potential benefits of bacteriophage-based biological control of plant pathogenic bacteria (for review see [25, 26]), few attempts have been made to test their efficacy by direct treatment of plants or planting material in large scale greenhouse or field experiments. One important difficulty is their well-known narrow host specificity [27]. Usually a particular bacteriophage is able to infect and kill only a limited range of strains of any one bacterial species [28]. This narrow specificity may hinder the value of bacteriophages in agriculture especially in situations in which more than one closely related pathogen is present. *Pectobacterium* spp. and *Dickeya* spp. can coexist in infected plant tissues [29–31], although there is little information on the relative importance of any synergistic effect in disease development [32]. It has been suggested that bacteria from both genera may interact in plants to facilitate infection [31, 33]. To our knowledge, there are no reports describing broad spectrum host lytic bacteriophages able to infect members of several species of soft rot *Enterobacteriaceae*.

The aim of this study was to isolate and characterize in detail broad host range lytic bacteriophages able to infect members of *Pectobacterium* spp. and *Dickeya* spp. especially against dominant soft rot and blackleg pathogens of potato in Europe *viz*. Pcc, Pwa and *D. solani*.

MATERIALS AND METHODS

Bacterial Strains and Media

Pectobacterium and *Dickeya* spp. isolates used in this study are listed in S1 Table. For routine tests, bacteria were grown at 28°C for 24–48 h on tryptone soya agar (TSA; Oxoid) or nutrient agar (NA; Oxoid) prior to use, unless stated otherwise. For liquid preparations, bacterial cultures were grown in tryptone soya broth (TSB; Oxoid) or in nutrient broth (NB; Oxoid) at 28°C with agitation at 200 rpm. For long-term storage, bacterial cultures were kept in sterile 40% (v/v) glycerol at −80°C.

Isolation of Bacteriophages from Environmental Samples and Enrichment of Bacteriophages in Their Host Bacterial Cultures

Soil, rhizosphere soil and potato plant and tuber samples were obtained from the Main Inspectorate of Plant Health and Seed Inspection, Poland. One hundred sixty four samples were collected between April 2013 and September 2013 in different regions in Poland. To isolate the bacteriophages from the environment the procedure described before [22] was used. *D. solani* strain IPO2222 [16], Pcc strain Ecc71 [34], Pba strain SCRI 1043 [35], Pwa strain SCC3193 [36], Pcb strain LMG 21371 [4] and *D. dianthicola* strain CFBP 1200 [37] were used to enrich lytic bacteriophages from soil samples as described previously [38]. After incubation, bacteria were removed by centrifugation (8000 ×*g* for 5 min) and supernatants were filter-sterilized with 0.22 μm syringe filters.

Purification of Individual Phage Particles

Purification of single phage particles was done using a soft top agar method [39] with several modifications [22]. Plaques obtained after the fourth purification were collected, resuspended in 2 ml sterile Ringer's buffer (Merck), shaken for 30 min at 200 rpm to release viral particles, filter-sterilized with a 0.22 μm membrane filter (VWR) and stored for further analysis at 4°C. Phage densities (plaque-forming units (pfu) ml^{-1}) were calculated for each bacteriophage.

Bacteriophage Morphology under Transmission Electron

Microscopy (TEM)

Transmission electron microscopy analysis of phage particles was performed in the Laboratory of Electron Microscopy, Faculty of Biology, University of Gdansk, Gdansk, Poland using negative staining of phage particles with uranyl acetate as described previously [22].

Determination of the Bacteriophages' Host Range

A host specificity assay was performed using 99 bacterial isolates: 41 isolates of *Dickeya* spp. belonging to six genomo-species [37] (eight isolates of *D. dadantii*, six isolates of *D. dianthicola*, five isolates of *D. zeae*, two isolates of *D. paradisiaca*, four isolates of *D. chrysanthemi*, 16 isolates of '*D. solani*') and 58 isolates of five *Pectobacterium* species and subspecies [3] (25 isolates of *P. atrosepticum*, 24 isolates of *P. carotovorum* subsp. *carotovorum*, 7 isolates of *P. wasabiae* and 2 isolates of *P. carotovorum* subsp. *brasiliense*) (S1 Table) as previously described before [22].

Determination of the Optimal Multiplicity of Infection (MOI)

Optimal MOI, defined as the optimal ratio between phage particles and bacterial host cells [40], was determined for phages ϕPD10.3 and ϕPD23.1. Bacterial culture (IPO2222, Ecc71 or SCC3193) was infected with phages at four different pfu/cfu ratios (MOI): 0.01, 0.1, 1 and 10. After an overnight incubation at 28°C with shaking (200 rpm), bacterial cultures were centrifuged (10 000 ×*g*, 10 min) and supernatants were assayed for phage titre as described above. The MOI resulting in the highest phage titre (the highest pfu ml^{-1}) was considered as optimal. The experiment was repeated three times and the results from all repetitions were averaged.

Phage Adsorption to Host Bacterial Cells

To determine the speed of phage adsorption to bacterial host cells, 1 ml of log-phase *Dickeya* spp. IPO2222, Pcc Ecc71 or Pwa SCC3193 cells (10^8 cfu ml^{-1}) was infected with a phage suspension (ϕPD10.3 or ϕPD23.1) to reach an MOI of 0.1 (*ca.* 10^7 pfu ml^{-1}) and incubated at 28°C for up to 20 min. After 0, 1, 2, 5, 10 and 20 min, two individual samples per phage were collected and centrifuged at 10 000 ×*g* for 5 min to sediment the bacteria together with the adsorbed bacteriophages. The resulting supernatants were filter-sterilized with a 0.22 μm syringe filter and assayed for free, unadsorbed phages. The experiment was repeated three times and the results were averaged. Phage adsorption was calculated as follows: percentage adsorption = ((control titre − residual titre) / control titre) × 100 [22].

One-Step Growth

To determine the latent period and the burst size of the ϕPD10.3 and ϕPD23.1 bacteriophages in bacterial hosts IPO2222, Ecc71 and SCC3193, a one-step growth experiment was conducted [41] with some modifications as described previously [22]. The number of phage particles was determined by the soft top agar method as described above. Viable bacterial counts were determined both before the bacteria were mixed with phages and at the end of the experiment. Burst size was estimated from three independent experiments (six independent measurements) by using the following equation: burst size = $\Delta V/\Delta B$, where ΔV—indicates changes in the number of phage particles and ΔB—represents the changes in the bacterial number during the experiment as described earlier [42].

Estimation of Bacteriophage Genome Size by Pulsed-Field Gel Electrophoresis (PFGE)

PFGE analysis of intact phage particles was prepared as previously described [43] with slight modifications. Briefly, 400 μl of purified phage suspensions (*ca.* 10^{14} pfu ml^{-1}; in PS buffer −0.1 M Tris, 0.1 M EDTA, pH = 8.0) were transferred to 2 ml-microcentrifuge tubes preheated to 50°C in a thermoblock and mixed with 400 μl of molten 1.2% plug agarose (1.2% SeaKem Gold Agarose (Cambrex Corp.) in 1×TE buffer). 250 μl of samples was transferred to wells of plug casting mold allowing them to solidify at room temperature for *ca.* 30 min. After the plugs had solidified, they were transferred to the 15 ml tubes (Sarstedt) containing 5 ml of PL buffer (50 mM Tris, 50 mM EDTA, 1% SDS) and 25 μl of proteinase K (Thermo Scientific, 20 mg ml^{-1}) and incubated at 54°C for 2 h to lyse phage capsids. After lysis, the plugs were washed four times with sterile TE buffer for 10 min. Prepared agar plugs containing the intact phage DNA were stored at 4°C in sterile TE buffer before analysis.

Phage DNA was run without digestion in a 1% SeaKem Gold Agarose gel in 0.5×TBE (45 mM Tris-borate, 1 ml EDTA, pH = 8.0) containing 100 mM thiourea (Sigma) with a CHEF-DR III chiller system (Bio-Rad Laboratories Inc.) using 0.5×TBE buffer with 100 mM thiourea as a running buffer. PFGE was performed at 6 V/cm field strength for 22 h at 14°C with the pulse time linearly increased from 2.2 to 54.2 s and a fixed reorientation angle of 120°. The gels were stained with 5 mg ml^{-1} ethidium bromide (Sigma), washed with water and analyzed under UV light in GelDoc (Biorad) imaging system. *Salmonella enterica* serotype Braenderup (strain H9812, ATCC) was used as a DNA marker [44].

Purification of Bacteriophage Genomic DNA for RFLP, RAPD-PCR and Whole Genome Sequencing

To obtain high phage titer (*ca.* 10^{13}–10^{14} pfu ml^{-1}) for genomic DNA purification, the ϕPD10.3 and ϕPD23.1 phages were enriched in *D. solani* IPO2222 culture as described previously [22]. After enrichment, bacterial cells were removed by centrifugation (8000 ×*g*, 20 min) and the resulting phage suspension (*ca.* 50 ml) was filtered through a 0.22 µm membrane filter (VWR) to remove bacterial debris. Phage suspension was treated with DNase I (Sigma-Aldrich) (final concentration: 0.5 mg ml^{-1}) for 60 min at 37°C with shaking (100 rpm) to digest the bacterial DNA. Phage particles were further purified and concentrated *via* centrifugation (15 000 ×g, 4°C) for 2 h. The supernatant was then removed and the pellet containing bacteriophage particles was resuspended in 500 µl of 5 mM $MgSO_4$ and washed two times with 5 mM $MgSO_4$ under the same conditions [45]. The phage genomic DNA was isolated as described previously [46].

Restriction Fragment Length Polymorphism (RFLP) of Phage Genomes

Purified phage genomic DNA (*ca.* 200–400 ng μl^{-1}) was subjected to a single-enzyme restriction analysis with *Nco*I, *Nde*I *Bam*HI, *Hind*III, *Kpn*I, *Sal*I, *Alu*I, *Xba*I, *Eco*RI, *Ksp*AI, *Alu*I, *Rsa*I, *Hpa*II and *Hin*6I (ThermoScientific) restriction endonucleases (RE) according to the manufacturer's protocol. Briefly, phage genomic DNA (*ca.* 200 ng per reaction) was digested for up to 24 h with 2.5 U of a single restriction endonuclease in separate digestions in a total volume of 10 μl. Digested DNA was electrophoresed in 1–2% agarose (0.5×TBE) gels. For the estimation of the size of the DNA fragments, λ genomic DNA digested with *Hind*III and *Eco*RI was used. Agarose gels were stained with 5 mg ml^{-1} of GelRed (Biotium) for visualization of DNA.

RAPD-PCR of Phage Genomes

The RAPD-PCR was executed according to Comeau *et al.* (2004) using primer R10D (5'-GTCASSWSSW-3', where S and W represent G/C and A/T, respectively) [47] with the following modifications. The phage DNA concentration was adjusted with Milipore water (MQ) to a final concentration of approximately 100 ng μl^{-1}. RAPD was performed in a total volume of 50 μl containing 1.5 U Platinum Taq DNA polymerase (Invitrogen), 4.5 mM $MgCl_2$, 0.6 mM of each deoxytriribonucleoside triphosphate (dNTPs), 2 μM of the R10D primer and 2–10 μl of phage DNA. PCR was carried out with the following program: initial denaturation at 95°C for 1.5 min., followed by 40 cycles of denaturation at 95°C for 45 s, annealing at 40°C for 3 min, extension at 72°C for 1 min and a final extension at 72°C for 10 min. As a control, instead of phage DNA, host bacterial DNA (*ca.* 100 ng μl^{-1}) was used. The amplified DNA was analyzed by electrophoresis in a 1.5% agarose gel in 0.5×TBE buffer stained with 5 mg ml^{-1} of ethidium bromide (Sigma). Gels were run for *ca.* 8 h at 100 V and at the room temperature (approx. 20–24°C). A 1 kb and 100 bp ladders (Fermentas) were used as size markers.

Phage genomes Sequencing, Annotation and in *Silico* Comparative Analysis

Phage ϕPD10.3 and ϕPD23.1 genomic DNA were sequenced using the Illumina technology and *de novo*re-assembled at Baseclear, The Netherlands (www.baseclear.com) using CLC Main Workbench program (http://www.clcbio.com/products/clc-main-workbench/). Structural and functional annotations were obtained from the IGS Annotation Service (Institute for Genome Sciences, University of Maryland School of Medicine automated pipeline http://ae.igs.umaryland.edu/cgi/index.cgi) and from RAST (Rapid Annotation using Subsystem Technology, accessed *via* the http://rast.nmpdr.org/ website). Data for phage genome assembly were subsampled from the Illumina paired reads. Fractions of

0.01, 0.03, 0.04, 0.05, 0.06, 0.07, 0.08, 0.09, 0.10, 0.20, 0.25, 0.33 0.50 were sampled from the reads with three different seeds (s10, s20, s30) using the tool seqtk (seqtk sample-s10 <file> <fraction>) (https://github.com/lh3/seqtk). The sampled fractions were assembled using Mira 4.9.3 (http://sourceforge.net/p/mira-assembler/wiki/Home/) and examined for complete phage contigs. The phage genomes were mapped and annotated using available genomic sequences deposited in GenBank (http://www.ncbi.nlm.nih.gov/genbank/). The *in silico* analysis of bacteriophages' genomes was done using Manatee (http://manatee.sourceforge.net/) accessed *via* website of Institute for Genome Sciences, University of Maryland School of Medicine. The lifestyle of phages ϕPD10.3 and ϕPD23.1 (temperate or lytic) was predicted using PHACTS [48]. Multiple genome alignment was performed using Mauve [49] and comparative genomics analyses were done using EDGAR [50]. We used a complete genome sequence of *Enterobacteriaceae* bacteriophage T4 (GenBank accession number: NC_000866.4) as outgroup for comparative bacteriophage genome analyses. To find potential genes acquired by the ϕPD10.3 and ϕPD23.1 coding for toxins and allergens bacteriophage genomes were subjected to bioinformatic analysis using Virulence Finder 1.2 (http://cge.cbs.dtu.dk/services/VirulenceFinder/) and VirulentPred (http://203.92.44.117/virulent/submit.html) accessed *via* the given websites.

Analyses of Bacteriophage ϕPD10.3 and ϕPD23.1 Proteins

SDS-PAGE Analysis of Phage Proteins

For SDS-PAGE electrophoresis, phages ϕPD10.3 and ϕPD23.1 (*ca.* 10^{10} pfu ml^{-1}) were diluted two times with sterile demineralized water. Laemmli buffer (4% SDS, 20% glycerol, 10% 2-mercaptoethanol, 0.004% bromophenol blue and 0.125 M Tris-HCl, pH 6.8, 6 × concentrated stock) was added to the samples to obtain 2 × concentrated working solution and the mixture was firstly frozen in liquid nitrogen for 1–2 min and then boiled for 5 min at 95°C. Twenty five µl of phage extracts were separated in 12% acrylamide SDS-PAGE gel (Rothiphorese Gel 30, 37.5:1) (ROTH) for 19 h at 50 V at room temperature (22°C). PageRuler Pre-stained Protein Ladder (Thermo Scientific), prepared according to instructions provided by the manufacturer, was used as a size marker. The gel was stained with PageBlue Coomasie Blue (Thermo Scientific) for 18 h and destained with sterile demineralized water for 6 h at room temperature (22°C) as suggested by the manufacturer. Protein bands were excised from the gel with a sterile scalpel and used for mass spectrometry analysis.

Mass Spectrometry Analysis of the Phage ϕPD10.3 and ϕPD23.1 *Proteins*

Electrospray ionization MS/MS analysis of phage proteins was performed at Mass Spectrometry Laboratory, Institute of Biochemistry and Biophysics, Polish Academy of Sciences in Warsaw, Poland.

Structural/Functional Assignment of Unknown Phage Proteins based on their 3D Protein Structures

In order to predict the molecular functions of the unknown structural proteins obtained from the SDS-PAGE and MS analysis of phage ϕPD10.3 and ϕPD23.1 proteins we used GeneSillico Protein Structure Prediction Meta-server (https://genesilico.pl/meta2/acl_users) containing known three-dimensional (3D) protein structures [51] and PSI-BLAST accessed *via* website of NCBI (http://www.ncbi.nlm.nih.gov/) [52]. The predicted proteins with the highest scores were considered as the most valid [51, 52].

Effect of Bacteriophages on Potato Tuber Tissue Maceration Caused by *D. solani* IPO2222, *P. wasabiae* SCC3139 and *P. carotovorum* subsp. *carotovorum* Ecc71

A Potato Slice Assay

A potato slice assay was used to assess if the ϕPD10.3 and ϕPD23.1 bacteriophages were able to protect potato tuber tissue when co-inoculated on potato slices with a mix of Dso IPO2222, Pcc Ecc71 and Pwa SCC3193 [22]. The bacteriophages' concentration was adjusted to 10^5 pfu ml^{-1} in sterile demineralized water and the bacterial density was adjusted to 10^7 cfu ml^{-1} in demineralized sterile water for each strain. Ware potato tubers of cultivar Bryza, obtained locally were surface-sterilized with 70% ethanol for 10 min, rinsed with tap water and dried with tissue paper. Potato tubers were cut into *ca.* 0.7 cm-thick transverse disks, using a sterile knife. For each slice, three wells (5 × 5 × 5 mm) were made using a sterile cork borer and these were filled with 50 μl of a mixture of containing 10^5 pfu ml^{-1} of one of the tested bacteriophages together with 10^7 cfu ml^{-1} of each bacterial strain tested (IPO2222 + Ecc71 + SCC3193). Three potato slices obtained from three different potato tubers were used per treatment. As a negative control, 50 μl aliquots of sterile demineralized water was used instead of bacterial and phage suspensions, while as a positive control, 50 μl of bacterial suspension containing 10^7 cfu ml^{-1} of three bacterial strains (IPO2222 + Ecc71 + SCC3193) was used. The protective effect of the phages on the potato tissue was measured after incubation for 72 h at 28°C in a humid box by calculating the ratio of the average diameter of rotten potato tissue around the wells co-inoculated with bacteria and bacteriophage to the average diameter of rotten tissue around wells inoculated with bacterial mixture only. Two duplications were made for each test and the entire experiment was independently repeated two times with the same setup. Results from the experiments were averaged.

A Whole Potato Tuber Assay

Whole potato tuber assay [23] was used to check whether application of ϕPD10.3 and ϕPD23.1 would protect potato tubers from vascular infections caused by a mix of Dso IPO2222, Pcc Ecc71 and Pwa SCC3193. Briefly, the bacteriophage and bacteria were prepared in the exact same way as described above for the potato slice assay. Ware potato tubers of cultivar Bryza, were surface-sterilized with 70% ethanol for 10 min, rinsed with tap water and dried with tissue paper. All tubers were weighted before the experiment. Per tuber, one 0.5 cm-thick transverse slice taken at a tuber rose end (end opposite to the stolon end) was incised and the cap was removed. 100 μl of the mixture containing 10^7 cfu ml^{-1} of each bacterial strain tested (IPO2222 + Ecc71 + SCC3193) in sterile demineralized water was pipetted on the surface of the capless tuber. Inoculated tubers were left for 30 min. allowing the bacteria to be absorbed into the (vascular) tuber tissue. As a control, 100 μl of sterile demineralized water was used instead of bacterial suspensions. To evaluate the protection effect of ϕPD10.3 and ϕPD23.1, 100 μl of phage suspension containing 10^5 pfu ml^{-1} of either individual or mixture of both phages was added onto the surface of infected or control tubers and these were once more left until the suspensions were absorbed inside vascular tissue. As a control, instead of bacteriophage suspensions, 100 μl of sterile demineralized water was used. The cut cap was re-attached on each tuber with a sterile toothpick. Treated tubers were left for 72 h at 28°C in a humid box. Ten potato tubers were used per treatment and the experiment was repeated two times with the same setup. The weight of the rotting tissue was measured for each tuber. The protective effect of the phage on the potato tubers was measured by calculating the percentage of rotting mass of each tuber. The results were averaged per treatment and per experiment.

RESULTS

Isolation of Bacteriophages

Between April and September 2013, 164 samples of potato rhizosphere soil, bulk soil, potato stem and tubers collected from arable potato fields in different regions in Poland were screened for the presence of lytic bacteriophages against different *Pectobacterium* and *Dickeya* species. After enrichment of potential bacteriophages in bacterial host cultures, 28 (17% of all tested samples) samples yielded lytic bacteriophages able to kill at least one of the six strains of bacterial species tested (*P. carotovorum* subsp. *carotovorum*, *P. wasabiae*, *P. atrosepticum*, *P. carotovorum* subsp. *brasiliense*, *D. solani* and *D. dianthicola*). From each positive sample one distinct plaque was isolated and further purified to obtain pure phage particles (Table 1). Enrichment of bacteriophages in their host bacterial cells resulted in phage suspensions with high titre averaging from 10^{12} to 10^{14} pfu ml^{-1} after an overnight incubation at 28°C.

Table 1: Bacteriophages isolated in this study.

No.	phage	Plague description [a]	Sample origin [b]	Geographical region in Poland
1	ϕPD 2.1	small, transparent	potato tuber	Warminsko-Mazurskie (Warmian-Masurian Province)
2	ϕPD 2.2	medium, semitransparent	bulk soil	Warminsko-Mazurskie (Warmian-Masurian Province)
3	ϕPD 3.1	big, transparent	bulk soil	Pomorskie (Pomeranian Province)
4	ϕPD 4.4	small, transparent	bulk soil	Lubelskie (Lublin Province)
5	ϕPD 4.6	medium, transparent	rhizosphere soil	Pomorskie (Pomeranian Province)
6	ϕPD 5.2	big, semitransparent	potato tuber	Mazowieckie (Masovian Province)
7	ϕPD 5.4	big, transparent	potato tuber	Podlaskie (Podlaskie Province)
8	ϕPD 7.1	big, transparent	bulk soil	Kujawsko-Pomorskie (Kuyavian-Pomeranian Province)
9	ϕPD 8.1	big, transparent	rhizosphere soil	Mazowieckie (Masovian Province)
10	ϕPD 8.5	medium	rhizosphere soil	Mazowieckie (Masovian Province)
11	ϕPD 9.1	big, transparent	potato tuber	Pomorskie (Pomeranian Province)
12	ϕPD 10.3	medium, transparent	potato stem	Mazowieckie (Masovian Province)
13	ϕPD 11.3	big, transparent	bulk soil	Mazowieckie (Masovian Province)
14	ϕPD 11.4	medium, semitransparent	bulk soil	Wielkopolskie (Greater Poland Province)
15	ϕPD 12.6	small, transparent	potato tuber	Pomorskie (Pomeranian Province)
16	ϕPD 15.3	small, transparent	bulk soil	Mazowieckie (Masovian Province)
17	ϕPD 17.1	big, transparent	potato stem	Pomorskie (Pomeranian Province)
18	ϕPD 18.2	small, transparent	potato stem	Pomorskie (Pomeranian Province)
19	ϕPD 20.2	medium, semitransparent	rhizosphere soil	Mazowieckie (Masovian Province)
20	ϕPD 22.1	big, transparent	bulk soil	Wielkopolskie (Greater Poland Province)
21	ϕPD 23.1	big, transparent	potato tuber	Wielkopolskie (Greater Poland Province)
22	ϕPD 25.1	big, transparent	potato tuber	Mazowieckie (Masovian Province)
23	ϕPD 26.2	medium, semitransparent	rhizosphere soil	Mazowieckie (Masovian Province)
24	ϕPD 27.2	medium, transparent	potato stem	Kujawsko-Pomorskie (Kuyavian-Pomeranian Province)
25	ϕPD 31.1	big, transparent	rhizosphere soil	Mazowieckie (Masovian Province)
26	ϕPD 31.2	small, transparent	potato tuber	Mazowieckie (Masovian Province)
27	ϕPD 32.3	big, semitransparent	potato tuber	Lubelskie (Lublin Province)
28	ϕPD 33.3	small, semitransparent	rhizosphere soil	Pomorskie (Pomeranian Province)

Bacteriophage Host Range

In order to find bacteriophages able to infect more than one species or subspecies of SRE, the above described 28 bacteriophages were tested for to lyse 99 *Dickeya* and *Pectobacterium* spp. strains including the type strains and environmental isolates; 8 isolates of *D. dadantii*, 6 isolates of *D. dianthicola*, 5 isolates of *D. zeae*, 2 isolates of *D. paradisiaca*, 4 isolates of *D. chrysanthemi*, 16 isolates of *D. solani*, 25 isolates of *P. atrosepticum*, 24 isolates of *P. carotovorum* subsp. *carotovorum*, 7 isolates of *P. wasabiae* and 2 isolates of *P. carotovorum* subsp. *brasiliense* were used (S1 Table). Of all bacteriophages tested, only two named ϕPD10.3 and ϕPD23.1 (abbreviations: ϕ—phage, P—*Pectobacterium*, D—*Dickeya*, number reflects the sample number) possessed the broadest host range from all the phages isolated in this study (were able to infect strains of Pcc, Pwa and Dso isolates) (S1 Table). Consequently, these two were selected for further analysis. None of the bacteriophage isolates tested in this study was able to infect strains of the remaining species (and subspecies) of soft rot *Enterobacteriaceae*, including Pba (S1 Table).

Transmission Electron Microscopy (TEM) Analysis

Transmission electron microscopy analysis performed for ϕPD10.3 and ϕPD23.1 revealed that both phages belonged to the order *Caudovirales*, family *Myoviridae* based on their morphology and presence of icosahedral head, neck and contractile tail (Fig. 1). The diameter of the heads was 85 and 86 nm, and the length of the tails was 117 nm and 121 nm for phages ϕPD10.3 and ϕPD23.1, respectively.

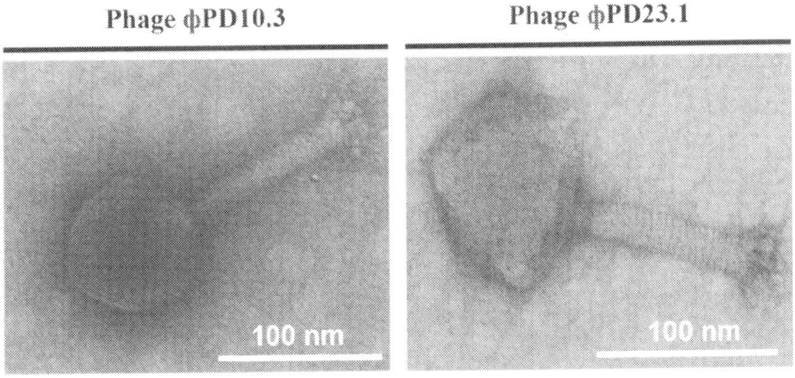

Figure 1: Transmission electron micrographs (TEM) of *Pectobacterium* **spp. and** *Dickeya* **spp. bacteriophages ϕPD10.3 and ϕPD23.1 stained with uranyl acetate.**

Phage Adsorption to Host Bacterial Cells, Optimal MOI, Latent Period and Burst Size

In order to better understand the interaction of bacteriophages with their bacterial hosts, ϕPD10.3 and ϕPD23.1 were assessed for their adsorption study to hosts, estimation of the optimal MOI and in one-step growth experiments in which the latent period and the burst size were measured. Optimal MOI, adsorption, latent period and burst size differed greatly for ϕPD10.3 and ϕPD23.1, for a given bacterial host. In general, for both phages, the optimal MOI was between 0.01 and 0.1, adsorption at 28°C after 20 min was between 70 and 83% and burst size was between 82 and 102 phages per host cell (Table 2).

Table 2: Interaction of bacteriophages ϕPD10.3 and ϕPD23.1 with their bacterial hosts *D. solani* **IPO2222,** *P. carotovorum* **subsp.** *carotovorum* **Ecc71 and** *P. wasabiae* **3193.**

Bacteriophage	Bacterial host [a]	Optimal MOI [b]	Adsorption [c]	Latent period [d]	Burst size [e]
ϕPD10.3	IPO 2222	0.01	70%	30 min.	95 ± 7
ϕPD10.3	Ecc71	0.1	75%	40 min.	82 ± 5
ϕPD10.3	3193	0.1	76%	30 min.	102 ± 3
ϕPD23.1	IPO2222	0.01	74%	30 min.	95 ± 5
ϕPD23.1	Ecc 71	0.1	83%	40 min.	91 ± 6
ϕPD23.1	3193	0.01	71%	20 min.	95 ± 5

RFLP, RAPD-PCR and PFGE Analysis

RFLP analysis

Phage ϕPD10.3 and ϕPD23.1 genomic DNA samples were insensitive to digestion with a majority of the restriction endonucleases tested (*Nco*I, *Nde*I, *Bam*HI, *Kpn*I, *Sal*I, *Alu*I, *Xba*I, *Eco*RI, *Ksp*AI, *Alu*I, *Rsa*I) except *Hin*6I, *Hind*III and *Hpa*II. The ϕPD10.3 and ϕPD23.1 exhibited different but closely related restriction endonuclease patterns with common fragments present in both phages (Fig. 2A).

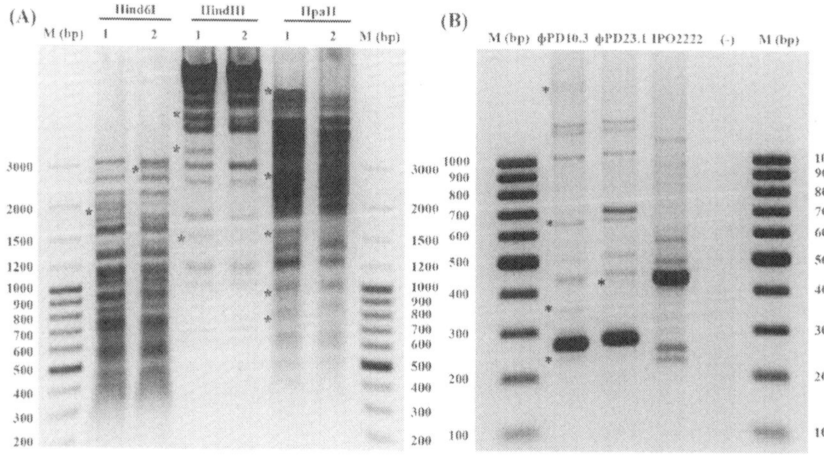

Figure 2: Fingerprinting of ϕPD10.3 and ϕPD23.1 bacteriophage genomic DNA with the use of restriction fragment length polymorphism (RFLP) (A) and Random Amplified Polymorphic DNA-PCR (RAPD-PCR) (B) analysis.

(A) Digestion of bacteriophage genomic DNA of ϕPD10.3 and ϕPD23.1 phages with endonucleases *Hin*6I, *Hind*III and *Hpa*II (all from

Fermentas). Phage DNA (*ca.* 200–400 ng ml^{-1}) was digested with 2.5 U of individual restriction endonuclease in single digestions for up to 24 h at 37°C according to protocol provided by the manufacturer. The digested DNA was ran on 1 or 1.5% agarose gel in 0.5 x TBE for up to 3 h and visualized under UV light by staining with 0.5 mg ml^{-1} **GelRed. GeneRuler 100 bp Plus DNA ladder (Fermentas) was used as a size marker. The asterisks (*) indicate DNA bands present either in ϕPD10.3 or ϕPD23.1 only. (B) RAPD-PCR band patterns obtained for ϕPD10.3 and ϕPD23.1 bacteriophages and *D. solani* IPO2222 (bacteriophage host) using ϕPD10.3 and ϕPD23.1 and IPO2222 genomic DNA as templates and primer R10D. The RAPD patterns were run on 1 or 1.5% agarose gel in 0.5 x TBE for up to 5 h and were visualized under UV light by staining with 0.5 mg ml^{-1}**GelRed. GeneRuler 100 bp DNA ladder (Fermentas) was used as a size marker. The asterisks (*) indicate bands present either in ϕPD10.3 or ϕPD23.1 only. For negative control (-), RAPD-PCR was performed with the same protocol, however, instead of DNA template an equal volume of sterile demineralized water was added to the reaction mixture.

RAPD-PCR Analysis

RAPD-PCR analyses were run in triplicate to assure consistent results. No amplification products were observed in reactions performed in the absence of DNA template. The RAPD profile of DNA isolated from *D. solani* IPO2222 was different that from ϕPD10.3 and ϕPD23.1. The ϕPD10.3 and ϕPD23.1 yielded different but closely related RAPD patterns (Fig. 2B).

PFGE Analysis

The PFGE analysis was performed on undigested genomic DNA with restriction endonucleases to estimate genome sizes of the ϕPD10.3 and ϕPD23.1. Both phages had genomes of similar size, *ca.* 180,000 to 190,000 bp (Fig. 3).

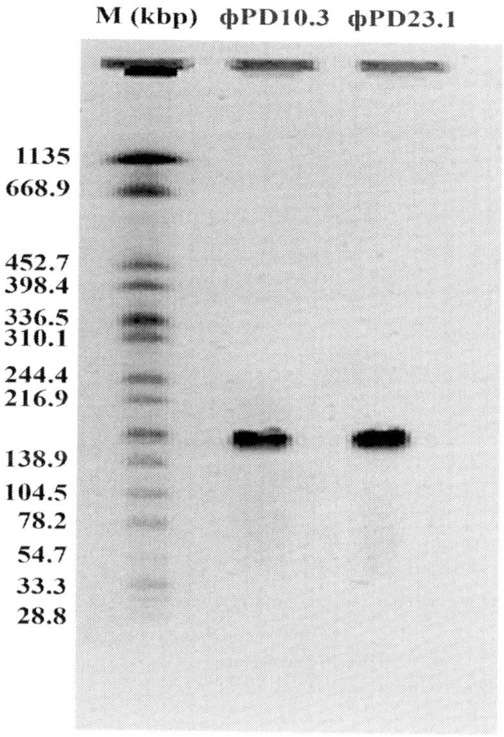

Figure 3: Estimation of bacteriophage ϕPD10.3 and ϕPD23.1 genome size by pulse-field gel electrophoresis (PFGE).

Indigested with restriction endonucleases genomic DNA from phage ϕPD10.3 and ϕPD23.1 particles was run on 1% SeaKem Gold agarose in 0.5 x TBE buffer with 100 mM thiourea for 22 h at 14°C with the pulse time linearly increased from 2.2 to 54.2 s and a fixed reorientation angle of 120°. The gels were stained with 5 mg ml^{-1} ethidium bromide (Sigma), washed with water for 10 min and analyzed under UV light in GelDoc imaging system. As a marker the *Salmonella enterica* serotype Braenderup (strain H9812, ATCC) was used.

Description of the ϕPD10.3 and ϕPD23.1 Draft Genomes and Comparative Genomic Analysis

Draft genome sequences of ϕPD10.3 and ϕPD23.1 were obtained using Illumina next generation sequencing technology with the average of

100-fold coverage and assembled *de novo* at BaseClear, the Netherlands (www.baseclear.com). The reads were assembled into 28 (ϕPD10.3) and 29 contigs (ϕPD23.1) with a total length of 192, 291 bp (ϕPD10.3) and 188,540 bp (ϕPD23.1). For each phage the contigs were concatenated into pseudomolecules with spacers between contigs that introduced start and stop codons in all six possible reading frames to assure proper translation from all 6 possible ORFs at a single locus. The draft genome sequences (contigs) were deposited at DDBJ/EMBL/GenBank under accession numbers: KM209229—KM209273 (ϕPD10.3) and KM209274—KM209320 (ϕPD23.1) (S1 Fig., S2 Table). The raw Illumina reads of ϕPD10.3 and ϕPD23.1 were subsampled and processed using Mira assembler in order to close their genomes. The largest scaffold obtained for ϕPD10.3 contained 156,113 bp. (S1 File), whereas the largest scaffold for ϕPD23.1 contained 153,365 bp. (S2 File). We were not able to close the genomes: 36,178 bp. (ϕPD10.3) and 35,175 bp. (ϕPD23.1) of the raw Illumina reads could not be scaffolded.

The general features of the ϕPD10.3 and ϕPD23.1 genomes are summarized in Table 3. A pairwise comparison of the nucleotide sequences of ϕPD10.3 and ϕPD23.1 revealed that they had considerable similarity suggesting a common origin. The common genome (core genome), estimated with EDGAR, consists of 207 PEGs (protein encoding genes), whereas only 19 and 16 PEGs are specific for phage ϕPD10.3 and ϕPD23.1, respectively. The majority of genes found in one but absent in the other phage were those coding for hypothetical proteins with unknown function. Genomes of ϕPD10.3 and ϕPD23.1 share significant homology with the genome of *Dickeya* spp. lytic bacteriophage ϕD5 isolated by us in the previous study [22]; 193 genes (ϕPD10.3) and 189 genes (ϕPD23.1) have homologs in the genome of ϕD5. Interestingly, the majority of ϕPD10.3 and ϕPD23.1 genes do not have homologs in of T4—one of the best characterized broad host range *Enterobacteriaceae* phage. Both phages share only two genes with T4, namely phage recombination protein and *regA*—translational repressor of early genes. The lifestyle of ϕPD10.3 and ϕPD23.1 predicted from PHACTS indicated that both may be lytic bacteriophages.

Table 3: General features of the bacteriophage ϕPD10.3 and ϕPD23.1 genomes.

Feature	Bacteriophage ϕPD10.3	Bacteriophage ϕPD23.1	Bacteriophage ϕD5(Czajkowski *et al.* 2014)
Genome size (bp)	192 291	188 540	155 346
% GC	48.6%	49.25%	49.7%
% of coding regions in the genome	79.25%	83.15%	89.9%
Number of predicted ORFs (PEGs)	226	223	196
% of PEGs with assigned functions	37.2%—with assigned functions 21.7%—unclassified with no assigned category 7.5%—have unknown functions	39.5%—with assigned functions 22.4%—unclassified with no assigned category 8%—have unknown functions	25.5%—with assigned functions 21.4%—unclassified with no assigned category 4.1%—have unknown functions
Average gene length (bp)	553–686	538–704	711
Number of PEGs in functional groups	2—nucleotide metabolism 2—energy metabolism 16—transport and binding 15—DNA metabolism 4—transcription 1—protein fate 2—regulatory functions 4—cell envelope 8—cellular processes 4—mobile or extrachromosomal elements 15—DNA metabolism	1—nucleotide metabolism 1—energy metabolism 17—transport and binding 15—DNA metabolism 5—transcription 7—cell envelope 9—cellular processes 4—mobile or Extrachromosomal elements	2—nucleotide metabolism 1—energy metabolism 7—transport and binding 9—DNA metabolism 3—transcription 1—regulatory functions 4—cell envelope 6—cellular processes 4—mobile or extrachromosomal elements
Transcription start codon (% of genes using this start codon)	ATG (87.3%) GTG (10.4%) TTG (2.3%)	ATG (85.3%) GTG (13.6%) TTG (1.1%)	ATG (94.4%) GTG (4.1%) TTG (1.5%)
tRNAs	tRNA-Met tRNA-Glu	tRNA-Met tRNA-Glu	tRNA-Met

The genomes of ϕPD10.3 and ϕPD23.1 did not contain any genes coding for toxins and allergens as tested by VirulenceFinder 1.2 and VirulentPred. A search in BLAST did not reveal the presence of any possible toxins, allergens and antibiotic resistance genes in the genomes of either bacteriophage.

Structural Proteome of ϕPD10.3 and ϕPD23.1 Bacteriophages

Electrospray ionization-MS/MS analysis of gel-separated ϕPD10.3 and ϕPD23.1 proteins led to the identified 13 and 12 phage proteins, respectively (Fig. 4). Of these, only have assigned functions based on sequence similarities with other phage proteins. The most abundant proteins in both samples were the major capsid protein (gp23) and tail sheath protein (gp18), as well as, in the ϕPD10.3 proteome tail tube protein (gp19) and the ϕPD23.1 proteome tail length tape measure protein and tailspike protein. Nine and 7 proteins present in ϕPD10.3 and ϕPD23.1 proteomes, respectively, were characterized as unknown structural proteins, to which function could not be assigned.

(A)

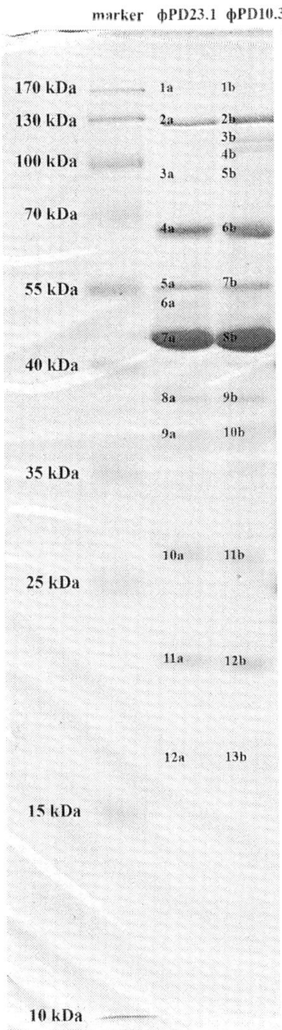

(B)

ϕPD23.1 protein in MS analysis	ϕPD23.1 unknown proteins in iterative database search
1a. unknown structural protein (177.8 kDa)	Tail-tube/spike protein
2a. unknown structural protein (128.2 kDa)	VrlC protein
3a. **tail length tape measure protein** (80.9 kDa)	
4a. **tail sheath protein gp18** (68.9 kDa)	
5a. unknown structural protein (55.8 kDa)	gp20 protein
6a. **tailspike protein** (53.7 kDa)	
7a. **major capsid protein gp23** (48.1 kDa)	
8a. unknown structural protein (38.7 kDa)	gp6 baseplate wedge
9a. **tail sheath protein gp18** (36.9 kDa)	
10a. unknown structural protein (28.3 kDa)	gp15 tail stabilizator
11a. unknown structural protein (20.6 kDa)	gp21 prohead protease
12a. unknown structural protein (18.4 kDa)	gp2 DNA protector

ϕPD10.3 protein in MS analysis	ϕPD10.3 unknown proteins in iterative database search
1b. unknown structural protein (177.8 kDa)	gp30 DNA ligase
2b. unknown structural protein (131.2 kDa)	VrlC protein
3b. unknown structural protein (129.5 kDa)	gp6 baseplase wedge
4b. unknown structural protein (110.7 kDa)	gp5 baseplate hub subunit
5b. unknown structural protein (88.7 kDa)	Tail length tape measure protein
6b. **tail sheath protein gp18** (68.9 kDa)	
7b. unknown structural protein (56.8 kDa)	gp20 protein
8b. **major capsid protein gp23** (48.1 kDa)	gp23 major capsid protein
9b. unknown structural protein (40.7 kDa)	gp6 baseplate wedge
10b. **tail sheath protein gp18** (28.9 kDa)	
11b. unknown structural protein (20.3 kDa)	gp15 tail stabilizator
12b. unknown structural protein (18.4 kDa)	gp21 prohead protease
13b. **tail tube protein gp19** (17.6 kDa)	

Figure 4: Bacteriophage ϕPD10.3 and ϕPD23.1 structural proteins separated in SDS-PAGE gels (A) and identification of bacteriophage ϕPD10.3 and ϕPD23.1 structural proteins with mass spectrometry (B).

(A) For SDS-PAGE electrophoresis *ca.* 5 x 10^9 pfu ml^{-1} were mixed with Laemmli buffer and frozen in liquid nitrogen for 1–2 min. following by boiling at 95°C for 5 min. The phage proteins were separated in 12% acrylamide SDS-PAGE gel for 19 h at 50 V at 22°C. PageRuler Pre-stained Protein Ladder (Thermo Scientific), prepared according to instructions provided by manufacturer, was used as a size marker. The gel was stained with PageBlue Coomasie Blue (Thermo Scientific) for 18 h and destained with sterile demineralized water for 6 h at room temperature (22°C) as suggested by the manufacturer. (B) For MS analysis of phage structural proteins, protein bands obtained for bacteriophage ϕPD10.3 and ϕPD23.1 were excised from gel with a sterile scalpel and sent for mass spectrometry analysis to Mass Spectrometry Laboratory, Institute of Biochemistry and Biophysics, Polish Academy of Sciences in Warsaw, Poland. Furthermore, possible molecular functions of the unknown structural proteins obtained from SDS-PAGE (A) phage ϕPD10.3 and ϕPD23.1 proteins against the sequence database of known three-dimensional (3D) protein structures was done using PSI-BLAST accessed *via* website of NCBI (http://www.ncbi.nlm.nih.gov/) [52] and using GeneSilico Protein Structure Prediction Meta-server (https://genesilico.pl/meta2/acl_users) [51]. The protein predictions with the highest scores were considered as the most valid [51, 52].

We were able to assign functions to all unknown structural proteins obtained in SDS-PAGE by comparison of protein 3D structure databases (Fig. 4). All the predicted proteins were phage-associated and functionally-involved in bacteriophage particle assembly.

Suppression of Soft Rot Development on Potato Slices and whole Tubers Co-inoculated with *P. carotovorum* subsp. *carotovorum* Ecc71, *D. solani* IPO2222, *P. wasabiae* SCC3193 and ϕPD10.3 and ϕPD23.1 Bacteriophages

The ϕPD10.3 and ϕPD23.1 were tested in a potato slice assay and whole tuber assay for their ability to reduce or prevent maceration of potato tuber tissue by *P. carotovorum* subsp. *carotovorum* Ecc71, *D. solani* IPO2222 or *P. wasabiae* SCC3193. We tested the bacteriophages in a worst case scenario in which we simulated infection with multiple soft rot *Enterobacteriaceae* pathogens. In replicated *in vitro*experiments, the two bacteriophages individually or co-inoculated with a combination of bacteria (Ecc71+IPO2222+SCC3193), were able to significantly reduce potato tuber tissue maceration by at least 80% of the control potato slices inoculated with a mixture of bacteria only and by at least 95% of the control whole tubers inoculated with a mixture of bacteria (Fig. 5)

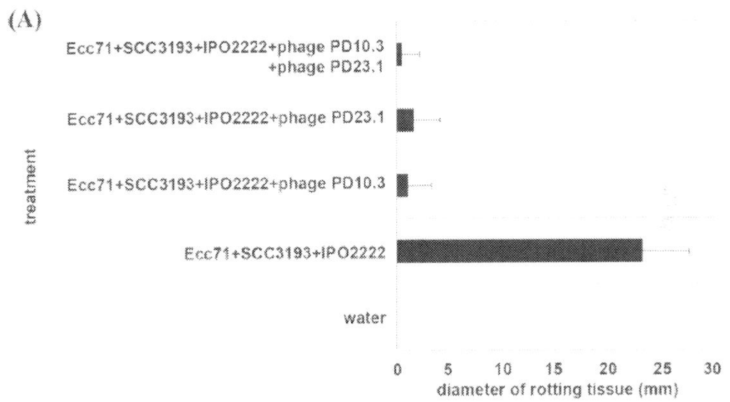

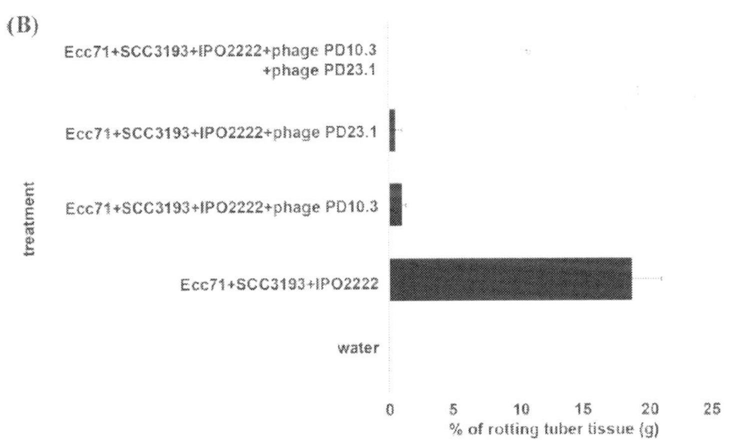

Figure 5: Protection effect on ϕPD10.3 and ϕPD23.1 on potato tuber slices (A) and whole tubers (B) against *D. solani* **IPO2222,** *P. carotovorum* **subsp.** *carotovorum* **Ecc71 and** *P. wasabiae* **3139.**

(A) Reduction of maceration ability of mixture of *D. solani* IPO2222, *P. carotovorum*subsp. *carotovorum* Ecc71 and *P. wasabiae* 3139 co-inoculated with bacteriophage ϕPD10.3 and ϕPD23.1 in potato tubers slices. The effect was determined by measuring the diameter of rotting tissue (in mm) after 72 h incubation at 28°C in a humid box. Wells of potato slices were filled up with 50 μl of sterile water (negative control), with 50 μl of bacterial suspension in water containing 10^7 cfu ml^{-1} of *D. solani* IPO2222, *P. carotovorum* subsp. *carotovorum* Ecc71 and *P. wasabiae* 3139 (positive control) or with 50 μl of bacterial suspension in water containing 10^7 cfu ml^{-1} of *D. solani* IPO2222, *P. carotovorum* subsp. *carotovorum* Ecc71 and *P. wasabiae* 3139 and 10^5 pfu ml^{-1} of ϕPD10.3 and/or ϕPD23.1 bacteriophage. Three potato slices containing 3 wells each and derived from three different tubers were used per experiment. The experiment was independently repeated one time and the results were averaged. (B) Reduction of infection ability of mixture of *D. solani* IPO2222, *P. carotovorum* subsp. *carotovorum*Ecc71 and *P. wasabiae* 3139 co-inoculated with bacteriophage ϕPD10.3 and ϕPD23.1 in the vascular tissue of potato tubers. The effect was determined by measuring the percentage of rotting tuber tissue (in g) after incubation the tubers at 28°C in a humid box. Caps of rose ends were removed from tubers with a sterile knife and resulting capless tubers were inoculated with 100 μl of bacterial suspension in water containing 10^7 cfu ml^{-1} of *D. solani* IPO2222, *P. carotovorum* subsp. *carotovorum* Ecc71 and *P. wasabiae* 3139 (positive control) or with 100 μl of bacterial suspension in water containing 10^7 cfu ml^{-1} of *D. solani* IPO2222, *P. carotovorum* subsp. *carotovorum* Ecc71 and *P. wasabiae* 3139 and co-inoculated with 100 μl of 10^5 pfu ml^{-1} of ϕPD10.3 and/or ϕPD23.1 bacteriophage. Ten potato tubers were used per treatment. The experiment was independently repeated one time and the results were averaged.

DISCUSSION

This study assessed the presence of lytic bacteriophages infecting blackleg and soft rot causing *Enterobacteriaceae* currently prevailing in potato in Europe *viz. D. solani*, *P. carotovorum* subsp. *carotovorum* and *P. wasabiae*. We focused on broad host range bacteriophages able to infect three SRE species most commonly associated with potato diseases in Europe. To our knowledge, this is the first study in which lytic bacteriophages able to infect more than one SRE species have been characterized in detail.

The frequency of isolation of lytic bacteriophages able to lyse the SRE bacteria in our study was relatively low, but was comparable to that in previous studies [21, 22, 53]. For example, lytic bacteriophages against *Dickeya* spp. were isolated from only 22% of soil samples [22] and Gross *et al.* (1991) reported that only 5% of potato rhizosphere soil samples contained bacteriophages against *P. carotovorum* subsp. *carotovorum* [53]. Of the 28 phages isolated in this study, 32.2% were isolated from (symptomatic) potato tubers (9 phages), 28.6% from bulk soil (8 phages), 25% from potato rhizosphere soil (7 phages) and 14.2% from (symptomatic) potato stems (4 phages) (Table 1). There is no information available in the literature addressing whether bacteriophages infecting *Pectobacterium* and *Dickeya* species are more frequently isolated from plants and soils infested with SRE than when the pathogens are absent. It can be speculated that the presence of viable host bacteria would favor the occurrence of bacteriophages. However, no correlation between the presence of SRE bacteria and bacteriophages was found in plant samples in this study (data not shown). As found previously (22), for all 164 samples tested here, an enrichment of phage particles in bacterial host cultures was required prior to phage isolation. However, the low recovery of bacteriophages from plant and soil samples this time cannot be explained by the seasonal differences. The phages were isolated during the spring/summer (April—September) when the highest populations of host bacteria can be expected, and from symptomatic plants harboring high bacterial populations [54].

All phages isolated in this study belonged to the order *Caudovirales* and family *Myoviridae* based on their morphology determined by transmission electron microscopy (Fig. 1). In addition, all of them have icosahedral heads and contractile tails of the diameter and sizes which would classify them as morphotype A1 [55]. Different studies have shown that more than 90% of lytic bacteriophages isolated and characterized with transmission electron microscopy belong to the order *Caudovirales* [55–57]. Previously [22] we reported the isolation of nine bacteriophages infecting *Dickeya* spp., belonging to the *Myoviridae*family [22]. Likewise, Adriaenssens and co-workers (2012) and Ravensdale *et al.* (2007) also isolated and characterized *Myoviridae* bacteriophages against *D. solani* and *P. carotovorum*, respectively [23, 24], suggesting that the SRE-infecting bacteriophages belonging to the family *Myoviridae* are the most abundant in the environment.

The large majority of phages (93%) were able to infect isolates of both *P. carotovorum* subsp. *carotovorum*and *D. solani* but only ϕPD10.3 and ϕPD23.1 (7% of isolated bacteriophages) were able to infect *D. solani*, *P. carotovorum* subsp. *carotovorum* and *P. wasabiae* strains as well. None of the bacteriophages was able to infect either *P. atrosepticum* or *P. carotovorum* subsp. *brasiliense* isolates. This may reflect recent changes in population dynamics of the host bacteria in the environment in Europe. In the past, *P. atrosepticum* was the main potato pathogen in Europe responsible for blackleg disease [58] and it is still the case in Scotland, however, since 2005 the incidence of disease caused by the pathogen has decreased significantly [29, 59]. There are also no reports of the occurrence of *P. carotovorum* subsp. *brasiliense* in Poland and an only limited number of reports describes the presence of *P. carotovorum* subsp. *brasiliense*in other European countries [60]. In contrast, since 2005, *D. solani* has become well established in the potato ecosystem in Europe [6] and that *P. carotovorum* subsp. *carotovorum* and *P. wasabiae* are known to be present in Europe for a long time in association with potato plants [1, 13, 61].

ϕPD10.3 and ϕPD23.1 were characterized in detail. They were indistinguishable from each other based on their morphology, their host range on 99 isolates of *Pectobacterium* and *Dickeya* species and their genome size estimated using PFGE. Restriction fragment length polymorphism (RFLP) analysis and randomly amplified polymorphic DNA fingerprinting (RAPD) showed that the ϕPD10.3 and ϕPD23.1 were genetically very similar although differences in RFLP and RAPD profiles could be observed. Comparison of the ϕPD10.3 and ϕPD23.1 genomes showed high homology in organization and composition of their genomes. Ninety two percent of ORFs in both bacteriophages were homologous as evidenced by comparative analysis using EDGAR. ϕPD10.3 and ϕPD23.1 bacteriophages also showed reasonably high similarity with the previously described *Dickeya* spp. bacteriophage LIMEstone1 [23] and ϕD5 [62]. Surprisingly, ϕPD10.3 and ϕPD23.1 do not share significant gene homology with *Enterobacteriaceae*bacteriophage T4 as seen for other *Myoviridae* T4-related bacteriophages [63]. There is no straightforward explanation of this phenomenon especially when considering the fact that the *Myoviridae* bacteriophages described to date have in majority very similar genome organizations [63]. It may be that the ϕ10.3 and ϕ23.1 phages are members of a new, T4-related family of viruses; however, to prove this, additional studies will be required and are now being conducted. Unfortunately, we were unable to close the genomes of ϕPD10.3 and ϕPD23.1 probably due to in part a lack of similar phage genome sequences available in international databases for use as scaffolds. The ordering of scaffolds based on the recently described ϕD5 and LimeStone1 bacteriophages was suboptimal.

The great similarity of the ϕPD10.3 and ϕPD23.1 genomes may indicate a common origin for these phages; they were isolated however from potato tuber and stem samples, respectively, originating from two different regions in Poland (Table 1). The ϕPD10.3 and ϕPD23.1 proteomes were relatively highly similar to that of LimeStone1 and ϕD5. Currently, there is only limited number of publications describing structural phage proteomes and, to our knowledge, except the work of Adriaenssens *et al.* (2012), there are no reports describing proteomes of bacteriophages infecting soft rot *Enterobacteriaceae*. Consequently, the majority of ϕPD10.3 and ϕPD23.1 proteins could only be characterized as unknown structural proteins for which no matches were found in protein databases. To further explore the possible functions of these unknown phage proteins we used GeneSillico Protein Structure Prediction Meta-server and PSI-BLAST as advised by others [51, 52]. Based on the three-dimensional protein structures we could predict their functions but with different levels of confidence. These predictions need to be supported with experimental data.

One-step growth, adsorption studies and analysis of the phage genomes supported the use of ϕPD10.3 and ϕPD23.1 bacteriophages as biological control agents. Both phages infect a number of SRE strains belonging to different species and isolated from different environments and countries. They both showed rapid adsorption (70–83% adsorption at 20 min depending on the host) to bacterial cells and large burst size of 82 to 102 phages per infected cell (Table 2). Furthermore, the bioinformatic analysis of the ϕPD10.3 and ϕPD23.1 genomes indicated that they lack known genes coding for toxins, potential allergens, transposases or integrases (S1 Fig. and S2 Table). The lifestyle of ϕPD10.3 and ϕPD23.1 predicted with PHACTS indicated that both phages are lytic and not lysogenic.

Moreover, we did not find any bacterial sequences in the genome nor any bacterial proteins in the phage proteomic analysis, and thus no generalized transduction of the bacterial hosts was not expected to occur. Bacteriophages are generally recognized as safe in the food industry and in veterinary medicine, and accordingly, products contains *Myoviridae* bacteriophages against *Listeria monocytogenes*, *Campylobacter jejuni*, *Staphylococcus aureus*are in use and are commercially available (for review see [64]).

In potato slice and whole tuber assays ϕPD10.3 and ϕPD23.1 were able to protect potato tuber tissue from maceration caused by a mixture of *P. carotovorum* subsp. *carotovorum*, *P. wasabiae* and *D. solani*. These bacteriophages were tested in the worst case scenario, in which bacterial isolates of different species were present in potato tuber tissue at relatively high numbers (5×10^5 cfu per inoculation) (10^7 cfu ml^{-1}) as it may occur during natural infections. Protection was conferred under warm (28°C) and humid (80% relative humidity) conditions favorable for development of soft rot on inoculated potato slices. The bacteriophages both individually and when applied together were able to reproducibly and significantly reduce soft rot infections by at least 80% to 95% in comparison to controls inoculated with a mixture of bacteria only. In the host range experiments, both ϕPD10.3 and ϕPD23.1 bacteriophages were able to infect a range of different strains of *P. carotovorum* subsp. *carotovorum*, *P. wasabiae* and *D. solani*. These results suggest that these phages would be valuable biological control agents under natural field conditions, where high variation in host bacterial strains would be expected. *P. carotovorum* subsp. *carotovorum* strains isolated from potato are genetically diverse [1, 65], whereas the majority of reports describe *P. wasabiae* and *D. solani* as more genetically homogenous [6, 66]. While *P. wasabiae* and *D. solani* are recognized in Europe as dominant blackleg pathogens in the field, *P. carotovorum* subsp. *carotovorum* is the main cause of soft rot of potato in storage and transit [7]. ϕPD10.3 and ϕPD23.1 being able to infect all these species might be used in all stages of potato production from planting in the field to harvest and tuber storage.

One of the main difficulties in applying bacteriophages to control bacterial phytopathogens in agriculture is their high specificity, infecting only the particular strains of the target bacterial species [20].

This problem has been partially overcome in food industry [67] and in veterinary medicine [68] by using bacteriophage cocktails consisting of several phages differing in host specificity to maximize the host range covered. In the case of soft rot *Enterobacteriaceae* there are, however, no well-characterized and readily available bacteriophage collections from which phages with desired host specificities might be selected, prepared and used in biological control under field conditions. To our knowledge, there is also little information on the use of lytic bacteriophages against the *Pectobacterium* and *Dickeya* species, e.g. there is less than twenty research publications worldwide describing lytic bacteriophages against *Pectobacterium* and *Dickeya* species. Additionally, all lytic bacteriophages against soft rot *Enterobacteriaceae* described to date have a restricted host range and have not been tested against a large panel of bacterial strains. In addition, for the majority of bacteriophages infecting *Pectobacterium* and *Dickeya* species there is no information about their genome organization, nor on their stability under a variety of environmental conditions. We postulate that use of broad host range lytic bacteriophages infecting the SRE dominant in potato in Europe may help in minimizing losses caused by these pathogens during potato production. In principle, broad host lytic bacteriophages may be further combined with other approaches to control blackleg and soft rot infections of potato in an integrated strategy [8].

To fully explore the practical potential of broad host lytic bacteriophages against the dominant soft rot *Enterobacteriaceae* in potato in Europe, additional studies are needed such as the long-term efficiency and consistency of control in the field, including bacteriophage population survival and dynamics in bulk soil, in potato rhizosphere soil, in and on potato plants and tubers, application method, timing of application, formulation of bacteriophages in order to prolong their stability under harsh conditions and ecotoxicological risks. Greenhouse studies are now being conducted to directly assess the biological control properties of bacteriophages ϕPD10.3 and ϕPD23.1 against *P. carotovorum* subsp. *carotovorum*, *P. wasabiae* and *D. solani* infections in potato.

ACKNOWLEDGMENTS

The authors would like to express their gratitude to Els Nijhuis (Plant Research International, Wageningen University and Research Centre,

Wageningen, The Netherlands) for guidance in RAPD-PCR analysis, Jacquie van der Waals (University of Pretoria, South Africa) for providing the *P. carotovorum* subsp. *brasiliense* JJ56 strain for the bacteriophages' host specificity assay, Iwona Cymerman (International Institute of Molecular and Cell Biology, Poland) and Sylwia Jafra and Magdalena Rajewska (Department of Biotechnology, Intercollegiate Faculty of Biotechnology, University of Gdansk and Medical University of Gdansk, Gdansk, Poland) for helpful discussion and M. Perombelon (ex. SCRI, Scotland, UK) and S. Lindow (University of California—Berkeley, USA) for critically reviving the manuscript before submission.

REFERENCES

1. Glasner JD, Marquez-Villavicencio M, Kim HS, Jahn CE, Ma B, Biehl BS, et al. Niche-specificity and the variable fraction of the Pectobacterium pan-genome. Mol Plant Microbe Interact. 2008;21: 1549–1560. pmid:18986251
2. Mansfield J, Genin S, Magori S, Citovsky V, Sriariyanum M, Ronald P, et al. Top 10 plant pathogenic bacteria in molecular plant pathology. Mol Plant Pathol. 2012;13: 614–629. pmid:22672649
3. Gardan L, Gouy C, Christen R, Samson R. Elevation of three subspecies of Pectobacterium carotovorum to species level: Pectobacterium atrosepticum sp. nov., Pectobacterium betavasculorum sp. nov. and Pectobacterium wasabiae sp. nov. Int J Syst Evol Microbiol. 2003;53: 381–391. pmid:12710602
4. Duarte V, De Boer SH, Ward LJ, De Oliveira AMR. Characterization of atypical Erwinia carotovora strains causing blackleg of potato in Brazil. J Appl Microbiol. 2004;96: 535–545. pmid:14962133
5. Pitman AR, Wright PJ, Galbraith MD, Harrow SA. Biochemical and genetic diversity of pectolytic enterobacteria causing soft rot disease of potatoes in New Zealand. Australas Plant Pathol. 2008;37: 559–568.
6. Toth IK, van der Wolf JM, Saddler G, Lojkowska E, Hélias V, Pirhonen M, et al. Dickeya species: an emerging problem for potato production in Europe. Plant Pathol. 2011;60: 385–399.
7. Pérombelon MCM. Potato diseases caused by soft rot erwinias: an overview of pathogenesis. Plant Pathol. 2002;51: 1–12.

8. Czajkowski R, Pérombelon MCM, van Veen JA, van der Wolf JM. Control of blackleg and tuber soft rot of potato caused by Pectobacterium and Dickeya species: a review. Plant Pathol. 2011;60: 999–1013.
9. Perombelon MCM, Kelman A. Ecology of the soft rot Erwinias. Annu Rev Phytopathol. 1980;18: 361–387.
10. Perombelon MCM. Blackleg risk potential of seed potatoes determined by quantification of tuber contamination by the causal agent and Erwinia carotovora subsp. atroseptica: a critical review. Bull. OEPP. 2000;30: 413–420.
11. van der Merwe J, Coutinho T, Korsten L, van Der Waals JE. Pectobacterium carotovorum subsp. brasiliensis causing blackleg on potatoes in South Africa. Eur J Plant Pathol. 2010;126: 175–185.
12. Goto M, Matsumoto K. Erwinia carotovora subsp. wasabiae subsp. nov. isolated from diseased rhizomes and fibrous roots of Japanese horseradish (Eutrema wasabi Maxim.). Int J Syst Bacteriol. 1987;37: 130–135.
13. Waleron M, Waleron K. Lojkowska E. Occurrence of Pectobacterium wasabiae in potato field samples. Eur J Plant Pathol. 2013;137: 149–158.
14. Nykyri J, Niemi O, Koskinen P, Nokso-Koivisto J, Pasanen M, Broberg M, et al. Revised phylogeny and novel horizontally acquired virulence determinants of the model soft rot phytopathogen Pectobacterium wasabiae SCC3193. PLoS Pathog. 2012;8: e1003013. pmid:23133391
15. Janse JD, Ruissen MA. Characterization and classification of Erwinia chrysanthemi strains from several hosts in the Netherlands. Phytopathology. 1988;78: 800–808.
16. Slawiak M, van Beckhoven JRCM, Speksnijder AGCL, Czajkowski R, Grabe G, van der Wolf JM. Biochemical and genetical analysis reveal a new clade of biovar 3 Dickeya spp. strains isolated from potato in Europe. Eur J Plant Pathol. 2009;125: 245–261.
17. van der Wolf JM, Nijhuis EH, Kowalewska MJ, Saddler GS, Parkinson N, Elphinstone JG, et al. Dickeya solani sp. nov., a pectinolytic plant pathogenic bacterium isolated from potato

(Solanum tuberosum). Int J Syst Evol Microbiol. 2014;64: 768–774. pmid:24225027

18. Prins H, Breukers A. In de puree? De gevolgen van aantasting door Erwinia voor de pootaardappelsector in kaart gebracht. Hague, the Netherlands: LEI Report; 2008. Dutch.
19. Pérombelon MCM, Salmond GPC. Bacterial soft rots. In: Singh US, Kohmoto K, editors. Pathogenesis and host specificity in plant diseases. Oxford: Pergamon Press Ltd.; 1995. vol.1,pp. 1–20.
20. Jones JB, Jackson LE, Balogh B, Obradovic A, Iriarte FB, Momol MT. Bacteriophages for plant disease control. Annu Rev Phytopathol. 2008;45: 245–262.
21. Eayre CG, Bartz JA, Concelmo DE. Bacteriophages of Erwinia carotovora and Erwinia ananas isolated from freshwater lakes. Plant Dis. 1995;79: 801–804.
22. Czajkowski R, Ozymko Z. Lojkowska E. Isolation and characterization of novel soilborne lytic bacteriophages infecting Dickeya spp. biovar 3 ('D. solani'). Plant Pathol. 2013;63: 758–772
23. Adriaenssens EM, Van Vaerenbergh J, Vandenheuvel D, Dunon V, Ceyssens PJ, De Proft M, et al. T4-Related bacteriophage LIMEstone isolates for the control of soft rot on potato Caused by 'Dickeya solani'. PLoS ONE. 2012;7: e33227. pmid:22413005
24. Ravensdale M, Blom TJ, Gracia-Garza JA, Svircev AM, Smith RJ. Bacteriophages and the control of Erwinia carotovora subsp. carotovora. Can J Plant Pathol. 2007;29: 121–130.
25. Frampton RA, Pitman AR. Fineran PC. Advances in bacteriophage-mediated control of plant pathogens. Int J Microbiol. 2012:326452.
26. Jones JB, Vallad GE, Iriarte FB, Obradović A, Wernsing MH, Jackson LE, et al. Considerations for using bacteriophages for plant disease control. Bacteriophage. 2012;2: 208–214. pmid:23531902
27. Thiel K. Old dogma, new tricks—21st Century phage therapy. Nat Biotechnol. 2004;22: 31–36. pmid:14704699
28. Koskella B, Meaden S. Understanding bacteriophage specificity in natural microbial communities. Viruses. 2013;5: 806–823. pmid:23478639
29. Czajkowski R, Grabe GJ. van der Wolf JM. Distribution of Dickeya

spp. and Pectobacterium carotovorum subsp. carotovorum in naturally infected seed potatoes. Eur J Plant Pathol. 2009;125: 263–275.

30. Degefu Y, Virtanen E. Väyrynen T. Pre-PCR processes in the molecular detection of blackleg and soft rot Erwiniae in seed potatoes. J Phytopathol. 2009;157: 370–378.

31. Ma B, Hibbing ME, Kim HS, Reedy RM, Yedidia I, Breuer J, et al. Host range and molecular phylogenies of the soft rot enterobacterial genera Pectobacterium and Dickeya. Phytopathology. 2007;97: 1150–1163. pmid:18944180

32. Czajkowski R, de Boer WJ, van der Zouwen PS, Kastelein P, Jafra S, de Haan EG, et al. Virulence of 'Dickeya solani' and Dickeya dianthicola biovar-1 and-7 strains on potato (Solanum tuberosum). Plant Pathology. 2012;62: 597–610. pmid:22520917

33. Garlant L, Nykyri J. Pirhonen M. A bacteriocin like molecule produced by soft rot pathogen Dickeya solani affects interactions with Pectobacterium spp. Proceedings of The 3rd International Erwinia Workshop; 2014 June 7–8; Shanghai, China. Available: http://www.icppb2014.org/page.asp?id=28

34. Willis JW, Engwall JK, Chatterjee AK. Cloning of genes for Erwinia carotovora subsp. carotovora pectolytic enzymes and further characterization of the polygalacturonases. Phytopathology. 1987;77: 1199–1205.

35. Bell KS, Sebaihia M, Pritchard L, Holden MTG, Hyman LJ, Holeva MC, et al. Genome sequence of the enterobacterial phytopathogen Erwinia carotovora subsp. atroseptica and characterization of virulence factors. Proc Natl Acad Sci U S A. 2004;101: 11105–11110. pmid:15263089

36. Pirhonen M, Saarilahti H, Karlsson MB, Palva ET. Identification of pathogenicity determinants of Erwinia carotovora subsp. carotovora by transposon mutagenesis. Mol Plant Microbe Interact. 1991;4: 276–283.

37. Samson R, Legendre JB, Christen R, Fischer-Le Saux M, Achouak W, Gardan L. Transfer of Pectobacterium chrysanthemi (Burkholder et al. 1953) Brenner et al. 1973 and Brenneria paradisiaca to the genus Dickeya gen. nov. as Dickeya chrysanthemi comb. nov.

and Dickeya paradisiaca comb. nov. and delineation of four novel species, Dickeya dadantii sp. nov., Dickeya dianthicola sp. nov., Dickeya dieffenbachiae sp. nov. and Dickeya zeae sp. nov. Int J Syst Evol Microbiol. 2005;55: 1415–1427. pmid:16014461

38. Twest R, Kropinski AM. Bacteriophage enrichment from water and soil. In: Clokie MJ, Kropinski A, editors. Bacteriophages. New York: Humana Press; 2009. pp. 15–21.

39. Bertani G. Studies on lysogenesis I: The mode of phage liberation by lysogenic Escherichia coli. J Bacteriol. 1951;62: 293–300. pmid:14888646

40. Gasic K, Ivanovic MM, Ignjatov M, Calic A, Obradovic A. Isolation and characterization of Xanthomonas euvesicatoria bacteriophages. J Plant Pathol. 2011;93: 415–423.

41. Ellis EL, Delbrück M. The growth of bacteriophage. J Gen Physiol. 1939;22: 365–384. pmid:19873108

42. Jiang S, Kellogg A, Paul JH. Characterization of marine temperate phage-host systems isolated from Mamala Bay, Oahu, Hawaii. Appl Environ Microbiol. 1998;64: 535–542. pmid:9464390

43. Lingohr E, Frost S, Johnson RP. Determination of bacteriophage genome size by pulsed-field gel electrophoresis. In: Clokie MJ, Kropinski A, editors. Bacteriophages. New York: Humana Press; 2009. pp. 19–25.

44. Hunter SB, Vauterin P, Lambert-Fair MA, Van Duyne MS, Kubota K, Graves L, et al. Establishment of a universal size standard strain for use with the PulseNet standardized pulsed-field gel electrophoresis protocols: converting the national databases to the new size standard. J Clin Microbiol. 2005;43: 1045–1050. pmid:15750058

45. Gill JJ, Svircev AM, Smith R, Castle AJ. Bacteriophages of Erwinia amylovora. Appl Environ Microbiol. 2003;69: 2133–2138. pmid:12676693

46. Clokie MRJ, Kropinski AM. Bacteriophages: methods and protocols, volume 1: isolation, characterization, and interactions. New York: Humana Press; 2009.

47. Comeau AM, Short S, Suttle CA. The use of degenerate-primed random amplification of polymorphic DNA (DP-RAPD) for strain-typing and inferring the genetic similarity among closely related

viruses. J Virol Methods. 2004;118: 95–100. pmid:15081604

48. McNair K, Bailey BA, Edwards RA. PHACTS, a computational approach to classifying the lifestyle of phages. Bioinformatics. 2012;28: 614–618. pmid:22238260
49. Darling AE, Mau B, Perna NT. progressiveMauve: multiple genome alignment with gene gain, loss and rearrangement. PLoS ONE. 2010;5: e11147. pmid:20593022
50. Blom J, Albaum S, Doppmeier D, Puhler A, Vorholter F-J, Zakrzewski M, et al. EDGAR: a software framework for the comparative analysis of prokaryotic genomes. BMC Bioinformatics. 2009;10: 154. pmid:19457249
51. Kurowski MA, Bujnicki JM. GeneSilico protein structure prediction meta-server. Nucleic Acids Res. 2003;31: 3305–3307. pmid:12824313
52. Kawabata T, Arisaka F, Nishikawa K. Structural/functional assignment of unknown bacteriophage T4 proteins by iterative database searches. Gene. 2000;259: 223–233. pmid:11163980
53. Gross DC, Powelson ML, Regner KM, Radamaker GK. A bacteriophage-typing system for surveying the diversity and distribution of strains of Erwinia carotovora in potato fields. Phytopathol. 1991;81: 220–226.
54. Pérombelon MCM. The impaired host and soft rot bacteria. Phytopathogenic Prokaryotes. 1982;2: 55–69.
55. Ackermann HW. Bacteriophage observations and evolution. Res Microbiol. 2003;154: 245–251. pmid:12798228
56. Ackermann HW. Tailed Bacteriophages: the Order Caudovirales. Adv Virus Res. 1998;51: 135–201. pmid:9891587
57. Ackermann HW. 5500 Phages examined in the electron microscope. Arch Virol. 2007;152: 227–243. pmid:17051420
58. de Boer SH, Li XZ, Ward LJ. Pectobacterium spp. associated with bacterial stem rot syndrome of potato in Canada. Phytopathol. 2012;102: 937–947.
59. Degefu Y, Potrykus M, Golanowska M, Virtanen E, Lojkowska E. A new clade of Dickeya spp. plays a major role in potato blackleg outbreaks in North Finland. Ann Appl Biol. 2013;162: 231–241.

60. Leite LN, de Haan EG, Krijger M, Kastelein P, van der Zouwen PS, van den Bovenkamp GW, et al. First report of potato blackleg caused by Pectobacterium carotovorum subsp. brasiliensis in the Netherlands. New Dis Rep. 2014;29: 24.
61. Nabhan S, De Boer SH, Maiss E, Wydra K. Taxonomic relatedness between Pectobacterium carotovorum subsp. carotovorum, Pectobacterium carotovorum subsp. odoriferum and Pectobacterium carotovorum subsp. brasiliense subsp. nov. J Appl Microbiol. 2012;113: 904–913. pmid:22747943
62. Czajkowski R, Ozymko Z, Zwirowski S, Lojkowska E. Complete genome sequence of a broad-host-range lytic Dickeya spp. bacteriophage ϕD5. Arch Virol. 2014;159: 3153–3155. pmid:25000903
63. Petrov VM, Ratnayaka S, Nolan JM, Miller ES, Karam JD. Genomes of the T 4-related bacteriophages as windows on microbial genome evolution. Virol J. 2010;7: 292–292. pmid:21029436
64. Karthik K, Muneeswaran NS, Manjunathachar HV, Gopi M, Elamurugan A, Kalaiyarasu S. Bacteriophages: effective alternative to antibiotics. Adv Anim Vet Sci. 2014;2: 1–7.
65. Smith C, Bartz JA. Variation in the pathogenicity and aggressiveness of strains of Erwinia carotovora subsp. carotovora isolated from different hosts. Plant Dis. 1990;74: 505–509.
66. Pitman AR, Harrow SA, Visnovsky SB. Genetic characterisation of Pectobacterium wasabiae causing soft rot disease of potato in New Zealand. Eur J Plant Pathol. 2010;126: 423–435.
67. García P, Martínez B, Obeso JM, Rodríguez A. Bacteriophages and their application in food safety. Lett Appl Microbiol. 2008;47: 479–485. pmid:19120914
68. Chan BK, Abedon ST, Loc-Carrillo C. Phage cocktails and the future of phage therapy. Future Microbiol. 2013;8: 769–783. pmid:23701332

CITATION

CHAPTER 2

Buttimer C, McAuliffe O, Ross RP, Hill C, O'Mahony J and Coffey A (2017) Bacteriophages and Bacterial Plant Diseases. Front. Microbiol. 8:34. doi: 10.3389/fmicb.2017.00034

CHAPTER 4

Alexandra Igorevna Klimenko, Yury Georgievich Matushkin, Nikolay Alexandrovich Kolchanov and Sergey Alexandrovich Lashin. BMC Microbiology, 16(Suppl 1):10 DOI 10.1186/s12866-015-0620-4

CHAPTER 6

Curtright AJ, Abedon ST (2011) Phage Therapy: Emergent Property Pharmacology. J Bioanal Biomed S6:002. doi:10.4172/1948-593X.S6-002

CHAPTER 9

Rebekah A. Frampton, Andrew R. Pitman, and Peter C. Fineran, "Advances in Bacteriophage-Mediated Control of Plant Pathogens," International Journal of Microbiology, vol. 2012, Article ID 326452, 11 pages, 2012. doi:10.1155/2012/326452

CHAPTER 10

Rombouts S, Volckaert A, Venneman S, Declercq B, Vandenheuvel D, Allonsius CN, Van Malderghem C, Jang HB, Briers Y, Noben JP, Klumpp J, Van Vaerenbergh J, Maes M and Lavigne R (2016) Characterization of Novel Bacteriophages for Biocontrol of Bacterial Blight in Leek Caused by *Pseudomonas syringae* pv. *porri*. *Front. Microbiol.* 7:279. doi: 10.3389/fmicb.2016.00279

CHAPTER 11

Das M, Bhowmick T S, Ahern S J, Young R, Gonzalez C F. PLoSONE 10(6):e0128902. doi:10.1371/ journal.pone.0128902

CHAPTER 12

Czajkowski R, Ozymko Z, de Jager V, Siwinska J, Smolarska A, Ossowicki A, et al. (2015) Genomic, Proteomic and Morphological Characterization of Two Novel Broad Host Lytic Bacteriophages ΦPD10.3 and ΦPD23.1 Infecting Pectinolytic *Pectobacterium* spp. and *Dickeya* spp. PLoS ONE 10(3): e0119812. https://doi.org/10.1371/journal.pone.0119812

INDEX

Q

R

S

T

U

X